D1214270

# CONTINUED FRACTIONS IN STATISTICAL APPLICATIONS

**STATISTICS: Textbooks and Monographs**

A Series Edited by

D. B. Owen, Coordinating Editor
*Department of Statistics*
*Southern Methodist University*
*Dallas, Texas*

R. G. Cornell, Associate Editor
for Biostatistics
*University of Michigan*

W. J. Kennedy, Associate Editor
for Statistical Computing
*Iowa State University*

A. M. Kshirsagar, Associate Editor
for Multivariate Analysis and
Experimental Design
*University of Michigan*

E. G. Schilling, Associate Editor
for Statistical Quality Control
*Rochester Institute of Technology*

**Vol. 1:** The Generalized Jackknife Statistic, *H. L. Gray and W. R. Schucany*
**Vol. 2:** Multivariate Analysis, *Anant M. Kshirsagar*
**Vol. 3:** Statistics and Society, *Walter T. Federer*
**Vol. 4:** Multivariate Analysis: A Selected and Abstracted Bibliography, 1957-1972, *Kocherlakota Subrahmaniam and Kathleen Subrahmaniam* (out of print)
**Vol. 5:** Design of Experiments: A Realistic Approach, *Virgil L. Anderson and Robert A. McLean*
**Vol. 6:** Statistical and Mathematical Aspects of Pollution Problems, *John W. Pratt*
**Vol. 7:** Introduction to Probability and Statistics (in two parts), Part I: Probability; Part II: Statistics, *Narayan C. Giri*
**Vol. 8:** Statistical Theory of the Analysis of Experimental Designs, *J. Ogawa*
**Vol. 9:** Statistical Techniques in Simulation (in two parts), *Jack P. C. Kleijnen*
**Vol. 10:** Data Quality Control and Editing, *Joseph I. Naus* (out of print)
**Vol. 11:** Cost of Living Index Numbers: Practice, Precision, and Theory, *Kali S. Banerjee*
**Vol. 12:** Weighing Designs: For Chemistry, Medicine, Economics, Operations Research, Statistics, *Kali S. Banerjee*
**Vol. 13:** The Search for Oil: Some Statistical Methods and Techniques, *edited by D. B. Owen*
**Vol. 14:** Sample Size Choice: Charts for Experiments with Linear Models, *Robert E. Odeh and Martin Fox*
**Vol. 15:** Statistical Methods for Engineers and Scientists, *Robert M. Bethea, Benjamin S. Duran, and Thomas L. Boullion*
**Vol. 16:** Statistical Quality Control Methods, *Irving W. Burr*
**Vol. 17:** On the History of Statistics and Probability, *edited by D. B. Owen*
**Vol. 18:** Econometrics, *Peter Schmidt*
**Vol. 19:** Sufficient Statistics: Selected Contributions, *Vasant S. Huzurbazar (edited by Anant M. Kshirsagar)*
**Vol. 20:** Handbook of Statistical Distributions, *Jagdish K. Patel, C. H. Kapadia, and D. B. Owen*
**Vol. 21:** Case Studies in Sample Design, *A. C. Rosander*
**Vol. 22:** Pocket Book of Statistical Tables, *compiled by R. E. Odeh, D. B. Owen, Z. W. Birnbaum, and L. Fisher*
**Vol. 23:** The Information in Contingency Tables, *D. V. Gokhale and Solomon Kullback*

**Vol. 24:** Statistical Analysis of Reliability and Life-Testing Models: Theory and Methods, *Lee J. Bain*
**Vol. 25:** Elementary Statistical Quality Control, *Irving W. Burr*
**Vol. 26:** An Introduction to Probability and Statistics Using BASIC, *Richard A. Groeneveld*
**Vol. 27:** Basic Applied Statistics, *B. L. Raktoe and J. J. Hubert*
**Vol. 28:** A Primer in Probability, *Kathleen Subrahmaniam*
**Vol. 29:** Random Processes: A First Look, *R. Syski*
**Vol. 30:** Regression Methods: A Tool for Data Analysis, *Rudolf J. Freund and Paul D. Minton*
**Vol. 31:** Randomization Tests, *Eugene S. Edgington*
**Vol. 32:** Tables for Normal Tolerance Limits, Sampling Plans, and Screening, *Robert E. Odeh and D. B. Owen*
**Vol. 33:** Statistical Computing, *William J. Kennedy, Jr. and James E. Gentle*
**Vol. 34:** Regression Analysis and Its Application: A Data-Oriented Approach, *Richard F. Gunst and Robert L. Mason*
**Vol. 35:** Scientific Strategies to Save Your Life, *I. D. J. Bross*
**Vol. 36:** Statistics in the Pharmaceutical Industry, *edited by C. Ralph Buncher and Jia-Yeong Tsay*
**Vol. 37:** Sampling from a Finite Population, *J. Hajek*
**Vol. 38:** Statistical Modeling Techniques, *S. S. Shapiro*
**Vol. 39:** Statistical Theory and Inference in Research, *T. A. Bancroft and C.-P. Han*
**Vol. 40:** Handbook of the Normal Distribution, *Jagdish K. Patel and Campbell B. Read*
**Vol. 41:** Recent Advances in Regression Methods, *Hrishikesh D. Vinod and Aman Ullah*
**Vol. 42:** Acceptance Sampling in Quality Control, *Edward G. Schilling*
**Vol. 43:** The Randomized Clinical Trial and Therapeutic Decisions, *edited by Niels Tygstrup, John M. Lachin, and Erik Juhl*
**Vol. 44:** Regression Analysis of Survival Data in Cancer Chemotherapy, *Walter H. Carter, Jr., Galen L. Wampler, and Donald M. Stablein*
**Vol. 45:** A Course in Linear Models, *Anant M. Kshirsagar*
**Vol. 46:** Clinical Trials: Issues and Approaches, *edited by Stanley H. Shapiro and Thomas H. Louis*
**Vol. 47:** Statistical Analysis of DNA Sequence Data, *edited by B. S. Weir*
**Vol. 48:** Nonlinear Regression Modeling: A Unified Practical Approach, *David A. Ratkowsky*
**Vol. 49:** Attribute Sampling Plans, Tables of Tests and Confidence Limits for Proportions, *Robert E. Odeh and D. B. Owen*
**Vol. 50:** Experimental Design, Statistical Models, and Genetic Statistics, *edited by Klaus Hinkelmann*
**Vol. 51:** Statistical Methods for Cancer Studies, *edited by Richard G. Cornell*
**Vol. 52:** Practical Statistical Sampling for Auditors, *Arthur J. Wilburn*
**Vol. 53:** Statistical Signal Processing, *edited by Edward J. Wegman and James G. Smith*
**Vol. 54:** Self-Organizing Methods in Modeling: GMDH Type Algorithms, *edited by Stanley J. Farlow*
**Vol. 55:** Applied Factorial and Fractional Designs, *Robert A. McLean and Virgil L. Anderson*
**Vol. 56:** Design of Experiments: Ranking and Selection, *edited by Thomas J. Santner and Ajit C. Tamhane*
**Vol. 57:** Statistical Methods for Engineers and Scientists. Second Edition, Revised and Expanded, *Robert M. Bethea, Benjamin S. Duran, and Thomas L. Boullion*
**Vol. 58:** Ensemble Modeling: Inference from Small-Scale Properties to Large-Scale Systems, *Alan E. Gelfand and Crayton C. Walker*
**Vol. 59:** Computer Modeling for Business and Industry, *Bruce L. Bowerman and Richard T. O'Connell*
**Vol. 60:** Bayesian Analysis of Linear Models, *Lyle D. Broemeling*

Vol. 61: Methodological Issues for Health Care Surveys, *Brenda Cox and Steven Cohen*

Vol. 62: Applied Regression Analysis and Experimental Design, *Richard J. Brook and Gregory C. Arnold*

Vol. 63: Statpal: A Statistical Package for Microcomputers – PC-DOS Version for the IBM PC and Compatibles, *Bruce J. Chalmer and David G. Whitmore*

Vol. 64: Statpal: A Statistical Package for Microcomputers – Apple Version for the II, II+, and IIe, *David G. Whitmore and Bruce J. Chalmer*

Vol. 65: Nonparametric Statistical Inference, Second Edition, Revised and Expanded, *Jean Dickinson Gibbons*

Vol. 66: Design and Analysis of Experiments, *Roger G. Petersen*

Vol. 67: Statistical Methods for Pharmaceutical Research Planning, *Sten W. Bergman and John C. Gittins*

Vol. 68: Goodness-of-Fit Techniques, *edited by Ralph B. D'Agostino and Michael A. Stephens*

Vol. 69: Statistical Methods in Discrimination Litigation, *edited by D. H. Kaye and Mikel Aickin*

Vol. 70: Truncated and Censored Samples from Normal Populations, *Helmut Schneider*

Vol. 71: Robust Inference, *M. L. Tiku, W. Y. Tan, and N. Balakrishnan*

Vol. 72: Statistical Image Processing and Graphics, *edited by Edward J. Wegman and Douglas J. DePriest*

Vol. 73: Assignment Methods in Combinatorial Data Analysis, *Lawrence J. Hubert*

Vol. 74: Econometrics and Structural Change, *Lyle D. Broemeling and Hiroki Tsurumi*

Vol. 75: Multivariate Interpretation of Clinical Laboratory Data, *Adelin Albert and Eugene K. Harris*

Vol. 76: Statistical Tools for Simulation Practitioners, *Jack P. C. Kleijnen*

Vol. 77: Randomization Tests, Second Edition, *Eugene S. Edgington*

Vol. 78: A Folio of Distributions: A Collection of Theoretical Quantile-Quantile Plots, *Edward B. Fowlkes*

Vol. 79: Applied Categorical Data Analysis, *Daniel H. Freeman, Jr.*

Vol. 80: Seemingly Unrelated Regression Equations Models : Estimation and Inference, *Virendra K. Srivastava and David E. A. Giles*

Vol. 81: Response Surfaces: Designs and Analyses, *Andre I. Khuri and John A. Cornell*

Vol. 82: Nonlinear Parameter Estimation: An Integrated System in BASIC, *John C. Nash and Mary Walker-Smith*

Vol. 83: Cancer Modeling, *edited by James R. Thompson and Barry W. Brown*

Vol. 84: Mixture Models: Inference and Applications to Clustering, *Geoffrey J. McLachlan and Kaye E. Basford*

Vol. 85: Randomized Response: Theory and Techniques, *Arijit Chaudhuri and Rahul Mukerjee*

Vol. 86: Biopharmaceutical Statistics for Drug Development, *edited by Karl E. Peace*

Vol. 87: Parts per Million Values for Estimating Quality Levels, *Robert E. Odeh and D. B. Owen*

Vol. 88: Lognormal Distributions: Theory and Applications, *edited by Edwin L. Crow and Kunio Shimizu*

Vol. 89: Properties of Estimators for the Gamma Distribution, *K. O. Bowman and L. R. Shenton*

Vol. 90: Spline Smoothing and Nonparametric Regression, *Randall L. Eubank*

Vol. 91: Linear Least Squares Computations, *R. W. Farebrother*

Vol. 92: Exploring Statistics, *Damaraju Raghavarao*

Vol. 93: Applied Time Series Analysis for Business and Economic Forecasting, *Sufi M. Nazem*

Vol. 94: Bayesian Analysis of Time Series and Dynamic Models, *edited by James C. Spall*

Vol. 95: The Inverse Gaussian Distribution: Theory, Methodology, and Applications, *Raj S. Chhikara and J. Leroy Folks*

Vol. 96: Parameter Estimation in Reliability and Life Span Models, *A. Clifford Cohen and Betty Jones Whitten*

Vol. 97: Pooled Cross-Sectional and Time Series Data Analysis, *Terry E. Dielman*

Vol. 98: Random Processes: A First Look, Second Edition, Revised and Expanded, *R. Syski*

Vol. 99: Generalized Poisson Distributions: Properties and Applications, *P.C. Consul*

Vol. 100: Nonlinear $L_p$-Norm Estimation, *René Gonin and Arthur H. Money*

Vol. 101: Model Discrimination for Nonlinear Regression Models, *Dale S. Borowiak*

Vol. 102: Applied Regression Analysis in Econometrics, *Howard E. Doran*

Vol. 103: Continued Fractions in Statistical Applications, *K.O. Bowman and L.R. Shenton*

Vol. 104: Statistical Methodology in the Pharmaceutical Sciences, *Donald A. Berry*

**ADDITIONAL VOLUMES IN PREPARATION**

# CONTINUED FRACTIONS IN STATISTICAL APPLICATIONS

K. O. BOWMAN
Oak Ridge National Laboratory
Oak Ridge, Tennessee

L. R. SHENTON
University of Georgia
Athens, Georgia

MARCEL DEKKER, INC. New York and Basel

ISBN 0-8247-8120-1



MARCEL DEKKER, INC.
270 Madison Avenue, New York, New York 10016

Current printing (last digit):
10 9 8 7 6 5 4 3 2 1

PRINTED IN THE UNITED STATES OF AMERICA

# PREFACE

This book is an account of the use of continued fractions and Padé sequences as tools in the interpretation of divergent or slowly convergent series occurring in theoretical statistics. In our early studies, an obstacle we had to overcome concerned the process of developing Taylor power series for moments of statistics, for example, the expected value of the sample standard deviation from some defined population. Traditionally one or two low-order terms sufficed because later terms involved increasingly heavy algebra that soon confounded the most ardent algebraist; the arrival of the digital computer in the 1950's had little impact on algorithmic approaches, which were generally not highly regarded, existing in a somewhat ill-defined gray area. With higher-order terms in series not available, the era of largeness of sample size was born as was the reliance from time to time on the uncertainties defined in the order of magnitude symbols (little o, big O).

Also, early in the present century, summation algorithms (or techniques) had few illustrations of divergent series, so a few classical cases received a lot of attention; many of them (by chance in the evolution of mathematical analysis) were such that the first few terms looked beguilingly "convergent" (the series part of $\ln \Gamma(x)$, for example, and the polygamma function series). So there appeared the notion of absolute error in stopping at a certain term being related to the first term omitted; whereas this rule is sometimes true, in general it has no basis.

Our treatment is partly mathematical but pivoted on computer oriented numerical analysis (or computer oriented algebra). The increased confidence in our main algorithm for the expectation of quotients of products of sample moments was the driving force for increased knowledge of summation processes for seemingly divergent Taylor series. The study, by no means complete, and no doubt involving unavoidable errors (hopefully few), has spanned more than two decades.

Multivariate situations are not considered for there are many outstanding problems in the univariate case. At the very least, we hope the reader will gain a new awareness of the many varieties of series and the corresponding summation problems.

In conclusion, let it be made clear that our development of the subject has been motivated by an attempt to clarify the interpretation of divergent series in practical situations. The classical theory of rational approximation is overloaded with a myriad of specialized theorems - albeit of some interest, and exhibiting acute mathematical expertise. But very few, if any, summation problems in practical situations can be embedded in known theorems - and the diversity of structure, only partially defined in those situations, makes "theorem proving" a hopeless objective in the present state of our knowledge. For the most part, our approach is really that of an experimental science and should appeal (with the help of modest computer facilities) to the curiosity of a wide spectrum of scientific workers (engineers, statisticians, and others).

We are indebted to Dr. Len Gray and Dr. Dennis Wolf for reading a drafted manuscript and their criticism. We also acknowledge the supercomputer time provided by the Department of Energy, Energy Sciences Advanced Computation Program. The research of K. O. Bowman was sponsored by the Applied Mathematical Sciences Research Program, Office of Energy Research, U.S. Department of Energy, under contract DE-AC05-84OR21400 with the Martin Marietta Energy Systems, Inc.; that of L. R. Shenton was supported by the Advanced Computational Methods Center, University of Georgia.

K. O. Bowman
L. R. Shenton

# CONTENTS

# CONTINUED FRACTIONS IN STATISTICAL APPLICATIONS

# INTRODUCTION

## 1. Genesis

It was in 1950's that one of us became interested in a statistical problem (the efficiency of the method of moments and the Gram-Charlier Type A distribution: Shenton, 1954) which required the evaluation of integrals of the form

$$\{P;Q\} = \int_{-\infty}^{\infty} \frac{e^{-x^2/2}P(x)}{Q(x)}\, dx \tag{1}$$

where $P$, $Q$ are polynomials in $x$ with $Q(x) > 0$ for $-\infty < x < \infty$. Typically $Q(\cdot)$ was a polynomial of degree four, whereas after reduction $P$ would be of degree three or less. A generalization to a Stieltjes integral involving a distribution function $\sigma(x)$, constant on $-\infty$ to $a$, and $b$ to $+\infty$ $(b > a)$ is clearly involved.

Looked at from the vantage point of present-day digital facilities ranging from programmable pocket calculators to micro or minicomputers to super computers such as the Cray series, the integrals involved appear rather trivial. One must also have in mind that numerical analysis, early in the present century, was scarcely accepted as a discipline in its own right -- it has an accepted notion, so it seemed, that quadrature problems could be dealt with efficiently, for were there not mathematically oriented remainder terms available? So why be concerned about the accuracy involved? Unfortunately remainder terms in the workshop of applied mathematics, although of value from the point of view of rigor, are frequently shelved because they involve more manipulation than the original problem. Pioneers like E. T. Whittaker (Whittaker and Robinson's "Calculus of Observations" appeared in the 1920's) and D. P. Hartree, Lanczos, and W. E. Milne slowly changed the climate of opinion on numerical work. Along the way accuracy was made to depend on diverse methods of approximation. In quadrature, for example, one could resort to increasing the number of intervals or trying out several different procedures (trapezoidal rule, Simpson's rule, Gaussian type formulas, and many others). Nonetheless, integrals involving an infinite range were time consuming problems on the desk calculators available (mainly Brunsviga mechanical machines) but could be readily evaluated on a "differential analyzer". Could there be a rough and ready (quick and dirty in modern terminology) approximate to $\{P;Q\}$? At this time, "least squares" was a popular process to deal with observations subject to errors, with a link to Fourier series, orthogonal functions, and orthogonal polynomials. It is not a great leap forward to link the integral in (1) with the Parseval formula

$$\sum_{\nu=0}^{\infty} |f_\nu|^2 = \int_{-\infty}^{\infty} |f(t)|^2 d\,\sigma(t) \tag{2}$$

where the Fourier coefficients with respect to the orthogonal set $\{\omega_s(t)\}$ are

$$f_\nu = \int_{-\infty}^{\infty} f(t)\omega_\nu(t)d\,\sigma(t).$$

In view of the structure of $\{P;Q\}$, the simplest case involving a linear form $Q(x) = x + z$, continued fractions of one form or another are going to play an important role.

## 2. Classical Work on Continued Fractions

A real positive number $n$ has as a first approximant $N$, the integer part; if the difference $n - N$ exceeds zero, then its reciprocal again has an integer part. For example

$$\pi = 3 + \cfrac{1}{7 + \cfrac{1}{15 + \cfrac{1}{1 + \cfrac{1}{292 + \cfrac{1}{1 + \cdots}}}}}$$

with the well-known approximants 3, 22/7, 333/106, 355/113, and 103993/33102, etc. A more convenient notation is

$$\pi = 3 + \frac{1}{7+}\,\frac{1}{15+}\,\frac{1}{1+}\,\frac{1}{292+}\,\frac{1}{1+}\,\cdots. \tag{3}$$

It looks as if it will be difficult to discern a pattern for the partial denominators 7, 15, $\cdots$; in addition to structure a continued fraction (c.f.), we need to know its decimal form. Of course we know $\pi$ to a very large number of decimal places (a recent computation on a super computer by Dr. Yasumasa Kanda (University of Tokyo) gives 134,217,728 decimal places of $\pi$; see New York Times, July 5, 1988, article by Malcom W. Browne), so the interest in (3) relates to how close $\pi$ can be approximated by a rational; there is also curiosity as to whether the partial denominators are bounded.

When a parameter is involved, there are various devices which lead to a c.f. development. Generally a power series expansion is looked for, say, $f(x) \sim a + bx + cx^2 + \cdots$. This is transformed using rational fractions in a c.f.

$$f(x) \sim \frac{a_0}{1+} \frac{a_1 x}{1+} \frac{a_2 x}{1+} \cdots, \tag{4}$$

with approximants $a_0$, $a_0/(1 + a_1 x)$, and so on. This class includes c.f.s for the hypergeometric function (in particular, $\ln(1+x)$, $\arcsin(x)$ arc tan $(x)$, $(1+x)^m$, and others), the error function, factorial series and many others. The Gauss c.f. for the hypergeometric function was introduced nearly 200 years ago. Note that the c.f. in (3) bears little relation to the Gauss form derivable from arctan 1, namely

$$\pi = \frac{4}{1+} \frac{1}{3+} \frac{4}{5+} \frac{9}{7+} \frac{16}{9+} \cdots .$$

There are many classical cases for which $a_0, a_1, \cdots$ in (4) can be found in simple closed form and many exceptions. For example, the series part of the logarithm of the factorial function has a c.f. form in which only the first few partial numerators are known. In general, of course, structures with regular sign and magnitude patterns have a high "appeal" index, quite apart from inherent merit.

Stieltjes (1918, Chapter 8) showed a tie-up between Stieltjes integral transforms and c.f.s. For example, under certain conditions one considers

$$\min_{P_n} \int_0^\infty \frac{P_n^2(t)d\,\sigma(t)}{t+x}, \qquad \min_{P_n} \int_0^\infty \frac{tP_n^2(t)d\,\sigma(t)}{t+x}$$

for $x > 0$, the minima being considered over all real polynomials of degree $\leqslant n$; here $\sigma(\cdot)$ is a nondecreasing bounded function -- a distribution function. An alternative formulation is to consider, for example,

$$\min_{P_n} \int_0^\infty (t+x)\{1/(t+x) - P_n(t)\}^2 d\,\sigma(t). \tag{5}$$

Especially suitable forms for $P_n(t)$ are the orthogonal set with respect to the weight function $d\,\sigma(t)$ or the related weight function $(t+x)d\,\sigma(t)$. This hints at the part played by orthogonal polynomials and the dominant role of second order recurrence relations in the structure of numerators and denominators of c.f.s. There is also a suggestion here of the reason why c.f.s often prove to be good approximators. In this connection, note that in going from a series to a c.f., the convergence region frequently changes from a circle for the series to the whole split plane for the c.f. However, by itself, this does not guarantee an improved rate of convergence, although it is possible for divergent series to translate to c.f. convergence (or divergence) and convergent series to translate to divergent (or convergent) c.f.s.

As an illustration of (5) consider

$$\min_{a_0, a_1} \int_0^1 (x+t)\{1/(x+t) - a_0 - a_1 t\}^2 dt \,. \qquad (x>0)$$

One readily finds the minimum values

$$\hat{a}_0 = (x+1)/(x^2+x+1/6), \qquad \hat{a}_1 = -1/(x^2+x+1/6),$$

leading to the approximation

$$\ln(1+x^{-1}) \sim (x+1/2)/(x^2+x+1/6), \tag{6}$$

which is the fourth convergent of the c.f.

$$\ln(1+x^{-1}) = \frac{1}{x+}\;\frac{1^2}{2+}\;\frac{1^2}{3x+}\;\frac{2^2}{4+}\;\frac{2^2}{5x+}\;\frac{3^2}{6+}\;\cdots. \tag{7}$$

Spectacular results are sometimes claimed for c.f.s, but this can be misleading. For example

$$\sqrt{2} = 1 + \frac{1}{2+}\;\frac{1}{2+}\;\frac{1}{2+}\;\cdots$$

and using 8 and 9 convergents leads to $1.4142132 < \sqrt{2} < 1.4142136$ far surpassing low order approximants derived from the binomial $\sqrt{(1+z)}$ when $z = 1$. On the other hand

$$e^{x^2/2}\int_x^\infty e^{-t^2/2}dt = \frac{1}{x+}\;\frac{1}{x+}\;\frac{2}{x+}\;\frac{3}{x+}\;\cdots$$

requires about 70 terms to pin the error down to $2.5 \times 10^{-7}$ when $x = 1$.

The main point of interest is transforming a divergent series to a convergent c.f.; more realistically, to transform an apparently divergent series to a more benign structure. For example, the factorial series $1 - 1!x + 2!x^2 - 3!x^3 + \cdots$ has historically been a prime exhibit in summation lore. It diverges unless $x = 0$. But the c.f. approximants form a bridge (as pointed out by Stieltjes) to a well behaved function, finite for all $x$ outside the split plane ($-\infty$ to 0). In fact, the series sums to

$$I_1(x) = \int_0^\infty \frac{e^{-t}\,dt}{1+xt}. \tag{8}$$

Euler studied the series for $I_1(1)$. According to Barbeau (1979) he tried four different approaches to the summation problem, believing that a correct answer had been found if several processes were in substantial agreement. One of his methods (as reported by Barbeau) is intriguing. It amounts to an unexpected use of finite difference calculus. Define a sequence $\{P_n\}$ by

$$P_{n+1} = nP_n + 1, \qquad (n = 1, 2, \cdots)$$

with $P_1 = 1$. Evidently

$$P_n = 1 + N + N(N-1) + N(N-1)(N-2) + \cdots, \qquad (N = n-1)$$

and $P_0 = I_1(1)$, stretching the meaning of the equality sign somewhat. But again, at least formally, using $E = 1 + \Delta$ for the incremental operator,

$$a_0 = E^{-1}a_1 = (1+\Delta)^{-1}a_1.$$

For example

$$P_0^{-1} = P_1^{-1} - \Delta P_1^{-1} + \Delta^2 P_1^{-1} - \cdots$$

with partial sums

| $r$ | $S_r$ | $r$ | $S_r$ |
|---|---|---|---|
| 1 | 1 | 7 | 1.614616 |
| 2 | 1.5 | 8 | 1.604664 |
| 3 | 1.7 | 10 | 1.665875 |
| 4 | 1.7375 | 16 | 1.888536 |
| 5 | 1.702885 | 20 | 1.592721 |
| 6 | 1.651740 | 30 | 0.641108 |

Similarly using $\ln(P_0) = (1+\Delta)^{-1}\ln(P_1)$ we have the partial sums

| $r$ | $T_r$ | $r$ | $T_r$ |
|---|---|---|---|
| 1 | 0.000000 | 6 | -0.537854 |
| 2 | -0.693147 | 7 | -0.533350 |
| 3 | -0.470004 | 8 | -0.520055 |
| 4 | -0.493720 | 9 | -0.504988 |
| 5 | -0.525649 | 10 | -0.493017 |

Readers will agree, we think, that there is no obvious sequence value to prefer, although roughly they all suggest a value for $I_1(1)$ near to 0.6, excepting $S_{30}$ in the first case (i.e., the algorithm for $1/P_0$). It is quite possible for the sequences to diverge, but the main point is that it shows how unusual devices can be constructed followed by false conclusions. There is the ever present risk that when the sum of a series is known, there will be a tendency to accept a value in a sequence without real justification, and for which, if the sequence were pursued (loss of accuracy being avoided), later values might indicate wild behavior.

Another aspect of c.f.s should be noted here concerning the relation to continuant determinants (determinants with elements in main diagonal, first superdiagonal and first subdiagonal only). For example, the c.f. in (8) can be expressed as the ratio of two continuants, namely, $N_r / D_r$, as $r \to \infty$, where $N_r$ is the principal minor of $D_r$, and $D_r$ is an $r \times r$ symmetric determinant with elements

$$
\begin{aligned}
P_{s,s} &= (2s-1)x + 1\,, && (s = 1, 2, \cdots, r)\\
P_{s,s+1} &= sx\,, && (s = 1, 2, \cdots, r-1)\\
P_{s,r} &= 0\,. && (r > s+1)
\end{aligned}
$$

For generalized c.f.s of higher order, continuant-type determinants with elements in 5, 7, $\cdots$ central diagonals occur naturally. Many examples of continuant determinant are given in Muir's *Contributions to the History of Determinants, 1900-1920* (Blackie), and the four previous volumes.

Basic properties of c.f.s are discussed in Chapter 1, including a number of new results for specialized forms. The point of view is the integral-square approach epitomized in (2).

## 3. Generalized Continued Fractions

In the early part of the present century, interest in methods of statistical estimation was heightened by the studies of the Karl Pearson school on moment methods, closely followed by concepts of more efficient maximum likelihood (m.l.) estimators introduced by R. A. Fisher. Efficient estimators were measured on a variance scale, and m.l. estimators provided a standard. As an illustration take the estimation of the parameter $k$ in the density

$$f(x;k) = e^{-x}(1+kx)/(1+k), \qquad (x>0;\ k>0)$$
$$= 0. \qquad (x>0)$$

From theory, the asymptotic variance of $k^*$, the m.l. estimator of $k$ is found from

$$n\ var(k^*) = 1/E\left(\frac{\partial \log(f)}{\partial k}\right)^2$$

for sample of size $n$. The denominator here reduces to

$$\alpha(k) = \int_0^\infty \frac{e^{-x}\,dx}{1+kx} - \frac{1}{1+k} \tag{9}$$

which involves a Stieltjes transform and a tie-up with c.f.s.

Gram-Charlier densities involve a basic density (normal, chi-square, etc.) modified by series (a perturbed normal density allows for changes in skewness and kurtosis by the introduction of third and fourth degree terms in the argument). It is natural then to consider extensions of examples such as (9); thus we might be interested in the simultaneous estimation of $k$ and $K$ from the density

$$f(x;k,K) = \frac{e^{-x}(1+kx+Kx^2/2)}{(1+k+K)}, \qquad (K>0,\ k^2<2K)$$

where the quadratic is nonnegative for $x>0$. Asymptotic efficiency would now basically involve integrals of the form

$$(k,K;A,B) = \int_0^\infty \frac{e^{-x}(A+Bx)dx}{1+kx+Kx^2/2}. \tag{10}$$

As mentioned earlier, these integrals appear to involve a simple exercise in quadrature via a computer (merely a desk calculator would suffice), although error analysis might not be trivial. But the 1950's only had early versions of computers, and computer expertise demanded more attention than anything else. So $(k,K;A,B)$ presented a nontrivial undertaking, as also similar examples involving third and fourth degree polynomials in the denominator of the integrand. Simple asymptotics in the parameters $k,K$ were looked for, the procedure being to generalize (10) and study

$$\{P,Q\} = \min_{\pi} \int_a^b P(x)\{1/Q(x) - \pi_r(x)\}^2 d\,\sigma(x)$$

where $P(\cdot)$, $Q(\cdot)$ are polynomials, $Q$ being nonnegative for $X \in (a,b)$, and the minimum taken with respect to polynomials of degree $\leqslant r$. In this way algebraic approximants could be set up for generalized Stieltjes integrals of the form

$$\int_a^b \frac{P_s(x)R_t(x)}{Q(x)} d\,\sigma(x). \tag{11}$$

Just as the usual c.f.s can be expressed as ratios of continuant determinants (determinants with elements only in the three central diagonals), so the generalized c.f.s appearing in approximants to (11) involve ratios of determinant with five, seven, $\cdots$ central diagonal elements only, depending on the degree of $Q(\cdot)$. Again, it is not surprising to find the orthogonal set which forms the basis of certain classes of c.f.s playing a fundamental role in generalized c.f.s. These in turn suggest the possibility of various recurrence relations between numerators (denominators) of the new c.f.s. Furthermore, a modified choice of the polynomial base leads to series developments analogous to the first order case

$$\int_a^b \frac{w(x)dx}{x+z} = \sum_{s=0}^{\infty} \frac{B_{2s+1}}{\omega_{2s}(z)\omega_{2s+2}(z)} \tag{12}$$

and its complementary odd part.

In response to a write-up of material on the recurrence situation, H. D. Ursell considered what would be the order of the recurrences for each function satisfying the set of equations

$$\begin{aligned} A_n &= d_n B_{n-1} - c_{n-1}D_n + a_{n-1}F_n\,, & D_n &= e_n C_{n-3}\,, \\ B_n &= d_n C_{n-2} - B_{n-1}D_n + a_{n-1}E_n\,, & E_n &= e_n B_{n-2}\,, \\ C_n &= c_n C_{n-1} - b_n B_n + a_n A_n\,, & F_n &= e_n D_{n-1}\,, \end{aligned}$$

the problem hinging on the variable coefficients $(a_n, b_n, \cdots)$; the structure of the set relates to the structure of 5-diagonal determinants appearing in "second order" c.f.s. Ursell showed that in this case each function satisfies a recurrence of order 6 (5 when the associated determinants are symmetric) and later (1958) generalized the case in a paper most readers find difficult to follow.

One fault with the generalized c.f.s developed here concerns "reversibility". Under certain conditions, a power series can be transformed into a c.f., or approximants set up in the form of a Padé table. The corresponding transformation for "second order" c.f.s, for example, has not been found; that one should exist is suggested by the fact that strongly divergent series (magnitude of coefficients increasing more rapidly than the "double" factorial series) apparently cannot be dealt with by the c.f. approach.

Second order c.f.s, those for which $Q(x)$ in (11) is a second degree polynomial, are discussed in Chapter 3. The general case along with recurrence relations for numerators and denominators is given in Chapter 4.

## 4. Divergent Series and Moments

There are many unsolved distributional problems in mathematical statistics. As an example, consider the distribution of the sample standard deviation (s.d.) in non-normal sampling, for a random sample $x_1, x_2, \cdots, x_n$, we define the s.d. to be

$$\sqrt{m_2} = \{\sum_{1}(x_j - \bar{x})^2/n\}^{1/2} \;;$$

the unbiased statistic $s^2 = nm_2/(n-1)$ has little advantage in the present context. The mean value is

$$E(\sqrt{m_2}) = \int_a^b \int_a^b \cdots \int_a^b \sqrt{m_2} \prod_{i=1}^{n} f(x_i)dx_i\,, \qquad (b > a)$$

when sampling from a population for which the density of a typical variate is $f(x)$. Now in general this multiple integral is intractable mathematically and out of reach of quadrature formulas for samples greater than ten or so. But we can expand $m_2$ in a series with argument $\epsilon = m_2 - E(m_2)$, $E$ referring to the expectation operator; here, however, difficulties arise in finding $E(\epsilon^s)$, $s = 1, 2, \cdots$. These difficulties are reduced if we use a bivariate expansion in the arguments $\epsilon_1 = m_1' - E(m_1')$, $\epsilon_2 = m_2' - E(m_2')$, where $m_s' = \sum x_j^s/n$. Expressions for

$$\nu_{r,s} = E(\epsilon_1^r \epsilon_2^s), \qquad (r, s = 0, 1, \cdots)$$

a typical product in the Taylor expansion, can be set up using interlaced recurrences (Shenton et al., 1971); the computer implementation of this process by Bowman was a turning point in the study and enabled significant advances to be made subsequently.

An early example concerns the moments of $\sqrt{m_2}$ in sampling from a gamma density

$$\begin{aligned} f(x) &= e^{-x} x^{\rho-1}/\Gamma(\rho), && (x > 0) \\ &= 0. && (x > 0;\ \rho > 0) \end{aligned}$$

Even moments of $\sqrt{m_2}$ present little difficulty for the low orders, but the odd moments become quite tiresome, complicated, and out of reach by hand calculations quite quickly. Craig (1929) found three moments, four terms for the mean, three terms for the variance, and two terms for the third central moment. He also found two or three terms in the first three moments of the skewness statistics $\sqrt{b_1}$ = (third central moment)/ $m_2^{3/2}$. If we can peer into the state of theoretical statistics or applied mathematics fifty years ago, and if a scientist at that time describes an algebraic undertaking as almost completely exhausting involving arduous work over many weeks, then in modern times the situation would be regarded either with awe or indifference. The computer age has diluted the appetite and curiosity for discovery by prolonged mathematical developments. However algebraic computer languages show much promise.

Our own computer extended version of the series (agreeing exactly with the numerical assessments of E. S. Pearson (1929) undertaken after seeing Craig's paper) has taken the first four moments of s.d. out to terms in $n^{-20}$ for values of the shape parameter $\rho = 20, 8, 4$, and 8/3, the values corresponding to Pearson's tabulations. We have also developed the four moment series for the skewness.

Studies of the first 4 - 6 moments of statistics such as the standard deviation, coefficient of variation, skewness, and kurtosis, as well as moments of parameter estimators have been carried out for a variety of non-normal populations whose moments exist -- altogether, several hundred cases.

The series have the appearance of divergence, with a variety of sign and magnitude patterns, generally the lower the moments involved in the statistic considered (mean and variance for example in preference to third and higher moments) the better the chance of some regularity in the coefficients. Of course, the population sampled plays an important role, there being a marked contrast between populations with a finite range (uniform, Pearson Type 1) and those with semi or doubly infinite ranges (chi-square, normal). Chapter 2 discuss some of the problems encountered, and emphasizes the summation problems for series.

## 5. Other Summation Techniques

Since consistency of approximation plays a dominant role in series sums, alternate algorithms are needed. We discuss (Chapter 5) a recent approach due to Levin (1973); this is a modification of a transformation introduced by Shanks (1955). Shanks expresses the partial sum of a series (or a sequence member) as a linear sum of transients; when the transients sweep off to infinity, the constant term in the approximation (Shanks calls this the antilimit) may account for the true asymptote of the sum. On the other hand, when all transients decay with time, the constant provides a limiting value. Levin modifies the transients to be basically related to the last term in the partial sum. It is possible that for divergent series Levin's algorithm diverges.

We also briefly describe a special summation technique relating to the Padé algorithm. In essence the coefficients in a series

$$S(n) \sim e_0 + e_1/n + e_2/n^2 + \cdots$$

are used tp determine the unknown parameters in the approximating function

$$F_{1,r}(n) = n\,\hat{\pi}_{r-1}(n) + \pi_r(n)\int_a^b \frac{d\,\sigma(t)}{1+t/n} \tag{13}$$

where $\hat{\pi}_{r-1}(\cdot)$, $\pi_r(\cdot)$ are polynomials of degree $r-1$ and $r$ and $\sigma(t)$ is known distribution function. The polynomial coefficients are determined by equating coefficients of powers of $n^{-1}$ in (13) with the corresponding ones in the series $S(n)$. An obvious extension is to use

$$F_{2,r}(n) = n\,\hat{\pi}_{r-1}(n) + \pi_r(n)\int_a^b \frac{d\,\sigma(t)}{1+t^2/n} + \pi_r'(n)\int_a^b \frac{t\,\sigma(t)}{1+t^2/n}$$

this being suitable for more rapidly divergent series. A number of detailed examples are given along with comparisons from other approaches.

## 6. Error Analysis

This aspect of the summation problem is not by any means completely resolved, and our remarks to some extent provisional. Series algorithmic sequences, to be useful, need to exhibit some stability; i.e., successive approximants should show a narrowing feature or some kind of monotonicity. The number of terms in the algorithm available is critical and very little can be done with, say, 2 to 5 terms in general. So there have to be comparative approximants: A unique feature of the series we have considered is the fact that in many cases exact (or nearly exact) sum values of the corresponding functions can be found for small values of $n$ ($n = 2, 3, 4, 5$). Difficulties may arise when the defining equations are implicit.

Again there is the question of how errors in the coefficients in the series affect sum sequence values. When very large numbers are involved, one recourse is to increase the precision, but this may prove uneconomical and demand considerable computer facilities.

## 7. History

Our main source of information on the divergency situations encountered has been the work of Stieltjes (Oeuvres, 1918), particularly volume 2. Supplementary information appears in the correspondence between Hermite and Stieltjes (1905). Most of these studies were based on a mathematical background, and there are few examples taken from applications such as occur in astronomy.

For present-day scientists, he is chiefly associated with the Stieltjes integral. But his studies in asymptotics in general, and in particular, his remarkable tie-up between the Stieltjes moment problem, rational fraction approximants, and Stieltjes integral transforms surely creates another landmark in the subject. The rational fraction approximants were expressed in terms of Stieltjes c.f.s, a remarkable notational achievement.

Stieltjes, for the most part, studied series generated by integrals or defined as specific functions. For example, he gave c.f.s for

$$\text{(a)}\quad \int_0^\infty e^{-xu}\,du\,/\,\{\cosh(u)+\sinh(u)\}^m\,,$$

$$\text{(b)}\quad \int_0^\infty \sinh(au)\sinh(bu)e^{-xu}\,du\,/\,\sinh(cu)\,,$$

$$\text{(c)}\quad \frac{\Gamma(x-a/2+1/4)\Gamma(x+a/2+1/4)}{\Gamma(x-a/2+3/4)\Gamma(x+a/2+3/4)}$$

$$\text{(d)}\quad \int_0^\infty e^{-xu}\,sn\,(u)du\,,$$

$$\text{(e)}\quad \int_0^\infty e^{-xu}\,sn^n\,(u)du\,/\,\int_{e^{-xu}} sn^{n-1}(u)du\,.$$

Stieltjes worked in a period when much of the controversy surrounding paradoxical sums had subsided. It was only 200 or so years earlier that conditions for the convergence of an alternating series were settled, along with a clear insight into the validity of a series. For example, the series $1+x+x^2+\cdots$ could yield $1-1+1-1+\cdots=½$ or $(1-1)+(1-1)+\cdots=0$, and several other values. At this time, Euler amused himself for a period with interpretations of $1-1!+2!-3!+\cdots$.

Emile Borel's lectures on divergent series were given at l'Ecole Normale in 1899-1900. The second edition (a considerable extension of the first) of this book appeared in 1928. The work sets out a fascinating history of the subject, including function theory. For our purpose, the chapter on the bridge supplied by the Stieltjes fractions is of considerable importance. An English translation (Charles L. Critchfield and Anna Vaker) appeared in 1975, indicating the vitality of the original lectures given 75 years earlier. Here is Borel's summary of the summation situation: ··· *The basic problem is the following: to make correspond to each numerical divergent series of as large a class as possible, a number such that substitution of that number for the series, in the usual calculations where it can appear, gives exact results or at least almost exact. This problem raises others and especially the following:*

*(1) to determine precisely the class of series one will consider;*

*(2) to enumerate the operations that are permitted upon these series while using their conventional sum:*

*The question being posed in this form, one readily conceives that it admits different solutions, or if one prefers, that their exist different processes of summation, that it to say, diverse ways of defining the conventional sum*".

The basic statement becomes simpler to digest for us because all our series (and there is an extraordinary rich field available, not to mention future potential) have a well-defined origin -- they relate in general to uniquely defined multiple integrals involving the parameter $n$ (sample size) about which asymptotes are to be constructed.

Treatments of continued fractions must include the two volumes of Perron (1957) and the function theory account of Wall (1948). The resurgence of interest in the last 20 or so years due mainly to workers in mathematical physics is described in Baker (1975) and Baker and Gammel (1970). One should also be aware of the detailed account of the subject in the bibliographies by Brezinski (1977, 1978, 1980a, 1981, 1985). Some of the present-day activity can be found in the symposiums edited by Graves-Morris (1973) and Saff and Varga (1977), and a recent treatise by Baker and Graves-Morris (1981).

# 1.

# BASIC CONCEPTS

## 1.1 Mean Square Approach

The intimate relation between continued fractions (c.f.s) and orthogonal polynomials is well known, and was brought out by Stieltjes, who made outstanding contributions to the subject. Consider for example the series

$$E(n) \sim 1/n - 1!/n^2 + 2!/n^3 - \cdots \tag{1.1}$$

which we may regard as a representation of

$$F(n) = \int_0^\infty \frac{e^{-t}\,dt}{t+n}. \qquad (n>0) \tag{1.2}$$

The series is divergent but rational fraction approximants to $F(n)$ may be derived from it by seeking the minimum of

$$S(n) = \int_0^\infty e^{-t}(t+n)\left\{\frac{1}{t+n} - \alpha_0\pi_0(t) - \alpha_1\pi_1(t) - \cdots - \alpha_s\pi_s(t)\right\}^2 dt \tag{1.3}$$

over the set $\{\alpha\}$; $\pi_0(t)$, $\pi_1(t)$, $\cdots$, being arbitrary real polynomials in $t$ of degrees 0, 1, $\cdots$. There are three interesting choices of these polynomials.

(a) $\pi_s(t) = t^s$

For simplicity, take $s = 2$ in (1.3). Then for the minimum set $\{\alpha^*\}$, we have

$$\alpha_0^*(n+1) + \alpha_1^*(n+2) + \alpha_2^*(2n+6) = 1$$

$$\alpha_0^*(n+2) + \alpha_1^*(2n+6) + \alpha_2^*(6n+24) = 1$$

$$\alpha_0^*(2n+6) + \alpha_1^*(6n+24) + \alpha_2^*(24n+6120) = 2$$

along with the equation for the minimum $S^*$ itself

$$\alpha_0^* + \alpha_1^* + 2\alpha_2^* = F(n) - S^*(n). \tag{1.4}$$

For the consistency of these equations, we must have

$$\begin{vmatrix} 1 & 1 & 2 & F(n)-S^*(n) \\ n+1 & n+2 & 2n+6 & 1 \\ n+2 & 2n+6 & 6n+24 & 1 \\ 2n+6 & 2n+24 & 24n+120 & 2 \end{vmatrix} = 0, \tag{1.5}$$

leading to the rational fraction approximant

$$P[2|3] = \frac{n^2 + 8n + 11}{n^3 + 9n^2 + 18n + 6}, \tag{1.6}$$

whose denominator clearly has no zeros on the positive real axis. Note that for general $s$ the determinant in (1.5) would be of order $s+2$ and awkward to evaluate, but leading to approximants $P[s|s+1]$ consisting of the ratio of polynomials of degrees $s$ and $s+1$.

(b) $\pi_s(t)$ and orthogonal polynomials

Let $\pi_s(t) = P_s(t)$ where $\{P_s(t)\}$ is the orthogonal set with respect to the weight function $W(t) = e^{-t}$ (this set is uniquely determined if we define them so that either the highest coefficient in $P_s(t)$ (that of $t^s$) is unity or if the integral square is taken to be unity; we shall in general prefer the first choice). The polynomials are those of Laguerre and for example

$$P_0(t) = 1, \quad P_1(t) = t - 1, \quad P_2(t) = t^2 - 4t + 2,$$

where

$$\int_0^\infty e^{-t} P_s(t) P_r(t) dt = 0, \quad r \neq s,$$

and if

$$\int_0^\infty e^{-t} P_s^2(t) = \phi_s, \quad \text{then} \quad \phi_0 = 1, \ \phi_1 = 1, \ \phi_2 = 4.$$

The minimal equations are now

$$\begin{aligned} \alpha_0^*(n+1) + \alpha_1^* \qquad\qquad\qquad &= 1, \\ \alpha_0^* \qquad + \alpha_1^*(n+3) + 4\alpha_2^* \qquad &= 0, \\ \alpha_0^* \qquad + 4\alpha_1^* \qquad + \alpha_2^*(4n+20) &= 0, \end{aligned} \tag{1.7}$$

which along with (1.4) lead to

$$P[2|3] = \begin{vmatrix} n+3 & 4 \\ 4 & 4n+20 \end{vmatrix} \div \begin{vmatrix} n+1 & 1 & 0 \\ 1 & n+3 & 4 \\ 0 & 4 & 4n+20 \end{vmatrix} \tag{1.8}$$

agreeing with (1.6). Actually carrying out this process for higher values of $s$, shows the approximants as the ratio of continuant determinants, each of which has zero elements except in the main diagonal, the sub-diagonal, and super-diagonal. Expanding a continuant by its last row brings out the property that it satisfies a linear recurrence relation of order 2 (three terms), immediately showing the relation to orthogonal systems. For the general approximant $P[s|s+1]$, note that both numerator and denominator satisfy exactly the same recurrence; they only differ in the initial values. If at this stage it is somewhat mystifying as to the origin of the three-term relation, we recall the simple Chebyshev system $\{\cos(s\theta)\}$ with $\cos(\theta) = t$ (orthogonal on the interval $[-1,1]$ with weight function $W(t) = 1/\sqrt{(1-t^2)}$), so that $\cos\{(s-1)\theta\} + \cos\{(s+1)\theta\} = 2\cos(\theta)\cos(s\theta)$, or $P_{s+1} = 2tP_s - P_{s-1}$.

### (c) $\pi_s(t)$ and the polynomials orthogonal to $W(t) = (t+n)e^{-t}$

Let $\{q_r(t)\}$ be orthogonal to $W(t)$, so that

$$\begin{aligned} q_0(t) &= 1, \\ q_1(t) &= t - (n+2)/(n-1), \\ q_2(t) &= t^2 - (4n^2+20n+12)t/\Delta + (2n^2+12n+12)/\Delta, \\ &\qquad (\Delta = n^2+4n+2) \end{aligned}$$

with

$$\begin{aligned} \int_0^\infty e^{-t}(t+n)q_s^2(t)\,dt &= n+1, && (s=0) \\ &= (n^2+4n+2)/(n+1), && (s=1) \\ &= (4n^3+36n^2+72n+24)/\Delta. && (s=2) \end{aligned}$$

For the minimum of (1.3) we now have

$$\begin{aligned} \alpha_0^*(n+1) &= 1, \\ \alpha_1^*(n^2+4n+2)/(n+1) &= -1/(n+1), \\ \alpha_2^*(4n^3+36n^2+72n+24)/\Delta &= 4/\Delta, \end{aligned}$$

leading to

$$P[2|3] = \frac{1}{(n+1)} + \frac{1}{(n+1)(n^2+4n+2)} + \frac{16}{(n^2+4n+2)(4n^3+36n^2+72n+24)} \tag{1.9}$$

which after simplification (addition from the left) leads again to (1.6). This form in the general case corresponds to the series expansion of a c.f. in terms of products of successive denominator convergents; we shall see that there is a corresponding expression for generalized c.f.s introduced in Chapter 3.

The polynomial base in (b) is the one most closely allied to the traditional development of c.f.s. For its implementation we need closed expressions for the orthogonal set $\{P_s(t)\}$, and in many cases these turn out to be the classical orthogonal polynomials. In situations where closed expressions are not known we may construct the polynomials by using the Gram-Schmidt process.

## 1.2 The Polynomial Base $\{P_s(t)\}$

Consider a non-negative weight function $W(x)$ for which $\int_a^b W(x)dx > 0$, where $b > a$, with the associated set of orthogonal polynomials $\{P_s(t)\}$, where

$$\begin{aligned} \int_a^b P_s(t)P_r(t)W(t)dt &= 0, \qquad s \neq r, \\ &= \phi_r, \qquad s = r. \qquad (r, s = 0, 1, \cdots) \end{aligned} \tag{1.10}$$

Let the recurrence satisfied by $\{P_s(t)\}$ be

$$P_s(t) = (t - a_s)P_{s-1}(t) - b_{s-1}P_{s-2}(t) \qquad (s = 1, 2, \cdots) \tag{1.11}$$

with $P_0 = 1$, $P_s(t) = 0$ if $s < 0$. For example,

$$P_1(t) = t - a_1, \quad P_2(t) = t^2 - t(a_1 + a_2) + a_1a_2 - b_1 .$$

From (1.11) note that by multiplying throughout by $P_{s-2}$ and integrating, using (1.10), we have

$$\begin{aligned} \phi_{s-1} &= b_{s-1}\phi_{s-2} \\ &= b_{s-1}b_{s-2} \cdots b_1b_0, \qquad (b_0 = \phi_0) \end{aligned} \tag{1.12}$$

so that in general $\{b_s\}$ is a positive sequence.

Now consider the minimum value, for $n > 0$, of

$$S_s(n) = \int_a^b (t+n)W(t)\{1/(t+n) - \sum_{r=0}^{s} \alpha_r P_r(t)\}^2 dt \tag{1.13}$$

over the set $\{\alpha_r\}$. Note that (1.11) may be written

$$(t+n)P_{r-1}(t) = P_r(t) + (n+a_r)P_{r-1}(t) + b_{r-1}P_{r-2}(t)$$

so that if

$$d_{r,s} = \int_a^b (t+n)P_r(t)P_s(t)W(t)dt$$

then $d_{r,s} = 0$ unless $s = r+1, r$, or $r-1$. Hence for the minimum

$$\begin{bmatrix} n+a_1 & b_1 & & & \\ 1 & n+a_2 & b_2 & & \\ & & \cdot & & \\ & & & \cdot & \\ & & & \cdot & b_s \\ & & & 1 & n+a_{s+1} \end{bmatrix} \begin{bmatrix} \alpha_0^* \\ \alpha_1^* \\ \cdot \\ \cdot \\ \cdot \\ \alpha_s^* \end{bmatrix} = \begin{bmatrix} 1 \\ 0 \\ \cdot \\ \cdot \\ \cdot \\ 0 \end{bmatrix} \tag{1.14}$$

where the minimum itself is given by

$$S_s^*(n) = F(n) - \alpha_0^* \phi_0 . \qquad (F(n) = \int_a^b \frac{W(t)}{t+n} dt)$$

But $\alpha_0^*$ is the leading element in the matrix inverse $A_{s+1}^{-1}(n)$, where (1.14) is equivalent to

$$A_{s+1}(n)\alpha_{s+1}^* = I$$

in an obvious notation. Hence an approximant to $F(n)/\phi_0$ is $P[s|s+1] = \chi_{s+1}(n)/\omega_{s+1}(n)$ where $\omega_{s+1}(n)$ refers to the continuant determinant $|A_{s+1}(n)|$, and $\chi_{s+1}(n)$ to the same determinant with its first row and column deleted. Each of these continuant determinants satisfies the same recurrence

$$\omega_{s+1}(n) = (n + a_{s+1})\omega_s(n) - b_s\omega_{s-1}(n) \tag{1.15}$$

with $\omega_0 = 1$, $\omega_1 = n + a_1$, $\chi_0 = 0$, $\chi_1 = 1$; the relation to the orthogonal system defined in (1.11) is obvious, and in fact

$$\omega_s(n) = (-1)^s P_s(-n).$$

It is now a short but important step to the usual formulation, i.e.,

$$\int_a^b \frac{W(t)dt}{t+n} = \frac{b_0}{n+a_1-}\ \frac{b_1}{n+a_2-}\ \frac{b_2}{n+a_3-}\ \cdots \tag{1.16}$$

in which the fractions may have an unlimited degree (we set aside for the moment questions of convergence) and where the notation means

$$\cfrac{b_0}{n+a_1-\cfrac{b_1}{n+a_2-\cfrac{b_2}{n+a_3-\cfrac{b_3}{n+a_4-\cdots}}}}.$$

However, the evaluation of the approximants is carried out much more efficiently using the recurrence (1.15) on the numerator and denominator components.

Note that if we apply (1.15) to the determinant

$$\Delta_s(n) = \omega_{s-1}(n)\chi_s(n) - \omega_s(n)\chi_{s-1}(n)$$

then for $s > 1$,

$$\Delta_s = b_{s-1}\Delta_{s-1} = b_{s-1}b_{s-2}\cdots b_1 b_0. \tag{1.17}$$

which is frequently called the *determinantal relation.* Hence

$$\frac{\chi_s(n)}{\omega_s(n)} = \frac{\chi_0}{\omega_0} + \left\{\frac{\chi_1}{\omega_1} - \frac{\chi_0}{\omega_0}\right\} + \cdots + \left\{\frac{\chi_s}{\omega_s} - \frac{\chi_{s-1}}{\omega_{s-1}}\right\} \tag{1.18a}$$

$$= \frac{b_0}{\omega_0\omega_1} + \frac{b_0 b_1}{\omega_1\omega_2} + \cdots + \frac{b_0 b_1 b_2 \cdots b_{s-1}}{\omega_{s-1}\omega_s}.$$

If there is convergence in some domain of $n$, then we may assert that

$$\int_a^b \frac{W(t)dt}{t+n} = \sum_{s=1}^{\infty} \frac{b_0 b_1 \cdots b_{s-1}}{\omega_{s-1}\omega_s}. \tag{1.18b}$$

Returning to (1.18a) we have already noted (see (1.12)) that $b_1$, $b_2$, $\cdots$, are positive; moreover, the zeros of $P_s(t)$ are known to be in the interval of orthogonality (see for example Szegö, 1939, p. 43) so that for the case $b > a \geqslant 0$, $\omega_s(n) = (-1)^s P_s(-n)$ is one-signed. Since $\omega_s(n) = n^s + \cdots$, it is clear that $\omega_s(n) > 0$ for $s = 0, 1, \cdots$, and $n > 0$. Hence the series in (1.18b) will be monotonic increasing in this case, and successive approximants from the c.f. (1.16) will have the same property. It is also evident that for large $n$, using (1.18a) in (1.18b), the remainder

$$F(n) - \chi_s(n)/\omega_s(n) = r_s(n), \tag{1.19}$$

is of order $(1/n^{2s+1})$ indicating that the rational fraction approximant to $F(n)$ may be determined by expanding in powers of $n^{-1}$ and equating the coefficients to the corresponding ones in $F(n)$.

Note that the series expansion of $F(n)$ in (1.18b) corresponds to the minimization problem stated in (1.3) using the polynomials orthogonal to $(t+n)W(t)$ as base.

**Example 1.1** Establish the c.f. for

$$F(n) = \int_0^\infty \frac{e^{-t}\,dt}{t+n}$$

using the fact that the elements of the orthogonal set $\{P_s(t)\}$ satisfy

$$P_s(t) = (t+2s-1)P_{s-1}(t) - (s-1)^2 P_{s-2}(t),$$

with $P_0 = 1$, $P_1 = t-1$. Show that

$$F(n) = \frac{1}{n+1-}\;\frac{1^2}{n+3-}\;\frac{2^2}{n+5-}\;\cdots. \qquad (n>0)$$

In particular, approximants to $F(1)$ are:

| $s$ | $\chi_s$ | $\omega_s$ | $\chi_s/\omega_s$ |
|---|---|---|---|
| 0 | 0 | 1 | 0 |
| 1 | 1 | 2 | 0.5 |
| 2 | 4 | 7 | 0.5714 |
| 3 | 20 | 34 | 0.5882 |
| 4 | 124 | 209 | 0.5933 |
| 5 | 920 | 1546 | 0.5951 |
| 6 | 7940 | 13327 | 0.59578 |
| 7 | 78040 | 130922 | 0.59608 |
| 8 | 859580 | 1441729 | 0.59621 |
| 9 | 10477880 | 17572114 | 0.59628 |
| 10 | 139931620 | 234662231 | 0.59631 |

In terms of the exponential integral

$$E_1(x) = \int_x^\infty \frac{e^{-t}\,dt}{t},$$

we have

$$F(n) = e^n E_1(n),$$

and from the Handbook of Mathematical Functions (1964, p. 240) $F(1) = 0.59635$. As a check, we have in the notation of (1.17) for the determinantal expression $\Delta_s$, $\Delta_{10} = 13168189440 = (9!)^2$ as it should. Also, the series expansion (1.18b) is

$$F(1) = \frac{1}{2} + \frac{1^2}{2.7} + \frac{1^2 \cdot 2^2}{7.34} + \frac{1^2 \cdot 2^2 \cdot 3^2}{34.209} + \frac{1^2 \cdot 2^2 \cdot 3^2 \cdot 4^2}{209.1546} + \cdots$$

## 1.3 Types of Continued Fractions

(a) $J$-fractions (J.f.s)

$$F(n) = \frac{b_0}{n + a_1 -}\ \frac{b_1}{n + a_2 -}\ \cdots \tag{1.20a}$$

If $a_1 = a_2 = \cdots = 0$, we have (Wall, 1948, p. 196 and p. 200)

(b) $S$-fractions (S.f.s)

$$F(n) = \frac{b_0}{n -}\ \frac{b_1}{n -}\ \cdots \tag{1.20b}$$

There is also a variant of these, namely

(c) Stieltjes'-fractions

$$F(n) = \frac{c_1}{n +}\ \frac{c_2}{1 +}\ \frac{c_3}{n +}\ \frac{c_4}{1 +}\ \cdots . \tag{1.21}$$

It is possible to express the Stieltjes form (1.21) as a J-fraction. Consider the recursions for the convergent $\chi_s / \omega_s$,

$$\begin{aligned} \omega_{2s} &= \omega_{2s-1} + c_{2s}\,\omega_{2s-2} \\ \omega_{2s+1} &= n\,\omega_{2s} + c_{2s+1}\omega_{2s-1}, \end{aligned}$$

so that

$$\begin{aligned} \omega_{2s} &= (n + c_{2s-1} + c_{2s})\omega_{2s-2} - c_{2s-1}c_{2s-2}\omega_{2s-4}, \qquad (s = 2, 3, \cdots) \\ \omega_2 &= (n + c_2)\omega_0 . \end{aligned}$$

Thus the *even-part* of (1.21) is

$$F(n) = \frac{c_1}{n+c_2-} \; \frac{c_2c_3}{n+c_3+c_4-} \; \frac{c_4c_5}{n+c_5+c_6-} \; \cdots \tag{1.22}$$

it being quite evident from (1.21) that $c_1$ cannot occur in the first partial denominator $n + c_2$.

There is also an *odd-part*, namely

$$F(n) = \frac{c_1}{n} - \frac{1}{n}\left[\frac{c_1c_2}{n+c_2+c_3-} \; \frac{c_3c_4}{n+c_4+c_5-} \; \frac{c_5c_6}{n+c_6+c_7-} \; \cdots\right] \tag{1.23}$$

derived in a similar manner, which we shall return to in the sequel. For many Stieltjes fractions, the partial numerators $c_1, c_2, \cdots$, are positive. This implies that the even and odd parts are monotonic increasing and decreasing sequences, respectively. (It is evident from (1.22) that $F(n) > c_1/(n+c_2)$ suggesting an increasing sequence; similarly $F(n) < c_1/n - c_1c_2/(n+c_2+c_3)$ from (1.23).)

Every Stieltjes fraction such as (1.21) has an even and odd part. However, it is not always possible to express a J.f. as an Stieltjes fraction such as (1.21) with positive partial numerators. Conditions for a solution when the $b$'s and $a$'s in (1.20a) are positive are given in Wall (1948, pp. 121-122).

A further operation on c.f.s needed from time to time consists of an *equivalence transformation*. Consider

$$\frac{c_1}{n+} \; \frac{kc_2}{k+} \; \frac{kc_3}{n+} \; \frac{c_4}{1+} \; \cdots$$

with convergents

| $s$ | $\chi_s$ | $\omega_s$ |
|---|---|---|
| 1 | $c_1$ | $n$ |
| 2 | $kc_1$ | $k(n+c_2)$ |
| 3 | $kc_1(n+c_3)$ | $k(n^2+nc_2+nc_3)$ |
| 4 | $kc_1(n+c_3+c_4)$ | $k\{n^2+n(c_2+c_3+c_4)+c_4c_2\}$ |

Evidently the introduction of $k\,(\neq 0)$ leaves the convergents unchanged. The operation can be repeated at any stage in the c.f. with any $k\,(\neq 0)$. It is useful since we may wish to express a Stieltjes fraction with all its partial numerators unity. Thus (1.21) becomes

$$F(n) = \frac{1}{k_1n+} \; \frac{1}{k_2+} \; \frac{1}{k_3n+} \; \frac{1}{k_4+} \; \cdots$$

where

$$k_1 = 1/c_1, \qquad k_2 = c_1/c_2,$$
$$k_3 = c_2/(c_1c_3), \qquad k_4 = c_1c_3/(c_2c_4),$$

and so on. We mention that Stieltjes showed that the Stieltjes moment problem (see Section **1.7**) has a unique solution if $k_1, k_2, \cdots$, are positive and $\sum k_s = \infty$.

There are other structures of c.f.s, some relating to solutions of differential equations (Perron, 1910).

## 1.4 The Odd-Part of a Stieltjes c.f. and Least Squares

It is well known that Stieltjes fractions are related to Stieltjes integral transforms and that under certain conditions (discussed in Chapter 3)

$$F(n) = \int_0^\infty \frac{d\,\psi(t)}{t+n} = \frac{c_1}{n+}\;\frac{c_2}{1+}\;\frac{c_3}{n+}\;\cdots \tag{1.24}$$

where $\psi(t)$ is a non-decreasing function with infinitely many points of increase. For example, (1.2) may be expressed as (1.24) by taking

$$\psi(t) = \int_0^t e^{-x}\,dx\,.$$

The form (1.24) is more general than the weight function version, and was introduced by Stieltjes to include step-functions. For example, a purely step function case is

$$F(n) = \sum_{x=0}^{\infty} \frac{m^x}{x!(x+n)}\,. \qquad (m>0)$$

In general we shall use the form (1.24).

Now consider

$$F(n) = \frac{c_1}{n} - \frac{1}{n}\int_0^\infty \frac{t\,d\,\psi(t)}{t+n} \tag{1.25}$$

and the least squares approximants to the second term. A formula due to Christoffel (Szegö, 1939, p. 28) asserts that the polynomials $\{q_r(t)\}$, orthogonal to $t\,d\,\psi(t)$, in terms of those $\{P_r(t)\}$ orthogonal to $d\,\psi(t)$, are given by

$$q_r(t) = \begin{bmatrix} P_r(t) & P_{r+1}(t) \\ P_r(0) & P_{r+1}(0) \end{bmatrix} \div \{-tP_r(0)\}, \tag{1.26}$$

$P_r(t)$ having highest coefficient unity. Since all orthogonal systems satisfy a second order recurrence, we may assume

$$q_r(t) = (t - A_r)q_{r-1}(t) - B_{r-1}q_{r-2}(t) \tag{1.27}$$

so that from (1.11),

$$\begin{cases} A_r = a_{r+1} + \dfrac{P_{r-1}(0)P_{r+1}(0) - P_r^2(0)}{P_{r-1}(0)P_r(0)}, & (r = 1, 2, \cdots) \\ B_{r-1} = b_{r-1}\dfrac{P_{r-1}(0)P_r(0)}{P_{r-1}^2(0)}. & (r = 2, 3, \cdots) \end{cases} \tag{1.28}$$

We now have the c.f. development

$$F(n) = \frac{c_1}{n} - \frac{c_1}{n}\left\{\frac{a_1}{n+A_1-}\ \frac{B_1}{n+A_2-}\ \frac{B_2}{n+A_3-}\ \cdots\right\}. \tag{1.29}$$

The first term in the c.f. arises from

$$\int_0^\infty t\,d\psi(t) = \int_0^\infty (t - a_1 + a_1)d\psi(t) = a_1c_1$$

using $P_1(t) = t - a_1$. We show that (1.29) is the odd part of the Stieltjes fraction for $F(n)$ given in (1.23). Write (for this paragraph only) $P_r = P_r(0)$. Then

$$\begin{aligned} P_0 &= 1, \\ P_1 &= -a_1 = -c_2, \\ P_2 &= -a_2P_1 - b_1P_0 = -(c_3+c_4)(-c_2) - c_2c_3 = c_2c_4, \end{aligned}$$

using in the notation of (1.16) and (1.22),

$$\begin{aligned} a_r &= c_{2r} + c_{2r-1}, & (r = 2, 3, \cdots) \\ b_r &= c_{2r}c_{2r+1}, & (r = 1, 2, \cdots) \end{aligned}$$

with $a_1 = c_1$. Using the recurrence

$$P_r = -a_rP_{r-1} - b_{r-1}P_{r-2}$$

one can prove inductively that

$$P_r = P_r(0) = (-1)^s c_2 c_4 \cdots c_{2r} . \tag{1.30}$$

Substituting in (1.28), we find

$$A_r = c_{2r} + c_{2r+1},$$
$$B_r = c_{2r+1} c_{2r+2},$$

and so (1.29) agrees with the odd-part of $F(n)$ defined in (1.23).

As an illustration, consider

$$F(n) = \frac{1}{\Gamma(a)} \int_0^\infty \frac{e^{-t} t^{a-1} dt}{t+n}. \qquad (a, n > 0) \tag{1.31}$$

The orthogonal set $\{P_r(t)\}$ are Laguerre polynomials (Szegö, pp. 96-101) and follow the recurrence

$$P_r(t) = (t - a - 2r + 2)P_{r-1}(t) - (r-1)(a + r - 2)P_{r-2}(t)$$
$$(r = 2, 3, \cdots)$$

with $P_0 = 1$, $P_1 = t - a$. Hence from (1.16),

$$F(n) = \frac{1}{n+a-} \; \frac{1 \cdot a}{n+a+2-} \; \frac{2 \cdot (a+1)}{n+a+4-} \; \cdots \tag{1.32}$$

as is known (the c.f. converges for all $n$ excluding the cut on the $n$-plane from 0 to $-\infty$). The odd-part is

$$\frac{1}{n} - \frac{1}{n} \left\{ \frac{a}{n+a+1-} \; \frac{1 \cdot (a+1)}{n+a+3-} \; \frac{2 \cdot (a+2)}{n+a+5-} \; \cdots \right\} \tag{1.33}$$

from the formulation in (1.25). Otherwise assume the Stieltjes fraction form in (1.21) and determine the coefficients by the equivalence of the even part and expression (1.32). For general $s$, we require

$$n + c_{2s} + c_{2s-1} = n + a + 2s - 2$$
$$c_{2s-1} c_{2s-2} = (s-1)(a + s - 2).$$

An obvious possibility is $c_{2s-2} = a + s - 2$, $c_{2s-1} = s - 1$ so that $c_{2s} + c_{2s-1} = a + 2s - 2$ as it should; for $s = 1$ we have $c_2 = a$ since we recall (see (1.22)) that the denominators do not contain $c_1$. Hence there is the Stieltjes fraction,

$$F(n) = \frac{1}{n+} \; \frac{a}{1+} \; \frac{1}{n+} \; \frac{a+1}{1+} \; \frac{2}{n+} \; \frac{a+2}{1+} \; \frac{3}{n+} \; \cdots \tag{1.34}$$

with, for $n > 0$,

$$\begin{aligned} P[0|1] &= 1/n > F(n), \\ \bar{P}[0|1] &= 1/(n+a) < F(n), \\ P[1|2] &= (n+1)/\{n^2+n(a+1)\} < P[0|1], \\ \bar{P}[1|2] &= (n+a+2)/\{n^2+2(a+1)n+a(a+1)\} > \bar{P}[0|1] \end{aligned}$$

and so on. As a numerical illustration there are the following upper bounds for the case $n = 1$, $a = 1$ (see Example 1.1);

| $s$ | $\chi_{2s+1}$ | $\omega_{2s+1}$ | $\chi_{2s+1}/\omega_{2s+1}$ |
|---|---|---|---|
| 0 | 1 | 1 | 1.0000 |
| 1 | 2 | 3 | 0.6667 |
| 2 | 8 | 13 | 0.6154 |
| 3 | 44 | 73 | 0.6027 |
| 4 | 300 | 501 | 0.5988 |
| 5 | 2420 | 4051 | 0.5974 |
| 6 | 22460 | 37633 | 0.5968 |
| 7 | 235260 | 394353 | 0.59657 |
| 8 | 2741660 | 4596553 | 0.59646 |
| 9 | 35152820 | 58941091 | 0.59641 |

As a check, one may show that

$$\Delta_s = \chi_{2s+1}\omega_{2s+3} - \chi_{2s+3}\omega_{2s+1} = s!\Gamma(a+s+1)/\Gamma(a),$$

and in particular $\Delta_8 = 2741660 \times 58941091 - 35152820 \times 4596553 = 14631321600$. Altogether then $0.59631 < F(1) < 0.59641$, the lower bound arising from $\chi_{20}/\omega_{20}$.

**Example 1.2 The Gauss c.f.**

The hypergeometric function

$$F(a,1,c;\theta) = 1 + \frac{a}{c}\theta + \frac{a(a+1)}{c(c+1)}\theta^2 + \cdots$$

may be expressed as an integral transform (assuming $c \neq 0, \neq -1, \cdots$)

$$F(a,1,c;\theta) = \frac{\Gamma(c)}{2^{c-2}\Gamma(a)\Gamma(c-a)} \int_{-1}^{1} \frac{(1+t)^{a-1}(1-t)^{c-a-1}}{2-\theta-\theta t} dt .$$

$$(c-a>0,\ a>0,\ |\theta|<1) \qquad (1.35)$$

Moreover, the orthogonal polynomials $\{P_r(t)\}$ with respect to the weight function $W(t) = (1+t)^{a-1}(1-t)^{c-a-1}$ are those of Jacobi (Szegö, chapter 4) and

$$W(t)P_r(t) = \frac{(-1)^r}{2^r r!}\left(\frac{d}{dt}\right)^r (1+t)^{a+r-1}(1-t)^{c-a+r-1}; \tag{1.36}$$

in particular $P_0 = 1$, $P_1 = (ct + c - 2a)/2$.

To set up the c.f. in the form (1.16) using (1.11) we require the recurrence for the orthogonal set when the highest coefficient is unity. Set $t+1 = 2y$ in (1.35) and (1.36) so that

$$F(a,1,c;\theta) = \frac{\Gamma(c)}{\Gamma(a)\Gamma(c-a)} \int_0^1 \frac{y^{a-1}(1-y)^{c-a-1}}{1-\theta y}\, dy$$

and the orthogonal set becomes

$$P_r(y) = r!P_r(t)/\{(c+2r-2)(c+2r-3)\cdots(c+r-1)\}, \tag{1.37}$$

with recurrence

$$P_r(y) = (y - a_r)P_{r-1}(y) - b_{r-1}P_{r-2}(y),$$

and $P_0 = 1$, $P_1 = y - a/c$, where

$$a_r = \frac{2(r-1)(c-2a)(c+r-2)}{c(c+2r-2)(c+2r-4)},$$
$$b_r = \frac{r(r+a-1)(r+c-a-1)(c+r-2)}{(c+2r-1)(c+2r-1)^2(c+2r-3)}.$$

Hence (1.16) leads to the c.f. for the Stieltjes integral transform

$$n^{-1}F(a,1,c;-n^{-1}) = \frac{\Gamma(c)}{\Gamma(a)\Gamma(c-a)} \int_0^1 \frac{y^{a-1}(1-y)^{c-a-1}}{y+n}\, dy$$
$$= \frac{1}{n+a_1-}\ \frac{b_1}{n+a_2-}\ \frac{b_2}{n+a_3-}\ \cdots. \tag{1.38}$$
$$(a>0,\quad c-a>0,\quad n>0)$$

Using Section 1.4 we derive the odd-part (the Gauss c.f.)

$$\frac{1}{n+}\ \frac{B_1}{1+}\ \frac{B_2}{n+}\ \frac{B_3}{1+}\ \cdots \tag{1.39}$$

where

$$\begin{cases} B_{2r} = r(c-a+r-1)/\{(c+2r6-2)(c+2r-1)\}, \\ B_{2r+1} = (r+a)(c+r-1)/\{(c+2r-1)(c+2r)\}. \end{cases}$$

Under certain restrictions the odd-part arises from the least-squares approach to the integral representation of

$$1 + \frac{a\theta}{c} F(a+1,1,c+1;\theta).$$

Is there anything to be gained from the use of the identity

$$F(a,1,c;\theta) = (1-\theta)^{-1}F(1,c-a,c;-\theta/(1-\theta)), \quad (|\theta| < 1, R(\theta) < \tfrac{1}{2})$$

followed by conversion to a c.f.? We find

$$F(a,1,c;\theta) = \frac{1-\theta}{1+}\ \frac{d_1\phi}{1+}\ \frac{d_2\phi}{1+}\ \cdots \tag{1.40}$$

where $\phi = \theta/(1-\theta)$ and

$$d_{2r} = r(a+r-1)/\{(c+2r-2)(c+2r-1)\},$$
$$d_{2r+1} = (c-a+r)(c+r-1)/\{(c+2r-1)(c+2r)\}.$$

After some algebra one can show that the even-convergents are the same as those of (1.39); however, the odd-convergents are

$$\frac{1}{1-\theta} - \frac{\theta(1-a)}{c(1-\theta)} \times (\text{Even convergent of (1.39)}).$$

A comprehensive discussion of the Gauss c.f. is given in Wall, chapter 18. He deals with cases involving complex coefficients, and also the ratio of two hypergeometric functions (see Further Example 4.8).

**Example 1.3** Determine the J.f. and Stieltjes fraction for

$$F(z) = \sum_{t=0}^{n-1} \frac{1}{z+t}, \qquad (n = 2, 3, \cdots; z > 0)$$

which is a Psi function $(\psi(x) = d\ln\Gamma(x)/dx)$ $\psi(n+z) - \psi(z)$ (Stieltjes (1918) gave a number of examples involving $\psi(z)$ using a series development based on a hyperbolic integral representation). The orthogonal set $\{P_r(t)\}$, if we can find it, leads directly to the c.f.s, and since $F(z)$ is a Stieltjes integral transform, a Stieltjes fraction with positive elements exists. The polynomials are in a discrete variable (note, if we write

$$F(z) = \int_0^\infty \frac{d\,\sigma(t)}{t+z}, \quad \text{then} \quad \begin{array}{ll} \sigma(t) = [t], & 0 \leqslant t \leqslant n-1; \\ \sigma(t) = n, & t \geqslant n \end{array} \quad ).$$

and (Aitken, 1932)

$$\sum_{t=0}^{n-1} T_r(t)T_s(t) = 0, \quad r \neq s, \quad (r, s = 0, 1, \cdots, n-1)$$

$$= \frac{n(n^2-1^2)(n^2-2^2)\cdots(n^2-r^2)}{(2r+1)r!r!}. \qquad (r = s)$$

The recurrence relation may now be found, using

$$T_r(t) = (a_r t + b_r)T_{r-1}(t) - c_r T_{r-2}(t).$$

It turns out that

$$r^2 T_r(t) = (2r-1)(2t-n+1)T_{r-1}(t) - (n-r+1)(n-r-1)T_{r-2}(t)$$

with $T_0 = 1$, $T_1 = 2t - n + 1$. Transforming to obtain a coefficient of unity for the highest term,

$$P_r(t) = \left(t - \frac{n-1}{2}\right)P_{r-1}(t) - \frac{(r-1)^2(n-r+1)(n-r-1)}{4(2r-1)(2r-3)} P_{r-2}(t)$$

$$(r = 1, 2, \cdots, n-1)$$

so that from (1.11) and (1.19),

$$F(z) = \frac{n}{z + \frac{(n-1)}{2} -} \;\frac{\frac{1^2(n^2-1)}{4\cdot 3\cdot 1}}{z + \frac{(n-1)}{2} -} \;\frac{\frac{2^2(n^2-4)}{4\cdot 5\cdot 3}}{z + \frac{(n-1)}{2} -} \;\frac{\frac{3^2(n^2-9)}{4\cdot 7\cdot 5}}{z + \frac{(n-1)}{2} -} \cdots$$

which is a terminating c.f. Now using the recurrence for the Stieltjes fraction, we find

$$F(z) = \frac{n}{z+} \;\frac{\frac{1(n-1)}{2}}{1+} \;\frac{\frac{1(n+1)}{6}}{z+} \;\frac{\frac{2(n-2)}{6}}{1+} \;\frac{\frac{2(n+2)}{10}}{z+}$$

$$\frac{\frac{3(n-3)}{10}}{1+} \;\frac{\frac{3(n+3)}{14}}{z+} \cdots \qquad (z > 0, n = 2, 3, \cdots)$$

**Example 1.4** For the binomial weight function

$$b(n, t; p) = \binom{n}{t} p^t q^{n-t}, \qquad \left(\begin{array}{l} t = 0, 1, \cdots, n; \\ q = 1-p, \quad n = 1, 2, \cdots \end{array}\right)$$

show that

$$F(z) = \sum_{t=0}^{n} \frac{b(n,t;p)}{z+t} \qquad (z>0)$$

has the Stieltjes fraction development

$$F(z) = \frac{1}{z+}\ \frac{np}{1+}\ \frac{q}{z+}\ \frac{(n-1)p}{1+}\ \frac{2q}{z+}\ \frac{(n-2)p}{1+}\ \frac{3q}{z+}\ \cdots \frac{nq}{z}.$$

*Hint.* The orthogonal set is defined by

$$P_r(t) = (t-r+1-(n-2r+2)p)P_{r-1}(t) - (r-1)(n-r+2)pqP_{r-2}(t)$$

with $P_0 = 1$, $P_1 = t - np$. They have been given by Aitken and Gonin (1934) and Krawtchouk (1929).

A more complicated case concerns the hypergeometric weight function (Aitken and Gonin, 1934)

$$h(t;N,m,n) = \binom{n}{t} m^{(t)}(N-m)^{(n-t)}/N(n) \qquad (t = 0, 1, \cdots, n)$$

where $m$, $n$, $N$ are positive integers with $n<N$, $m<N$, and $x^{(l)} \equiv x(x-1)\cdots(x-l+1)$.

In the notation of (1.11) the orthogonal set $\{P_r(t)\}$ is given by

$$a_r = r - 1 + \frac{r(n-r+1)(m-r+1)}{N-2r+2} - \frac{(r-1)(n-r+2)(m-r+2)}{N-2r+4},$$

$$b_r = \frac{r(n-r+1)(m-r+1)(N-n-r+1)(N-m-r+1)(N-r+2)}{(N-2r+3)(N-2r+2)(N-2r+2)(N-2r+1)},$$

with $P_0 = 1$, $P_1 = t - nm/N$.

The Stieltjes fraction for $F(z) = \sum_{t=0}^{n} h(t;N,m,n)/(z+t)$ is, setting it up from the corresponding J.f.,

$$F(z) = \frac{c_1}{z+}\ \frac{c_2}{1+}\ \frac{c_3}{z+}\ \cdots \qquad (z>0)$$

where

$$c_{2s} = \frac{(n-s+1)(N-s+2)(m-s+1)}{(N-2s+2)(N-2s+3)},$$

$$c_{2s+1} = \frac{s(N-n-s+1)(N-m-s+1)}{(N-2s+1)(N-2s+2)}. \qquad (s = 1, 2, \cdots;\ c_1 = 1)$$

**Example 1.5**

$$F(\frac{1}{2},1,\frac{3}{2};-\frac{1}{3}) = \frac{3}{2}\int_0^1 \frac{1}{(t+3)}\frac{dt}{\sqrt{t}} = \frac{\pi}{2\sqrt{3}}.$$

Hence

$$\frac{\pi}{3\sqrt{3}} = \frac{2}{3+}\;\frac{\frac{1^2}{1\cdot 3}}{1+}\;\frac{\frac{2^2}{3\cdot 5}}{3+}\;\frac{\frac{3^2}{5\cdot 7}}{1+}\;\frac{\frac{4^2}{7\cdot 9}}{1+}\;\cdots,$$

| $s$ | $\chi_s$ | $\omega_s$ | Approximant to $\pi$ |
|---|---|---|---|
| 1 | 2 | 3 | 3.46 |
| 2 | 2 | 10/3 | 3.118 |
| 3 | 98/3·5 | 162/3·5 | 3.1434 |
| 4 | 740/3·5·7 | 1224/3·5·7 | 3.14146 |
| 5 | 21548/3·5·7·9 | 35650/3·5·7·9 | 3.141602 |
| 6 | 255528/3·5·7·9·11 | 422640/3·5·7·9·11 | 3.1415920 |
| 7 | 10741320/3·5·7·9·11·13 | 17766000/3·5·7·9·11·13 | 3.141592702 |
| 8 | 173640672/3·5·7·9·11·13·15 | 287199360/3·5·7·9·11·13·15 | 3.141592651 |

**Example 1.6**

$$F(n,m) = \sum_{t=0}^{\infty} \frac{\psi(t,m)}{t+n}, \tag{1.41}$$

where

$$\psi(t,m) = e^{-m}m^t/t!. \qquad (m,n>0)$$

Here the Stieltjes transform is a step-function with infinitely many points of increase and relates to the moment problem associated with a Poisson probability function $\psi(t,m)$; actually, there is little interest in this moment problem, for any doubts can be settled by simpler methods than setting up the Stieltjes c.f. Nonetheless, the c.f. is of some interest, and, for the present, let it be said that the moment problem in this case relates to finding a $\phi(t)$ such that

$$\int_0^\infty t(t-1)\cdots(t-s+1)d\,\phi(t) = m^s. \qquad (s=1,2,\cdots;\ m>0)$$

Further discussion is given later in the sequel. The orthogonal set $\{P_r(t)\}$ has been treated by Szegö (2.81, p. 33) and Aitken and Gonin (1934). In fact

$$P_r(t) = \sum_{s=0}^{r} (-1)^s \binom{r}{s} t^{(r-s)} m^s , \qquad (t^{(k)} = t(t-1)\cdots(t-k+1))$$

with $P_0 = 1$, $P_1 = t - m$, $P_2 = t(t-1) - 2tm + m^2$ for example, and recurrence

$$P_r(t) = (t - r + 1 - m)P_{r-1}(t) - (r-1)mP_{r-2}(t). \tag{1.42}$$

Using (1.11) and (1.16),

$$F(n,m) = \frac{1}{n+m-}\;\frac{m}{n+m+1-}\;\frac{2m}{n+m+2-}\;\frac{3m}{n+m+3-}\;\cdots,$$

with Stieltjes form

$$F(n,m) = \frac{1}{n+}\;\frac{m}{1+}\;\frac{1}{n+}\;\frac{m}{1+}\;\frac{2}{n+}\;\frac{m}{1+}\;\frac{3}{n+}\;\frac{m}{1+}\;\cdots. \tag{1.43}$$

A c.f. equivalent to this is due to Padé (see Perron, 1957, p. 280).

In particular when $n = 1$,

$$e^{-m} = 1 - \frac{m}{1+}\;\frac{m}{1+}\;\frac{1}{1+}\;\frac{m}{1+}\;\frac{2}{1+}\;\frac{m}{1+}\;\frac{3}{1+}\;\cdots,$$

or

$$e^{m} = \frac{1}{1-}\;\frac{m}{1+}\;\frac{m}{1+}\;\frac{1}{1+}\;\frac{m}{1+}\;\frac{2}{1+}\;\frac{m}{1+}\;\frac{3}{1+}\;\cdots. \tag{1.44}$$

If the $s$ th convergent of (1.44) is $\chi_s / \omega_s$, then one may prove

$$\chi_{2s+1} = s!(1 + m/1! + \cdots + m^s/s!),$$
$$\omega_{2s+1} = s!,$$

and using

$$\omega_{2s+1}\chi_{2s} - \omega_{2s}\chi_{2s+1} = m^{s+1}(s-1)!$$

deduce

$$\frac{\chi_{2s}}{\omega_{2s}} = 1 + \frac{m}{1!} + \cdots + \frac{m^s}{s!} + \frac{m^{s+1}}{(s-m)s!}.$$

$$(s = 1, 2, \cdots;\ m \neq s, m > 0)$$

## 1.5 Formula for a Numerator

Consider

$$P_s^*(n) = (-1)^{s-1} \int_a^b \frac{P_s(t) - P_s(-n)}{t+n} \, d\psi(t) \tag{1.45}$$

where $\{P_s(t)\}$ are orthogonal with respect to $d\psi(t)$. But

$$P_0^*(n) = 0, \quad P_1^*(n) = b_0, \quad P_2^*(n) = (n + a_2)b_0,$$

and $P_s^*(n)$ clearly follows the recurrence of the numerators of the c.f. in (1.16). Hence $P_s^*(n) = \chi_s(n)$. It follows that the remainder

$$r_s(n) = F(n) - \chi_s(n)/\omega_s(n) = \frac{1}{P_s(-n)} \int_a^b \frac{P_s(t)}{t+n} \, d\psi(t), \tag{1.46}$$

a formula implied in Szegö (p. 54). It may be used to assess the rate of convergence of the c.f. but requires a closed form for $P_s(t)$, $s$ large, for its implementation (see Allen, et al., 1975). As in convergence aspects of c.f.s (a unified theory is still lacking), assessments of remainders present many difficulties and may depend on special procedures for particular structures. Some cases are now considered.

## 1.6 Illustrations

A recurring problem associated with summation algorithms (c.f.s, Borel, Shanks, for example) is the rate of convergence when an infinity of terms is available, or the apparent rate of convergence for finite sums. Rather than attempt to deal with this aspect, once and for all, we shall refer to it from time to time when appropriate. The assimilation of myriads of convergence theorems can still leave a gap in one's appreciation of numerical assessment of quickness of approach to a limit (or optimum value). A crude assessment can be quite valuable whereas refined bounds may involve more sophistication than the original problem.

The simple c.f.

$$f = \frac{1}{a_1+} \; \frac{1}{a_2+} \; \frac{1}{a_3+} \; \frac{1}{a_4+} \; \cdots, \qquad (a_2, a_2, \cdots > 0)$$

for which

$$\chi_s/\omega_s - \chi_{s-1}/\omega_{s-1} = (-1)^{s-1}/(\omega_{s-1}\omega_s)$$

suggests fast convergence for large $a$'s and slow convergence for small $a$'s. This is consistent with the known fact (Khovanskii, 1963 p. 42) that the *divergence of the series* $\Sigma a_s$ *is necessary* and *sufficient* for *convergence* -- note that the relaxation of the positivity of the partial denominators creates a completely different and more complex convergence problem, as does the convergence of the series with positive elements. One may now turn to the c.f.s in (1.34) and (1.43) and reach some appraisal of the rate of convergence.

**Example 1.7**

$$f = \frac{1}{1^2+}\;\frac{1}{2^2+}\;\frac{1}{3^2+}\;\frac{1}{4^2+}\;\cdots$$

| $s$ | $\chi_s$ | $\omega_s$ | $\chi_s / \omega_s$ |
|---|---|---|---|
| 0 | 0 | 1 | 0.000000 |
| 1 | 1 | 1 | 1.000000 |
| 2 | 4 | 5 | 0.800000 |
| 3 | 37 | 46 | 0.804348 |
| 4 | 596 | 741 | 0.804318 |
| 5 | 14937 | 18571 | 0.804319 |

$$f = \frac{1}{1+}\;\frac{1}{1/2!+}\;\frac{1}{1/3!+}\;\frac{1}{1/4!+}\;\cdots$$

| $s$ | $\chi_s$ | $\omega_s$ | $\chi_s / \omega_s$ |
|---|---|---|---|
| 0 | 0 | 1 | 0.0000 |
| 1 | 1 | 1 | 1.0000 |
| 2 | 0.5 | 1.5 | 0.3333 |
| 3 | $1.08\dot{3}$ | 1.25 | 0.8667 |
| 4 | 0.545139 | 1.552083 | 0.3512 |
| 5 | 1.087876 | 1.262934 | 0.8614 |
| 6 | 0.546650 | 1.553837 | 0.3518 |

### 1.6.1 The Mills' Ratio

Let the density of a standard normal deviate be

$$g(t) = \{1/\sqrt{(2\pi)}\}\exp(-t^2/2). \qquad (-\infty < t < \infty)$$

Then the probability of a deviate $T$ exceeding $t\,(>0)$ is

$$Pr(T > t) = \int_t^\infty g(x)dx\,,$$

and the Mills' ratio

$$R(t) = \int_t^\infty g(x)dx \,/\, g(t) \qquad (1.47)$$

has been used for computational purposes and is well known in statistical literature. (Applied mathematicians are more accustomed to the error function notation which merely omits the factor ½ in the exponential.) We shall also refer to the complementary ratio

$$\bar{R}(t) = \int_0^t g(x)dx / g(t) \tag{1.48}$$

so that

$$R(t) + \bar{R}(t) = 1/\{2g(t)\}. \tag{1.49}$$

There are the series expansions

$$\begin{aligned} R(t) &\sim \frac{1}{t} - \frac{1}{t^3} + \frac{1\cdot 3}{t^5} - \cdots \\ \bar{R}(t) &\sim \frac{t}{1} + \frac{t^3}{1\cdot 3} + \frac{t^5}{1\cdot 3\cdot 5} + \cdots, \end{aligned} \tag{1.50}$$

the first asymptotic and useful for large $t$ (the error committed in using $s$ terms being less in magnitude than the first term omitted), the second convergent and only useful for small $t$. C.f.s are available for both. For the asymptotic series,

$$R(t) = t\int_{-\infty}^{\infty} \frac{g(x)dx}{x^2+t^2} = \frac{t}{2\sqrt{\pi}} \int_0^{\infty} \frac{e^{-x}x^{-1/2}dx}{x+t^2/2} \qquad (t>0)$$

so using (1.31) and (1.34)

$$\begin{aligned} R(t) &= \frac{t/2}{t^2/2+}\;\frac{1/2}{1+}\;\frac{1}{t^2/2+}\;\frac{3/2}{1+}\;\frac{2}{t^2/2+}\;\frac{5/2}{1+}\;\frac{3}{t^2/2+}\;\cdots \\ &= \frac{1}{t+}\;\frac{1}{t+}\;\frac{2}{t+}\;\frac{3}{t+}\;\cdots \end{aligned} \tag{1.51}$$

by an equivalence transformation.

For $\bar{R}(t)$ we use the Gauss c.f. for the confluent hypergeometric function

$${}_1F_1(a;b;x) = 1 + \frac{a}{b}\frac{x}{1!} + \frac{a(a+1)}{b(b+1)}\frac{x^2}{2!} + \cdots$$

so that

$$\bar{R}(t) = t\cdot {}_1F_1(1;3/2;t^2/2)$$

and from Wall (p. 348)

$$\bar{R}(t) = \frac{t}{1-}\;\frac{t^2}{3+}\;\frac{2t^2}{5-}\;\frac{3t^2}{7+}\;\frac{4t^2}{9-}\;\cdots . \tag{1.52}$$

From Example 1.7 on rate of convergence, it can be seen that (1.51) will be useful for large $t$ and (1.52) for small $t$. The case for $R(t)$ can be dealt with approximately by elementary methods. First show (Shenton, 1954a, and also 1954b)

$$R(t) = \int_0^\infty \exp(-x^2/2 - xt)dx ,$$

which after differentiation and manipulation of the integral leads to

$$\begin{aligned} R(t)\omega_s - \chi_s &= (-1)^s \int_t^\infty (x-t)^s g(x)dx / g(t) \\ &= (-1)^s \int_0^\infty u^s \exp(-u^2/2 - ut)du , \end{aligned} \tag{1.53}$$

where the $s$th convergent of (1.51) is $\chi_s / \omega_s$. Now identify $\omega_s$ with the $s$th Hermite polynomial; for example, $\omega_1 = H_1(t) = t$, $\omega_2 = H_2(t) = t^2 + 1$. For large $s$, one considers the differential equation for $W_s = \omega_s e^{-t^2/4}$ namely

$$\frac{d^2W_s}{dt^2} = (\frac{t^2}{4} + s + \frac{1}{2})W_s ,$$

to show that

$$\omega_{2s} \sim \frac{1}{2}\exp\{-\frac{t^2}{2} + t\sqrt{(2s)} + s\ln(s) + (s + \frac{1}{2})\ln(2) - s\}$$

with a similar expression for $\omega_{2s+1}$. Again the integral in (1.53) for $R(t)\omega_s - \chi_s$ is of order

$$(-1)^s \sqrt{\pi} \exp\{-s/2 - t\sqrt{s} + s/2 \ln(s)\} .$$

Finally then

$$R(t) - \chi_s/\omega_s \sim (-1)^s \sqrt{(2\pi)} \exp(t^2/2 - 2t\sqrt{s}), \qquad (s \to \infty) \tag{1.54}$$

which decreases moderately fast with $s$, with some assistance for large $t$. To obtain an absolute accuracy of $\epsilon$, the approximate number of terms is $s = S(\epsilon)$, where

$$S(\epsilon) = \left\{ \frac{1}{2t}\ln\left[\frac{\epsilon}{\sqrt{(2\pi)}}\right] - \frac{t}{4} \right\}^2 . \tag{1.55}$$

| | Values of $S(\epsilon)$ | | |
|---|---|---|---|
| $t$ | $\epsilon = 2.5E{-}07$ | $\epsilon = 2.5E{-}13$ | $\epsilon = 2.5E{-}19$ |
| 0.1 | 6501 (4) | 22412 (6) | 47866 |
| 0.5 | 264 (6) | 904 (10) | 1925 |
| 1.0 | 69 (8) | 232 (12) | 490 |
| 5.0 | 8 | 18 | 32 |
| 10.0 | 11 | 16 | 22 |

(For $t = 10$, more refined methods show $S(\epsilon) = 4$ and 9 instead of 11 and 16.)

The parenthetic entries refer to the number of terms required for a given accuracy for $R(t)$ using the Gauss c.f. (1.52) together with (1.49); they were calculated numerically, the analytical approach being complicated. We conclude that the Laplace c.f. works well for $t \geqslant 5$ whereas the Gauss does well for $0 < t < 1$. From another point of view, suppose a practical problem supplied a finite set of terms for the series $R(t) \sim 1/t - 1/t^3 + \cdots$; then the c.f. approach would be in a parlous state for small $t$.

**Example 1.8**

(i) R(10) by the Laplace c.f. ($c_s = \chi_s / \omega_s$)

$$c_{14} = \frac{22366536329850}{225859369179635} = 0.099028596471731 9216,$$

$$c_{15} = \frac{251693089438620}{2541620283495250} = 0.0990285964717319214.$$

(ii) $\bar{R}(t)$ by Gauss c.f.

$$c_6 = \frac{11088}{7860} = 1.410687, \qquad c_7 = \frac{1504942}{106680} = 1.410686,$$

$$c_{11} = \frac{15236333520}{10800654480} = 1.410686134642463,$$

$$c_{12} = \frac{342655143480}{242899632360} = 1.410686134642448.$$

(iii) $\bar{R}(1/4)$ by Gauss c.f.

$$c_2 = \frac{12}{47} = 0.25532, \qquad c_3 = \frac{242}{948} = 0.25527,$$

$$c_7 = \frac{557084352}{2182299420} = 0.2552740228469657,$$

$$c_8 = \frac{33350265756}{130644964905} = 0.25527402284696568.$$

### 1.6.2 The "Gamma" Weight c.f.

$$F(n) = \frac{1}{\Gamma(a)} \int_0^\infty \frac{e^{-t} t^{a-1} dt}{t+n}. \quad \text{(see Example 1.1)}$$

Let the $s$th convergent of the c.f. in (1.32) be $\chi_s / \omega_s$. Then if $r_s(n) = F(n)\omega_s - \chi_s$, integration by parts shows

$$r_s(n) = n^{a-1} \frac{\Gamma(s+a)}{\Gamma(a)} \int_0^\infty \frac{e^{-t} t^s \, dt}{(t+n)^{s+a}}. \tag{1.56}$$

(A result in Wall (p. 355) shows the alternative

$$r_s(n) = \frac{s!}{\Gamma(a)} \int_0^\infty \frac{e^{-t} t^{s+a-1} dt}{(t+n)^{s+1}}.)$$

Moreover, the denominators $\omega_s$ are Laguerre polynomials which for large $s$, from Perron's formula (Szegö, p. 193), have the asymptotic

$$\omega_s(n) \sim \frac{s!}{2\sqrt{\pi}} n^{(1-2a)/4} s^{(2a-3)/4} \exp\{-\frac{n}{2} + 2\sqrt{(sn)}\}. \tag{1.57}$$

Using the saddle point technique on (1.56), we find

$$r_s(n) \sim n^{a-1} \sqrt{\pi} \frac{\Gamma(s+a)}{\Gamma(a)} (ns)^{(1/4-a/2)} \exp\{\frac{n}{2} - 2\sqrt{(sn)}\}, \tag{1.58}$$

so that for the remainder for large $s$

$$F(n) - \frac{\chi_s}{\omega_s} \sim \frac{2\pi (ns)^{(a-1)/2}}{\Gamma(a)} \exp\{n - 4\sqrt{(sn)}\}. \tag{1.59}$$

| $n$ | $a$ | $s$ | $F(n) - \chi_s / \omega_s$ |
|---|---|---|---|
| 1 | 1 | 20 | 2.9E-07 |
| 10 | 1 | 20 | 3.7E-20 |
| 1 | 5 | 20 | 4.8E-06 |
| 10 | 5 | 20 | 6.2E-17 |
| 50 | 1 | 10 | 4.6E-17 |
| 0.5 | 0.5 | 20 | 1.1E-05 |

Convergence is speeded up for large $s$ and $n$ and retarded for increases in $a$.

## 1.7 The Moment Problem

As mentioned in Example 1.6, the moment problem concerns the existence of bounded non-decreasing functions $\psi(t)$ satisfying the equations

$$\mu_s' = \int_a^b t^s \, d\,\psi(t), \qquad (s = 0, 1, \cdots) \tag{1.60}$$

where $\{\mu_s'\}$ is a given real sequence of values, and the limits of integration are known. It was originally a problem in mechanics, and addressed itself to deciding whether a knowledge of the moments of a set of masses on a line determined the mass distribution itself. Surprisingly enough, it was found that there could be no solutions, a unique solution, or an infinity of solutions. Again it was natural to discuss three cases: the line mass (i) finite, (ii) semi-finite, and (iii) infinite in both direction. In terms of (1.60) this translates to $(a, b)$ finite (sometimes called the reduced (or little) moment problem (*rmp*)), or $a = 0$, $b = \infty$ corresponding to the Stieltjes moment problem (*Smp*), and lastly $a = -\infty$, $b = \infty$ corresponding to the Hamburger moment problem (*Hmp*; to banish commercial innuendos, one should note that this one manufactured moment solutions some half century ago).

The problem has an obvious statistical or probability interpretation, the terminology shifting to distributions and their moments about a point, and usually four moments are commonly conceptualized (the standard deviation corresponding to the radius of gyration in mechanics), and there is rarely the necessity to ponder the possibility of an infinite set not being adequate.

For our present purpose special attention is given to the *Smp*. Here the Stieltjes integral transformation and Stieltjes fraction are fundamental entities. Thus

$$F(t, \psi) = \int_0^\infty \frac{d\,\psi(x)}{t + x} \tag{1.61}$$

is the formal representation of the moment generating function

$$\mu_0'/t - \mu_1'/t^2 + \mu_2'/t^3 - \cdots, \tag{1.62}$$

which itself may be associated with

$$\phi(t) = \frac{1}{\alpha_1 t +}\;\frac{1}{\alpha_2 +}\;\frac{1}{\alpha_3 t +}\;\frac{1}{\alpha_4 +}\;\cdots, \tag{1.63a}$$

or

$$\phi(t) = \frac{c_1}{t+} \frac{c_2}{1+} \frac{c_3}{t+} \frac{c_4}{1+} \cdots, \tag{1.63b}$$

where

$$\begin{aligned} c_{2s} &= A_{s-1}B_s/(A_s B_{s-1}), \quad (A_0 = B_0 = 1) \\ c_{2s+1} &= A_{s+1}B_{s-1}/(A_s B_s), \end{aligned}$$

and

$$A_s = |\mu'_{i+j}|_{i,j=0}^{s-1}, \qquad B_s = |\mu'_{i+j+1}|_{i,j=0}^{s-1}. \tag{1.64}$$

Stieltjes showed that the positivity of the $\alpha$'s in (1.63a) is a necessary condition for the existence of solutions, and that if in addition $\Sigma\alpha_s = \infty$, then there is a unique bounded non-decreasing $\psi(t)$ with infinitely many points of increase for $0 \leqslant t < \infty$.

The existence of orthogonal polynomials is of prime importance, and this depends on the nature of the weight function. For example, it is obvious that its moments must exist and it must be one signed. For example, if

$$\begin{aligned} W(t) &= (t-1)\exp(-t), \qquad (0 \leqslant t < \infty) \\ &= 0, \qquad (t < 0) \end{aligned}$$

then $P_1(t)$ does not exist.

The question of uniqueness relates to the possibility of skeleton functions not detectable by moments. Thus (as Shohat and Tamarkin (1950) point out), does

$$\int_0^\infty W(t)t^s\,dt = \int_0^\infty \{\exp(-kt^\alpha)\}t^s\,dt, \qquad (s = 0, 1, \cdots;\ k, \alpha > 0) \tag{1.65}$$

imply $W(t) = \exp(-kt^\alpha)$? Shohat and Tamarkin show that there is a unique solution if $\alpha > 1/2$, but many solutions if $0 < \alpha < 1/2$ (the latter corresponding to a density function which tends to zero very slowly as $t \to \infty$; in statistical jargon, a long "tailed" distribution with extremely large moments). Thus the skeleton density component $\lambda\{\exp(-kt^\alpha)\}\sin\{kt^\alpha\tan(\alpha\pi)\}$ has zero moments, as can be seen by using

$$\int_0^\infty y^{c-1}\exp(-by)\,dy = b^{-c}\Gamma(c), \qquad (b, c > 0)$$

with $c = (s+1)/\alpha$, $b = k + il$, $l/k = \tan(\alpha\pi)$, $0 < \alpha < 1/2$. Note that $\alpha = 1/2$ corresponds to $\mu'_s$ increasing as fast as $(2s)!$.

## 1.8 Carleman's Theorem

If the moment determinants in (1.64) are positive, then the *Smp* is determined (has a unique solution) provided

$$\sum (1/\mu_s')^{1/(2s)} = \infty. \tag{1.66a}$$

Similarly, the *Hmp* is determined if $A_s > 0$, $s = 1, 2, \cdots$, and

$$\sum (1/\mu_{2s}')^{1/(2s)} = \infty. \tag{1.66b}$$

The positivity of the determinants in both cases is ensured by the existence of the orthogonal system of polynomials (in the Stieltjes case the existence of those associated with $d\psi(t)$ ensures the existence of those associated with $t d\psi(t)$). Refer to Wall, p. 330.

**Example 1.9**

For the gamma density integral (1.31) and (1.34), the orthogonal polynomials are those of Laguerre and in (1.65), the parameter $\alpha = 1$. Moreover for the partial denominators in the Stieltjes fraction (1.63a)

$$\sum_2^\infty \alpha_s = \sum_1^\infty \frac{\Gamma(a)\Gamma(s)}{\Gamma(a+s)} + \sum_1^\infty \frac{\Gamma(a+s)}{\Gamma(a)\Gamma(s+1)}$$

so that divergence is evident for $a > 0$; in fact, a typical term in the second sum is $s^{a-1}/\Gamma(a)$ approximately for $s \to \infty$. Moreover $\mu_s'^{-s/2} \sim \sqrt{(e/s)}$ so the moment series diverges.

Similarly for Example 1.6, concerning the weight function $\psi(t,m) = (m^t/t!)\exp(-m)$, the Stieltjes integral $F(n,m)$ has Stieltjes fraction for which

$$\sum_1^\infty \alpha_s = \sum_1^\infty \Gamma(s)/m^s + \sum_0^\infty m^s/\Gamma(s+1), \qquad (m > 0)$$

so that the even series diverges, and the odd series converges.

## 1.9 Stieltjes Summability

If we are given an infinite series $\sum e_s/n^s$ which diverges (or converges slowly) then the c.f., if it exists, may be constructed, convergence in a domain and to a given function demonstrated. For a finite set of terms, as may be the case when further coefficients are extremely complicated (even allowing for modern computer facilities), the c.f. in its truncated form may still throw some light on numerical assessments. Examples of this type of situation will be given in later chapters, but for the present we remark that in general it is seemingly impossible to derive more than the $n^{-1}$ and $n^{-2}$ coefficients in the low-order moments of maximum likelihood estimators in simultaneous estimation; in single parameter estimation, one or two higher terms may be found (Shenton and Bowman, 1977a). To appreciate the situation, suppose the probability of a discrete random variate taking a value $x$ is $P(x \mid m_1, m_2)$ where

$$\sum_{x=0}^{\infty} t^x P(x \mid m_1, m_2) = \exp\{m_1(e^{m_2(t-1)} - 1)\}. \qquad (m_1, m_2 > 0)$$

Then for a random sample $(x_1, x_2, \cdots, x_n)$ from this population we set up the maximum likelihood estimators $(\hat{m}_1, \hat{m}_2)$ of $(m_1, m_2)$ from the equations

$$\sum_{j=1}^{n} \frac{\partial}{\partial \hat{m}_1} \log P(x_j \mid \hat{m}_1, \hat{m}_2) = 0$$

$$\sum_{j=1}^{n} \frac{\partial}{\partial \hat{m}_2} \log P(x_j \mid \hat{m}_1, \hat{m}_2) = 0,$$

so that $\hat{m}_j = f_j(x_1, x_2, \cdots, x_n)$, $j = 1, 2$. We seek from these the mean values of $\hat{m}_j$ and the product $\hat{m}_1\hat{m}_2$, to give minimum information on their joint distribution. Clearly the problem is complicated. There are many similar examples but involving more terms (see for example Milton Van Dyke (1974, 1975), Gaunt and Guttman (1973), Baker (1975), Bender and Wu (1969) for cases in critical phenomena and other aspects of statistical physics.)

### 1.9.1 The Psi Function and Its Derivatives

Starting with the representation

$$\ln\Gamma(z) = (z - ½)\ln(z) - z + \frac{1}{2}\ln(2\pi) + 2\int_0^{\infty} \frac{\arctan(t / z)dt}{e^{2\pi t} - 1}, \qquad (\mathrm{Re}(z) > 0) \qquad (1.67)$$

we find for the Psi function $\psi(z) = d\{\ln\Gamma(z)\}/ dz = \Gamma'(z)/ \Gamma(z)$,

$$\psi(z) = \ln(z) - \frac{1}{2z} - \int_0^{\infty} \frac{dx}{(x + z^2)(y - 1)}, \qquad (y = \exp(2\pi\sqrt{x})) \qquad (1.68a)$$

and its derivatives, after integration by parts

$$\psi_1(z) = \frac{1}{z} + \frac{1}{2z^2} + \frac{2\pi}{z}\int_0^\infty \frac{y\sqrt{x}\,dx}{(x+z^2)(y-1)^2}, \tag{1.68b}$$

$$\psi_2(z) = -\frac{1}{z^2} - \frac{1}{z^3} - \left(\frac{2\pi}{z}\right)^2 \int_0^\infty \frac{(y+y^2)x\,dx}{(x+z^2)(y-1)^3}, \tag{1.68c}$$

and so on.

Stieltjes (Oeuvres, 1918) by noticing some special properties of integrals of the form

$$I = \int_0^\infty \frac{c\sinh(au)\sinh(bu)e^{-xu}\,du}{\sinh(cu)}$$

gave the results

$$\psi_1(z) = \frac{1}{z-\tfrac{1}{2}+}\;\frac{a_1}{z-\tfrac{1}{2}+}\;\frac{a_2}{z-\tfrac{1}{2}+}\;\cdots, \tag{1.69a}$$

$$\psi_2(z) = \frac{1}{z^2} - \frac{1}{z^3} - \frac{1}{2z^2}\left(\frac{1}{z^2+}\;\frac{p_1}{1+}\;\frac{q_1}{z^2+}\;\frac{p_2}{1+}\;\frac{q_2}{z^2+}\;\cdots\right) \tag{1.69b}$$

where

$$a_s = s^4/\{4(2s-1)(2s+1)\},$$
$$p_s = s^2(s+1)/(4s+2),$$
$$q_s = s(s+1)^2/(4s+2).$$

We can derive c.f.s from the Stieltjes transforms in (1.68) which are different from (1.69), and which generalize to all the polygamma functions. Define the integrals (omitting the factors before the integral signs) in (1.68a), (1.68b), and (1.68c) as $g_0(z)$, $g_1(z)$, and $g_2(z)$. Note that

$$g_m(z) = \int_0^\infty \frac{dF_m(x)}{x+z^2} \tag{1.70}$$

where

$$F_m(u) = \int_0^u \frac{x^{m/2}\theta_m(y)\,dy}{(y-1)^{m+1}}, \qquad (m = 0, 1, 2)$$

with $\theta_0(y) = 1$, $\theta_1(y) = y$, $\theta_2(y) = y + y^2$, and $y = \exp(2\pi\sqrt{x})$. Clearly $F_m(u)$ is a bounded non-decreasing function with infinitely many points of increase; the integrand is of order $(1/\sqrt{x})$ for small $x$, and of order $(e^{-2\pi\sqrt{x}}x^{m/2})$ for $x \to \infty$. To finalize the existence of the Stieltjes fractions we need the series for $g_m(z)$. But the integral $J(z)$ in (1.67) has the asymptotic expansion

$$J(z) \sim \sum_{s=1}^{\infty} \frac{(-1)^s B_{2s+2}}{(2s+1)(2s+2)z^{2s+1}} \tag{1.71}$$

where $B_0 = 1$, $B_2 = 1/6$, $B_4 = 1/30$, $\cdots$, are the Bernoulli numbers. Hence since asymptotic series may be differentiated,

$$g_m(z) \sim \frac{1}{(2\pi)^m} \sum_{s=0}^{\infty} \frac{(-1)^s \Gamma(2s+m+2) B_{2s+2}}{z^{2s+2}\Gamma(2s+3)} \tag{1.72}$$

$$= \sum_{s=0}^{\infty} \frac{(-1)^s \mu_s^{(m)}}{z^{s+2}}.$$

For Carleman's criterion (Section 1.8), using the asymptote $B_{2s} \sim 2(2s)!/(2\pi)^{2s}$, and Stirling's formula for the factorial, we have

$$\frac{1}{2s} \ln \mu_s^{(m)} \sim (1 + \frac{2m+3}{s}) \ln(s) \tag{1.73}$$

so that the series behaves asymptotically as

$$\sum_{s=1}^{\infty} \frac{1}{s^{1+c/s}}$$

where $c > 0$. Writing $a_s$ for the general term, we have for sufficiently large $s$,

$$s \left\{ \frac{a_{s+1}}{a_s} - 1 + \frac{1}{s} \right\} \sim \frac{c \ln(s)}{s}$$

so that from Raabe's convergence test (Knopp, 1928, p. 285) the series $\Sigma a_s = \infty$. For each of the integrals in (1.68) we have established the existence of $F_m(u)$ and the fulfillment of Carleman's criterion, so that Stieltjes c.f.s exist for the polygamma functions mentioned.

In particular,

$$\psi(z) = \ln(z) - \frac{1}{2z} - \int_0^{\infty} \frac{dF_1(u)}{u+z^2} \tag{1.74}$$

where

$$F(u) = \int_0^u \frac{dx}{e^{2\pi\sqrt{x}} - 1},$$

and

$$\int_0^\infty \frac{dF(u)}{u+z^2} \sim \sum_{s=0}^\infty \frac{(-1)^s B_{2s+2}}{(2s+2)z^{2s+2}}.$$

Using Wall's algorithm for converting a series into a c.f. (the use of the determinant form in (1.63) and (1.64) is complicated, as also the construction of the orthogonal polynomials by first principles), we find

$$\psi(z) - \ln(z) + \frac{1}{2z} = -\frac{c_1}{z_2+}\ \frac{c_2}{1+}\ \frac{c_3}{z_2+}\ \cdots \tag{1.75}$$

$$= -\frac{b_0}{z^2+a_1-}\ \frac{b_1}{z^2+a_2-}\ \cdots, \qquad (\mathrm{Re}(z)>0)$$

where

$$c_0 = \frac{1}{12}, \quad c_1 = \frac{1}{10}, \quad c_2 = \frac{79}{210},$$

$$c_3 = \frac{1205}{1659}, \quad c_4 = \frac{262445}{209429}, \quad c_5 = \frac{2643428414511}{1429053441530};$$

$$b_0 = \frac{1}{12}, \quad b_1 = \frac{79}{2100} \quad b_2 = \frac{1312225}{1441671};$$

$$a_1 = 10, \quad a_2 = \frac{871}{90}, \quad a_3 = \frac{1672667011}{539062030}.$$

Similarly, for $\mathrm{Re}(z)>0$,

$$z\left\{\psi_1(z) - \frac{1}{z} - \frac{1}{2z^2}\right\} = \frac{1/6}{z^2+1/5-}\ \frac{18/175}{z^2+22/15-}\ \frac{1000/693}{z^2+145/39-}\ \cdots$$

$$= \frac{1/6}{z^2+}\ \frac{1/5}{1+}\ \frac{18/35}{z^2+}\ \frac{20/21}{1+}\ \frac{50/33}{z^2+}\ \frac{315/143}{1+}\ \cdots, \tag{1.76}$$

$$z^3\left\{\psi_3(z) - \frac{2}{z^3} - \frac{3}{z^4}\right\} = \frac{2}{z^2+½-}\ \frac{5/12}{z^2+23/10-}\ \frac{232/75}{z^2+734/145-}\ \cdots$$

$$= \frac{2}{z^2+}\ \frac{1/2}{1+}\ \frac{5/6}{z^2+}\ \frac{22/15}{1+}\ \frac{116/55}{z^2+}\ \frac{942/319}{1+}\ \cdots; \tag{1.77}$$

we have omitted $\psi_2(z)$ since the few terms evaluated agree with those given by Stieltjes, as they should.

### 1.9.2 A Logarithmic Integral

Define

$$F(z,\alpha) = -\int_0^1 t^{\alpha-1}\frac{\ln\{1-(1-t)/(z+1)\}dt}{1-t}. \quad (\alpha, z > 0) \tag{1.78}$$

Does $F(z,\alpha)$ have a Stieltjes c.f. development? Before studying this question, a brief background is in order. In sampling from the so-called negative binomial distribution, with probability function

$$P(x) = \frac{\Gamma(\alpha+x)}{\Gamma(\alpha)\Gamma(x+1)}\frac{\lambda^x\alpha^\alpha}{(\lambda+\alpha)^{\alpha+x}}, \qquad (\lambda, \alpha > 0)$$

for $x = 0, 1, \cdots$, the maximum likelihood estimator $\hat{\alpha}$ of $\alpha$ for a sample size $n$, has variance $Var(\hat{\alpha})$ given by

$$\{n\ Var(\hat{\alpha})\}^{-1} = k\sum_{s=2}^{\infty} u_s$$

where

$$k = \frac{1}{2}\left(\frac{\lambda}{\lambda+\alpha}\right)^2\frac{1}{\alpha(\alpha+1)}, \qquad u_s = \frac{2}{s}\left(\frac{\lambda}{\lambda+\alpha}\right)^{s-2}\frac{\Gamma(s)\Gamma(\alpha+2)}{\Gamma(s+\alpha)}.$$

Note that the variance is an asymptotic one for large $n$; moreover with $z = \alpha/\lambda$,

$$\sum_{s=2}^{\infty} u_s = 1 + \frac{2\cdot 2!}{3(z+1)(\alpha+2)} + \frac{2\cdot 3!}{4(z+1)^2(\alpha+3)(\alpha+2)} + \cdots$$

and

$$F(z,\alpha) = \frac{1}{(z+1)\alpha} + \frac{1}{2(z+1)^2\alpha(\alpha+1)} + \frac{2!}{3(z+1)^3\alpha(\alpha+1)(\alpha+2)} + \cdots, \tag{1.79}$$

so that

$$\sum_2^{\infty} u_s = 2\alpha(\alpha+1)(z+1)^2[F(z,\alpha) - 1/\{\alpha(z+1)\}].$$

The Stieltjes integral arises as follows. We have

$$F(z,\alpha) - F(z,\alpha+1) = \frac{1}{\alpha(z+1)}\int_0^1 \frac{(1-t)^\alpha\, dt}{1-t/(z+1)} = \frac{1}{\alpha}\int_0^1 \frac{x^\alpha\, dx}{x+z},$$

$$F(z,\alpha) = \int_0^1 \frac{x^\alpha f(x;\alpha)\,dx}{x+z} \sim \mu_0'/z - \mu_1'/z^2 + \cdots \tag{1.80}$$

where

$$f(x;\alpha) = \sum_{s=0}^{\infty} \frac{x^s}{\alpha+s}, \qquad \mu_s' = \frac{1}{s+1}\sum_{r=0}^{s} \frac{1}{r+\alpha}.$$

Note that

$$xf(x;1) = \ln(1-x)^{-1},$$

$$x^m f(x;m) = \ln(1-x)^{-1} - \sum_{s=1}^{m-1} x^s/s. \qquad (m=2,3,\cdots)$$

Now $F(z,\alpha)$ from (1.80) can be written

$$F(z,\alpha) = \int_0^\infty \frac{d\,\phi(t)}{z+t},$$

where

$$\phi(t) = \alpha\int_0^t \frac{(t-u)u^{\alpha-1}du}{1-u}, \qquad 0 \leqslant t \leqslant 1; \tag{1.81}$$

$$\phi(t) = 1,\ t>1; \qquad \phi(t) = 0,\ t<0.$$

Evidently $\phi(t)$ is bounded and non-decreasing has infinitely many points of increase for $t>0$. As for the moments

$$\mu_s' = \frac{1}{s+1}\{\psi(s+\alpha+1) - \psi(\alpha)\}, \qquad (\psi(\cdot) \text{ the Psi function})$$

$$\sim \frac{1}{s+1}\ln(s+\alpha+1), \qquad (s \gg 0) \tag{1.82}$$

so that $(\mu_s')^{1/(2s)} \to 1$ and Carleman's criterion is satisfied. Hence

$$\alpha F(z,\alpha) = \frac{1}{z+}\;\frac{c_2}{1+}\;\frac{c_3}{z+}\;\cdots \tag{1.83}$$

and using the determinantal formulae in (1.64)

$$c_2 = \frac{2\alpha+1}{2\alpha+2}, \qquad c_3 = \frac{5\alpha+2}{6(\alpha+1)(\alpha+2)(2\alpha+1)},$$

$$c_4 = \frac{(\alpha+1)R}{3(\alpha+2)(\alpha+3)(2\alpha+1)(5\alpha+2)}, \qquad c_5 = \frac{4(2\alpha+1)S}{5(\alpha+3)(\alpha+4)(5\alpha+2)R},$$

$$c_6 = \frac{3(\alpha+2)(5\alpha+2)T}{10(\alpha+4)(\alpha+5)RS}, \qquad c_7 = \frac{27RU}{14(\alpha+5)(\alpha+6)ST},$$

where

$$\begin{aligned} R &= 30\alpha^3 + 78\alpha^2 + 58\alpha + 12, \\ S &= 245\alpha^3 + 587\alpha^2 + 390\alpha + 72, \\ T &= 4900\alpha^6 + 36070\alpha^5 + 102608\alpha^4 + 143084\alpha^3 \\ &\qquad + 101958\alpha^2 + 34704\alpha + 4320, \\ U &= 140161\alpha^6 + 990172\alpha^5 + 2683877\alpha^4 + 3536882\alpha^3 \\ &\qquad + 2360376\alpha^2 + 747360\alpha + 86400. \end{aligned}$$

### Example 1.10

In Example 1.4, take $n = -k$, $p = -p^*$, so that $q = 1 + p^*$. The binomial function now becomes

$$B(k, t; p^*) = \frac{\Gamma(k+t)}{\Gamma(k)t!} p^{*t}(1+p^*)^{-k-t},$$

$$(k > 0, \ p^* > 0; \ t = 0, 1, \cdots)$$

and Stieltjes integral

$$\sum_{t=0}^{\infty} \frac{B(k, t; p^*)}{z+t} = \frac{1}{z+} \frac{kp^*}{1+} \frac{q}{z+} \frac{(k+1)p^*}{1+} \frac{2q}{z+} \frac{(k+2)p^*}{1+} \cdots .$$

This establishes the Stieltjes fraction for the moment problem for the negative binomial function $B(k, t; p^*)$.

### Example 1.11 A Bessel Function Transform

The modified Bessel function may be defined as

$$K_0(t) = \int_0^\infty \exp\{-t \cosh(\theta)\} d\theta, \qquad (t > 0)$$

where (Watson, 1962),

$$2\int_0^\infty t^s K_0(2\sqrt{t})dt = \Gamma^2(s+1). \quad (s = 0, 1, \cdots)$$

Hence $2K_0(2\sqrt{t}) > 0$ for $t > 0$ and the moments exist (for Carleman, $\mu_s'^{-1/(2s)} \sim 1/s$ ensuring divergence of the series), and so the Stieltjes fraction exists. In fact

$$\int_0^\infty \frac{2K_0(2\sqrt{t})dt}{t+z} = \frac{a_0}{z+b_1-}\ \frac{a_1}{z+b_2-}\ \cdots$$

where, for example,

$$\begin{aligned}
a_0 &= 1, & b_1 &= 1,\\
a_1 &= 3, & b_2 &= 9.666667,\\
a_2 &= 7.288889\text{E}01, & b_3 &= 9.818699\text{E}01,\\
a_3 &= 4.106859\text{E}02, & b_4 &= 5.657190\text{E}01,\\
a_4 &= 1.363471\text{E}03, & b_5 &= 9.482393\text{E}01,\\
a_5 &= 3.424421\text{E}03, & b_6 &= 1.429441\text{E}02,
\end{aligned}$$

to seven digits. The Stieltjes fraction is readily set up.

### Example 1.12 A Mills' Ratio for a Trigonometric Integral

If

$$H(\theta,k,r) = \frac{\int_\theta^{\pi/2} e^{-k\phi}\cos^r(\phi)d\phi}{e^{-k\theta}\cos^r(\theta)}, \quad (r > 1, -\frac{\pi}{2} < \theta < \frac{\pi}{2})$$

then

$$H(\theta,k,r) = \frac{1/(r+1)}{t+a_1+}\ \frac{b_1}{t+a_2+}\ \frac{b_2}{t+a_3+}\ \cdots, \quad (t = \tan\theta)$$

where

$$\begin{aligned}
a_s &= \frac{kr}{(r+2s-2)(r+2s)},\\
b_s &= \frac{s(r+s)\{k^2+(r+2s)^2\}}{(r+2s-1)(r+2s)^2(r+2s+1)}.
\end{aligned}$$

In particular,

$$H(\theta, 0, r) = \frac{1}{(r+1)t+} \; \frac{1(r+1)}{(r+3)t+} \; \frac{2(r+2)}{(r+5)t+} \; \cdots ,$$

and

$$H(\theta, k, 0) = \frac{1}{t+k/2+} \; \frac{(k^2+2^2)/4}{3t+} \; \frac{(k^2+4^2)/4}{5t+} \; \cdots ,$$

the latter being added by us recently to the results in Shenton and Carpenter (1964). The results in the paper can be shown to hold for $r > 0$. By taking $k = -2K$, the reader may show

$$e^{2K(\pi/2-\theta)} = 1 + \frac{2K}{t-K+} \; \frac{K^2+1}{3t+} \; \frac{K^2+4}{5t+} \; \cdots ,$$

a result given by Wall(p. 346) and attributed to Laguerre.

### 1.10 Padé Approximants

J. fractions are approximants to series $F(n) = e_0 + e_1/n + \cdots$ of the form

$$P[s \mid s] = \pi_s^*(n)/\pi_s(n), \quad (s = 1, 2, \cdots)$$

where $\pi_s^*$, $\pi_s$ are polynomials in $n$ of degree $s$, the coefficients being determined from the asymptotic equivalence ($n$ large) of the fraction and series to the necessary number of terms. The Stieltjes fraction has a similar structure (see (1.21)). A series may be truncated and the same procedure followed. For example

$$F(n) \sim e_0 + \frac{1}{n} P^*[s \mid s], \qquad (s = 1, 2, \cdots)$$

yields another sequence of approximants $n^{-1}P_1[s+1 \mid s]$ with numerator one degree higher than the denominator. Similarly truncating terms up to $e_{r-1}/n^{s-1}$ leads to sequences $n^{-r}P[s+r \mid s]$. Exactly the same procedure can be followed in general with the reciprocal series, new approximants arising for one, two, etc., truncations. Padé's thesis was published in 1892.

For example $1/n - 1/n^2 + 2!/n^3 - 3!/n^4 + \cdots$ has the c.f. expansion ((1.32) with $a = 1$)

$$\frac{1}{n+1-} \; \frac{1^2}{n+3-} \; \frac{2^2}{n+5-} \; \cdots ; \tag{1.84}$$

$\frac{1!}{n^2} - \frac{2!}{n^3} + \cdots$ has the c.f. ((1.32) with $a = 2$)

$$\frac{1}{n}\left[\frac{1}{n+2-}\;\frac{1\cdot 2}{n+4-}\;\frac{2\cdot 3}{n+6-}\;\cdots\right], \tag{1.85}$$

and so on.

**Example 1.13**

For the Mills' ratio c.f. (see (1.47), (1.51)), show that

$$R(t) = \frac{1}{t} - \frac{1}{t^3} + \cdots + (-1)^s \frac{1\cdot 3\cdot \cdots (2s-1)}{t^{2s+1}} \tag{1.86}$$

$$+(-1)^{s+1}\frac{1\cdot 3\cdots(2s+1)}{t^{2s+2}}\left[\frac{1}{t+}\;\frac{2s+3}{t+}\;\frac{2}{t+}\;\frac{2s+5}{t+}\;\frac{4}{t+}\;\cdots\right].$$

$$(s = 0, 1, \cdots;\; t > 0)$$

Similarly for the complementary function $\bar{R}(t)$ ((1.48), (1.52)), show that

$$\bar{R}(t) = \sum_{r=0}^{s-1} \frac{t^{2r+1}}{1\cdot 3\cdots(2r+1)} + \frac{t^{2s}}{1\cdot 3\cdots(2s+1)} \tag{1.87}$$

$$\times\left[\frac{t}{2s+1-}\;\frac{(2s+1)t^2}{2s+3+}\;\frac{2t^2}{2s+5-}\;\frac{(2s+3)t^2}{2s+7+}\;\frac{4t^2}{2s+9-}\;\cdots\right].$$

**Example 1.14**

In sampling from a normal distribution, the statistic $\sqrt{b_1} = m_3/m_2^{3/2}$ is a measure of the sample skewness and an estimate of the population skewness $\sqrt{\beta_1} = 0$ (for the sample $(x_1, x_2, \cdots, x_n)$ the central moments are defined by $m_r = \sum_{j=1}^{n}(x_j - \bar{x})^2/n$ where $\bar{x} = \sum x_j/n$). Since the mean value of $\sqrt{b_1}$ is zero, the variance reduced to the mean value of $m_3^2/m_2^3$. Prior to work of R. A. Fisher on the independence of $\sqrt{b_1}$ and the variance, E. S. Pearson (1930) by some non-trivial algebra derived for $x = (n-1)\sqrt{[b_1/\{6(n-2)\}]}$

$$Var(x) \sim 1 - 6/n + 22/n^2 - 70/n^3 + 214/n^4 - 646/n^5, \tag{1.88}$$

excepting the last two terms which we insert for our present purposes.

Padé approximants to the true value $Var(x) = (n-1)^2/\{(n+1)(n+3)\}$ are with $y = 1/n$:

$$\begin{cases} P[4|0] = 1 - 6y + 22y^2 - 70y^2 + 214y^4, \\ P[3|1] = (1 - 2.9429y + 3.6571y^2 - 2.7429y^3)/(1 + 3.057y), \\ P[2|2] = (1-y)^2/\{(1+y)(1+3y)\}, \\ P[1|3] = (1 - 1.3637y)/(1 + 4.6364y + 5.8182y^2 + 2.9091y^2), \\ P[0|4] = 1/(1 + 6y + 14y^2 + 22y^3 + 30y^4). \quad (0<y<1) \end{cases} \tag{1.89}$$

Notice that $P[2|2]$ is exact, also $P[s|s]$, $s>2$ from elementary considerations. What happens if we set up, for example $P[s|1]$ or $P[1|s]$ for large $s$? Set $P[s|1] = \pi_s(y)/(1+b_s y)$; then it turns out by solving the linear equations

$$b_s = (4{\cdot}3^s - 1)/(4{\cdot}3^{s-1} - 1)$$

which $\to 3$ as $s \to \infty$. Also

$$(4{\cdot}3^{s-1} - 1)\pi_s(y) = 4{\cdot}3^{s-1} - 1 + (5 - 4{\cdot}3^s)y + 16{\cdot}3^{s-1}X_s - 16y^2 Y_s,$$

where

$$X_s = y^2 - y^3 + \cdots + (-1)^s y^s,$$
$$Y_s = 1 - 3y + \cdots + (-3y)^{s-2},$$

so that $\pi_s \to (1-y)^2/(1+y)$. Hence $P[s|1] \to (1-y)^2/\{(1+y)(1+3y)\}$ since $0<y<1$.

Similarly $P[1|s] = (1+B_s y)/\pi_s^*(y)$, where $B_s = -(4s+3)(4s-1)$ and $\to -1$ as $s \to \infty$. Also

$$(4s-1)\pi_s^* = 4s - 1 + (20s-9)y + 32sy^2(1-y^{s-1})/(1-y)$$
$$- 32y^2(1-sy^{s-1})/(1-y) - 32y^3(1-y^{s-1})/(1-y)^2$$

as $\pi_s^* \to (1+y)(1+3y)/(1-y)$, so $P[1|s]$ tends to the true value as $s \to \infty$.

**Example 1.15**

In Section 2.8.3 we consider the series expansion for

$$y(n) = \frac{(n-1)\Gamma(n/2)}{\sqrt{(2n)}\Gamma(n/2+½)}, \qquad (n>1)$$

which is $E(\sqrt{m_2})$ in normal sampling. Show that the first few Padé approximants $P_s[s|s+1]$, where $x = 1/n$, are (to four digits):

$$P[1\,|\,2] = (1 - 1.05x)/(1 - 0.3x - 0.0063x^2),$$

$$P[1\,|\,3] = (1 - 0.9875x)/(1 - 0.2375x + 0.0406x^2 + 0.0488x^3),$$

$$P[1\,|\,4] = \frac{(1 - 0.9399x)}{(1 - 0.1899x + 0.0763x^2 + 0.0860x^3 + 0.0391x^4)}.$$

For $n = 1, 2$, the Padé tables are:

| $n = 1$ $y = 0$ | | | | | |
|---|---|---|---|---|---|
| 0.2500 | -0.0588 | 0.0721 | 0.0147 | .0594 | -0.0253 |
| 0.0313 | −0.0724 | -0.0603 | 0.1519 | 0.0342 | 0.0948 |
| -0.0391 | 0.0186 | 0.0869 | -0.0169 | -0.0340 | 0.0053 |
| -0.0102 | -0.0666 | 0.0057 | −0.0351 | -0.0196 | 0.1020 |
| 0.0487 | 0.0266 | -0.0240 | -0.0056 | 0.0552 | -0.0079 |
| 0.0133 | 0.0585 | -0.0032 | -0.0203 | 0.0023 | −0.0227 |
| $n = 2$ $y = 0.564190$ | | | | | |
| 0.6250 | 0.5610 | 0.5599 | 0.5641 | 0.5655 | 0.5644 |
| 0.5703 | 0.5598 | 0.5607 | 0.5667 | 0.5649 | 0.5701 |
| 0.5615 | 0.5630 | 0.5659 | 0.5640 | 0.5636 | 0.5642 |
| 0.5633 | 0.4843 | 0.5644 | 0.5636 | 0.5639 | 0.5647 |
| 0.5652 | 0.5647 | 0.5639 | 0.5641 | 0.5646 | 0.5642 |
| 0.5646 | 0.5656 | 0.5641 | 0.5634 | 0.5642 | 0.5640 |

More terms yield for $n = 1$

$$p[13\,|\,14] = -0.0087, \qquad P[14\,|\,14] = -0.0005,$$

and for $n = 2$,

$$P[13\,|\,14] = 0.564176, \qquad P[14\,|\,14] = 0.564189,$$

$$(1/\sqrt{\pi} = 0.5641896).$$

Since our main interest is in the $P[s\,|\,s+1]$, and contiguous sequences, we do not pursue the general Padé summation technique further. Convergence questions, and error analysis, as in most other algorithms, present unsolved problems. A great amount of investigation has occurred in the last two decades; the state of the art then and now can be gleaned from the treatment in the last chapter of Wall (it being mainly theoretically oriented) and the recent account, with many applications, by Baker (1975). A Baker, et al (1961) conjecture is worth noting, the flavor of it being--"*a power series representation of a function (regular within the unit circle) is such that at least a subsequence of* $P[s\,|\,s]$ *approximants converges to the function, excepting isolated points in the unit circle.*"

## 1.10 Remarks and References

We make no claim that this brief sketch does anything more than cover aspects of the subject germane to our specific requirements. The subject was very much alive a century or so ago, and simultaneous with the development of computer facilities became rejuvenated to a vigorous discipline. A recent bibliography of Brezinski (1977-1981) gave over 100 references to Padé approximants, twice as many for c.f.s, and about the same number for applications.

In addition to the references already given in this chapter, recent developments in physics related applications are to be found in "Padé approximant method and its application in mechanics" (H. Cabannes, ed., 1976) and "Padé approximants" (P. R. Graves-Morris, ed., 1973). Survey articles on series analysis (mainly relating to Padé techniques) are those of Hunter and Baker (1975), Baker and Hunter (1973); a similar study is that of Gaunt and Guttman (1973), and this covers various aspects of the subject. There are the classical sources including principally Stieltjes collected works, Perron's two volume treatise, and Wall's function theory account; a recent classical treatment is given by Jones and Thron (1980). Some unusual aspects of the subject are to be found in Khovanskii (1963), translated from the Russian by P. Wynn. A concise account of analytic theory is given in Henrici (1977).

Further information on the Mills' ratio in 1.6.1 is given in Shenton (1954a); the ranking of the approximants as stated is not quite correct, and presents a tricky problem. Actually since the c.f. in (1.52) for $\overline{R}(t)$ is intended for small $t > 0$, we now state for $0 < t < \sqrt{3}$, in the terminology of the paper,

$$r_0 < r_1 < r_4 < r_5 < r_8 < r_9 < \cdots < \overline{R} < \cdots < r_{11} < r_{10} < r_7 < r_6 < r_3 < r_2 .$$

Fuller details for the c.f.s for polygamma functions in Section 1.9.1, including a few terms for the general case, are given in Shenton and Bowman (1971). Similarly for the logarithmic integral in Section 1.9.2, see Shenton (1963), and for the Mills' ratio in Example 1.12 see Shenton and Carpenter (1964).

### Further Examples

**Further Example 1.1**

Show that the c.f.

$$\arctan(z) = \frac{z}{1+}\ \frac{z^2}{3+}\ \frac{4z^2}{5+}\ \frac{9z^2}{7+}\ \frac{16z^2}{9+}\ \frac{25z^2}{11+}\ \cdots$$

may be written as

$$\arctan(z) = \frac{z}{1+}\;\frac{\frac{z^2}{3}}{1+}\;\frac{\frac{4z^2}{3\cdot5}}{1+}\;\frac{\frac{9z^2}{5\cdot7}}{1+}\;\frac{\frac{16z^2}{7\cdot9}}{1+}\;\frac{\frac{25z^2}{9\cdot11}}{1+}\;\cdots$$

so that the $s$ th partial numerator $a_s z^2$ is given by

$$a_s = s^2/\{(2s-1)(2s+1)\}.$$

Deduce $a_s \to 1/4$ as $s \to \infty$.

Note that the c.f.

$$\arctan(z) = \frac{z}{1+}\;\frac{1^2z^2}{3-z^2+}\;\frac{3^2z^2}{5-3z^2+}\;\frac{5^2z^2}{7-5z^2+}\;\cdots$$

is merely equivalent to the series

$$\arctan(z) = z - z^3/3 + z^5/5 - z^7/7 + \cdots .$$

Compare the rate of convergence of the two c.f.s when $z = 1$, in which case the equivalent c.f. takes the form

$$\frac{\pi}{4} = \frac{1}{1+}\;\frac{1^2/2}{1+}\;\frac{3^2/4}{1+}\;\frac{5^2/4}{1+}\;\cdots$$

so that $a_s = (2s-1)^2/4$, $s = 2, 3, \cdots$ and $a_s \to s^2$. Actually the 29th and 30th convergents are 0.794 and 0.777 and do not pin down even the second decimal digit.

**Further Example 1.2**

Consider

$$F(x;p,q) = \int_0^x \frac{t^p\,dt}{N+t^q} \qquad \text{(Khovanskii,1963)}$$

which can be expressed as

$$\frac{x^{p+1}}{qx^2}\int_0^1 \frac{y^{(p+1-q)/q}\,dy}{n+y}. \qquad (n = N/x^2)$$

Using properties of the Jacobi orthogonal polynomial $\{q_r(t)\}$, orthogonal with respect to the weight function $W(t) = (1-t)^\alpha(1+t)^\beta$ $(-1 \leqslant t \leqslant 1)$ show that

$$\int_0^x \frac{t^p\,dt}{n+t^2} = \frac{x^{p+1}}{p+1}\left(\frac{1}{N+}\;\frac{P_1}{1+}\;\frac{Q_1}{N+}\;\cdots\right)$$

where

$$P_s = \frac{x^q(\beta+s)^2}{(\beta+2s-1)(\beta+2s)}, \qquad Q_s = \frac{s^2x^q}{(\beta+2s)(\beta+2s+1)},$$

$\beta = (P+1)/q - 1$ and $x > 0,\ q > 0,\ p > -1,\ N > 0$. In particular,

$$\int_0^1 \frac{x^2dx}{1+x^4} = \frac{1}{3+}\ \frac{9}{7+}\ \frac{16}{11+}\ \frac{49}{15+}\ \frac{64}{19+}\ \frac{121}{23+}\ \cdots$$

the partial numerators being $16s^2$, $(4s+3)^2$.

By elementary methods show that the integral equals $\{\pi/2 - \ln(\sqrt{2}+1)\}/(2\sqrt{2})$.

A trapezoidal formula on a desk calculator gives 0.24374775 for 50 and 100 step formulas. Double-double precision arithmetic on IBM gave the lower bound (28th convergent) 0.24374774719968O524179 with upper bound the same digits except 180 replacing the last three.

The c.f. converges quickly; note that the $s$th partial numerator tends to 1/4 (the partial denominators being transformed to unity).

**Further Example 1.3**

Show that

$$\sqrt[3]{\left(1+\frac{1}{n}\right)} \sim 1 + \frac{1/3}{n+1/3-}\ \frac{2/27}{n+\frac{1}{2}-}\ \frac{7/108}{n+\frac{1}{2}-}\ \frac{4/63}{n+\frac{1}{2}-}\ \frac{143/2268}{n+\frac{1}{2}-}\ \cdots.$$

Evaluate the first few approximants for $n = 5, 10$, and -2, and comment on the nature of the approximation sequences.

Lagrange gave the expansion

$$\left(1+\frac{1}{n}\right)^k = \frac{n}{n-}\ \frac{k}{1+}\ \frac{\frac{1(1+k)}{1\cdot2}}{n+}\ \frac{\frac{1(1-k)}{2\cdot3}}{1+}\ \frac{\frac{2(2+k)}{3\cdot4}}{n+}\ \frac{\frac{2(2-k)}{4\cdot5}}{1+}\ \cdots$$

providing bounds for the binomial expression when $|k| < 1,\ n > 0$. Using the notation

$$\left(1+\frac{1}{n}\right)^k = \frac{n}{n-}\;\frac{p_s}{1+}\;\frac{q_1}{n+}\;\frac{p_2}{1+}\;\frac{q_2}{n+}\;\cdots$$

$$= \frac{n}{n+}\;\frac{1}{\alpha_0+}\;\frac{1}{\beta_1 n+}\;\frac{1}{\alpha_1+}\;\frac{1}{\beta_2 n+}\;\cdots,$$

show that

$$\alpha_s = \frac{2s+1}{k}\prod_{m=1}^{s}\left(\frac{m+k}{m-k}\right), \qquad \beta_s = \frac{2k}{s+k}\prod_{m=1}^{s-1}\left(\frac{m-k}{m+k}\right).$$

$$(s = 1, 2, \cdots)$$

Derive the form

$$1-\left(1-\frac{1}{n}\right)^{-k} = \frac{k}{n+}\;\frac{\frac{1(1+k)}{1\cdot 2}}{1+}\;\frac{\frac{1(1-k)}{2\cdot 3}}{n+}\;\frac{\frac{2(2+k)}{3\cdot 4}}{1+}\;\frac{\frac{2(2-k)}{4\cdot 5}}{n+}\;\cdots.$$

**Further Example 1.4**

Show that

$$\frac{1}{2\pi}\int_0^4 \frac{\sqrt{(4/t-1)}dt}{(x+t)\{\alpha t+(\alpha-1)^2\}\{\beta t+(\beta-1)^2\}\{\gamma t+(\gamma-1)^2\}\{\delta t+(\delta-1)^2\}}$$

$$= \frac{q_0}{x+}\;\frac{p_1}{1+}\;\frac{q_1}{x+}\;\frac{p_2}{1+\tfrac{1}{2}q_2\{\sqrt{(4/x+1)}-1\}} = I(x;\alpha,\beta,\gamma,\delta)$$

where

$q_0 = \Delta_3/(\Delta_1\Delta_2)$, $p_1 = \Delta_1\Delta_4/\Delta_3$, $q_1 = \Delta_2/(\Delta_3\Delta_4)$, $p_2 = \Delta_3/\Delta_4$, $q_2 = \Delta_4$;

$\Delta_1 = (1-\alpha)(1-\beta)(1-\gamma)(1-\delta)$,

$\Delta_2 = (1-\alpha\beta)(1-\alpha\gamma)(1-\alpha\delta)(1-\beta\gamma)(1-\beta\delta)(1-\gamma\delta)$,

$\Delta_3 = 1-(\alpha\beta\gamma+\alpha\beta\delta+\alpha\gamma\delta+\beta\gamma\delta)+\alpha\beta\gamma\delta(\alpha+\beta+\gamma+\delta)-(\alpha\beta\gamma\delta)^2$,

$\Delta_4 = (1-\alpha\beta\gamma\delta)$.

$(0 \leqslant \alpha, \beta, \gamma, \delta < 1;\ x > 0$ or complex and not in $(0, -4))$

Moreover

$$I(x;\alpha,\beta,\gamma,\delta) = \frac{q_0\{(x+q_1)(2+q_2K)+2p_2\}}{(2+q_2K)\{x^2+x(p_1+q_1)\}+2p_2(x+p_1)}.$$

$$(K = \sqrt{(4/x+1)}-1)$$

# 2.

# SERIES AND CONTINUED FRACTIONS

## 2.1 Background

Series expressed as c.f.s have a long history: the subject reaches back to around 1800. Perron (1957) quotes a relevant result of Kausler (1803) in which a determinantal form is given for the conversion. Perron considers the series $1+c_1x+c_2x^2+\cdots$ and its c.f. form

$$1+\frac{a_1x}{1+}\;\frac{a_2x}{1+}\;\cdots \tag{2.1}$$

where

$$a_{2s}=-\frac{\psi_{s+1}\phi_{s-1}}{\phi_s\psi_s},\qquad a_{2s+1}=-\frac{\phi_{s+1}\psi_s}{\psi_{s+1}\phi_s},\qquad (s=1,2,\cdots)$$

$$a_1=\phi_1,\quad \phi_0=1,\quad \psi_1=1,$$

$$\phi_s=|c_1c_3\cdots c_{2s-1}|,\ \psi_s=|c_2c_4\cdots c_{2s-2}|.$$

There are several algorithms available to-day to determine the c.f. partial numerators in (2.1) when they exist (see for example, Wall (1948, p198), Brezinski (1980b, Appendix), Bowman and Shenton (1988, p208). These algorithms generally avoid the evaluation of determinants. Various algorithms associated with Padé approximants are given in Brezinski (1979).

We note that even with a desk calculator algorithms can be set up to include 12 or so coefficients. (Loss of accuracy may arise even in early numerators if large numbers are involved, as for example occurs in the double factorial; this loss is not always easy to detect.) An example is worth recording (readers may like to compare results derived from a desk calculator):

**Example 2.1**

$$I_n = \frac{1}{2\pi}\int_0^4 \frac{\sqrt{(4/t-1)}}{n+t}(t-1)^6 dt \,,$$

$$\sim \frac{15}{n} - \frac{51}{n^2} + \frac{178}{n^3} - \frac{628}{n^4} + \frac{2236}{n^5} - \frac{8025}{n^6} + \frac{29004}{n^7} - \frac{105477}{n^8}$$

$$+ \frac{385698}{n^9} - \frac{1417341}{n^{10}} + \frac{5231460}{n^{11}} - \frac{19386859}{n^{12}} + \cdots ,$$

$$\sim \frac{15}{n+}\;\frac{17/5}{1+}\;\frac{23/75}{n+}\;\frac{179/115}{1+}\;\frac{3700/1587}{n+}\;\frac{6335/3404}{1+}$$

$$\frac{9039/21904}{n+}\;\frac{54597/19388}{1+}\;\frac{44548/85805}{n+}$$

$$\frac{10579/5633}{1+}\;\frac{16113/9245}{n+}\;\frac{16113/9245}{1+}\;\cdots .$$

The integral can be computed by quadrature (taking note of the singularity at the origin; desk algebraic computers do not always succeed with this type of integral) or evaluated as

$$I_n = \frac{(n+1)^6}{2}\{-1+\sqrt{(\frac{4}{n}+1)}\} + 2 - 9n - 12n^2 - 11n^3 - 5n^4 - n^5 .$$

The answer for $n = 4/3$ is $4765/1458 = 3.2681756$.

| Approximants ($A_s$) | | |
|---|---|---|
| $s$ | $A_{2s+1}$ | $A_{2s+2}$ |
| 0 | 11.25 | 3.169014 |
| 1 | 3.320122 | 3.241691 |
| 2 | 3.270278 | 3.267007 |
| 3 | 3.268652 | 3.268117 |
| 4 | 3.268207 | 3.268164 |
| 5 | 3.268177 | |

One can always check for loss of accuracy by testing the algorithm against a classical example, such as the c.f. for the factorial series $1/n - 1!/n^2 + 2!/n^3 \cdots$.

We recall that Stieltjes (1918, p425-429) in considering the c.f. for

$$\frac{e_0}{n} - \frac{e_1}{n^2} + \frac{e_2}{n^3} \quad \cdots \tag{2.2}$$

in the form

$$\frac{1}{a_1 n +}\ \frac{1}{a_2 +}\ \frac{1}{a_3 n +}\ \cdots,$$

showed that

$$a_{2s} = A_s^2 / (B_s B_{s-1}), \qquad a_{2s+1} = B_s^2 / (A_s A_{s+1}), \tag{2.3}$$

where

$$A_0 = B_0 = 1,$$

$$A_s = | e_0 e_2 \cdots e_{2s-2} |, \quad B_s = | e_1 e_3 \cdots e_{2s-1} |.$$

In addition he showed that

$$a_1 + a_3 + \cdots + a_{2s+1} = C_s / A_{s+1}. \qquad (C_s = | e_2 e_4 \cdots e_{2s} |) \tag{2.4}$$

One of our main interests is the information that can be obtained from series which arise from more or less intractable functions (multiple integrals of high dimensionality, for example). So we consider the effects of slight modifications in series on the rational fraction summation algorithm. Whereas Taylor series are scarcely less manageable because of the modification of a particular coefficient, or the addition of a new term and other elementary algebraic processes (addition, differentiation, etc.), this is certainly not the case with the c.f. representation. Stieltjes addressed this problem in the context of the mechanical moment problem. For example, what happens to the c.f. structure when the mass at the origin is increased, noting that a negative mass at the origin or other location is inadmissible in the moment problem. For statistical applications the moments are non-central and

$$\mu_s' = \int_0^\infty t^s \, d\sigma(t), \qquad (s = 0, 1, \cdots) \tag{2.5}$$

so that (if $\mu_0' = 1$) $\mu_1'$ is the mean, $\mu_2' - \mu_1'^2$ the variance, and so on; in mechanics these relate to the mass center and radius of gyration. Clearly an arbitrary set $\{\mu_s'\}$ may not satisfy the many restraints implied in (2.5) when $\sigma(\cdot)$ is in fact a distribution function.

The chapter concludes with several examples of different types of moment series occurring in statistics. A rich field of examples of divergent (or slowly convergent) series is brought to light for commonly occurring means and variances of statistics in applied science. In the past such examples have been rare so that classical (idealistic) series have been deeply researched and many elegant properties highlighted; we remark that the c.f. associated with $\ln \Gamma(x)$ (the series part) does not have much appeal because the partial numerators are not available in simple closed form, but the c.f. for the single factorial series has remarkable simplicity.

In general statistical series rarely conform to the latter, and we have to escape the bondage of the aesthetically pleasing and enjoy the rewards of mathematical consistency.

Note, an unusual series with surges is briefly described in Appendix I.

## 2.2 Convergency

Series and their c.f. forms may both converge in some domain, converge and diverge, diverge and converge, or diverge and diverge. Our applications often are concerned with apparently divergent series and the possibility of transformation to convergent c.f. equivalents.

Perron (1957, p146) considers the function

$$R(x) = \{1 + 6x + \sqrt{(1 - 24x^3)}\} / \{2(1 + x)\} \tag{2.6}$$

with c.f. development (periodic)

$$R_c(x) = 1 + \frac{2x}{1+}\ \frac{x}{1-}\ \frac{3x}{1+}\ \frac{2x}{1+}\ \frac{x}{1-}\ \frac{3x}{1+}\ \cdots \tag{2.7}$$

and series

$$\begin{aligned} R_s(x) = {} & 1 + 2x - 2x^2 - 4x^3 + 4x^4 - 4x^5 - 32x^6 + 32x^7 - 32x^8 \\ & - 400x^9 + 400x^{10} - 400x^{11} - 6080x^{12} + 6080x^{13} - 6080x^{14} \\ & - 102784x^{15} + 102784x^{16} - 102784x^{17} - 1856768x^{18} \\ & + 1856768x^{19} - 1856768x^{20} - 35094784x^{21} + 35094784x^{22} \\ & - 35094784x^{23} - 685460480x^{24} + 685460480x^{25} \\ & - 685460480x^{26} - 13725644800x^{27} + 13725644800x^{28} \\ & - 13725644800x^{29} + \cdots . \end{aligned} \tag{2.8}$$

Perron gives the following convergence comparisons:

| Argument Value | Series | c.f. |
|---|---|---|
| $-\frac{1}{\sqrt[3]{24}} \leqslant x < \frac{1}{3}$ | Convergent | Convergent |
| $x = 1/3$ | Convergent | Divergent |
| $\frac{1}{3} < x \leqslant \frac{1}{\sqrt[3]{24}}$ | Convergent | Convergent |
| $x = -1$ | Divergent | Divergent |
| $x < -1$ | Divergent | Convergent |
| $x > \frac{1}{\sqrt[3]{24}}$ | Divergent | Divergent |

When $x = 1/3$ the c.f. becomes

$$R_c(\tfrac{1}{3}) = 1 + \frac{2/3}{1+}\,\frac{1/3}{1-}\,\frac{1}{1+}\,\frac{2/3}{1+}\,\frac{1/3}{1-}\,\frac{1}{1+}\,\cdots.$$

Approximants are $C_s^* = 1 + C_s$ where $C_1 = 2/3$, $C_2 = 1/2$ etc.

| $s$ | $C_s$ |
|---|---|
| 2 | 0.5 |
| 3 | 0.0 |
| 4 | $0.36\dot{3}\dot{6}$ |
| 5 | $0.333\dot{3}$ |
| 6 | 0.0 |
| 7 | 0.2963 |

Summing the c.f. backwards it is easy to see that $C_{3s} = 0$, $s = 1, 2, \cdots$. Similarly writing

$$p = \frac{1/3}{1-}\,\frac{1}{1+}\,\frac{2/3}{1+}\,\frac{1/3}{1-}\,\frac{1}{1+}\,\frac{2/3}{1+}\,\cdots$$

we have

$$p = \frac{1/3}{1-}\,\frac{1}{1+}\,\frac{2/3}{1+p},$$

so $p = 5/3$ and $C_{3s+1} \to 1/4$ as $s \to \infty$; similarly $C_{3s+2} \to 1/4$. Hence the c.f. diverges by oscillation. The series (2.8) does converge although convergence is slow. Writing $S_r(1/3)$ for the partial sum of $r$ terms we have

| $r$ | $S_r(1/3)$ | $r$ | $S_r(1/3)$ |
|---|---|---|---|
| 4 | 1.296 | 31 | 1.2538 |
| 7 | 1.285 | 32 | 1.2543 |
| 10 | 1.275 | 33 | 1.2541 |
| 13 | 1.268 | 34 | 1.2531 |
| 19 | 1.260 | 35 | 1.2534 |
| 22 | 1.258 | 36 | 1.25331 |
| 25 | 1.256 | 37 | 1.25249 |
| 28 | 1.2548 | 38 | 1.25276 |
| 29 | 1.2554 | 39 | 1.252672 |
| 30 | 1.2552 | 40 | 1.252031 |
| | | $\infty$ | 1.250000 |

As an example of a diverging series but convergent c.f. take $x = -2$; the series now has a sign pattern of period 6, and clearly oscillates between $-\infty$ and $\infty$. The c.f. progresses towards its limit -1.446222 slowly. For example,

| $s$ | $C_s^*$ | $s$ | $C_s^*$ |
|---|---|---|---|
| 30 | -0.778 | 138 | -1.442 |
| 33 | -2.082 | 141 | -1.450 |
| 45 | -1.796 | 300 | -1.446220 |
| 48 | -1.157 | 301 | -1.446223 |
| 78 | -1.377 | 303 | -1.446223 |
| 81 | -1.507 | 304 | -1.446221 |

## 2.3 Modification of Series: Changes of Mass at the Origin

Suppose the moments $(\mu_0', \mu_1', \cdots)$ are replaced by $(\mu_0' + \mu, \mu_1', \cdots)$; the fact that $\mu_0' + \mu > 0$ may be different from unity is of no consequence here. If the Stieltjes moment problem is determined for the original set of moments, then it is determined for the enhanced set of moments. Hence the determinantal form in (2.3) yields the modified c.f.

$$\frac{1}{\delta_1' n +} \; \frac{1}{\delta_2' +} \; \frac{1}{\delta_3' n +} \; \cdots \tag{2.9}$$

where

$$\delta_{2s+1}' = \frac{1}{\mu}\left\{\frac{1}{1+\mu\sum_{m=1}^{s}\delta_{2m-1}} - \frac{1}{1+\mu\sum_{m=1}^{s+1}\delta_{2m-1}}\right\}. \qquad (\mu > 0)$$

From this

$$\delta_1' + \delta_3' + \cdots + \delta_{2s+1}' = \frac{\sum_{m=1}^{s+1}\delta_{2m-1}}{1+\mu\sum_{m=1}^{s+1}\delta_{2m-1}}. \tag{2.10}$$

Moreover

$$\delta_{2s}' = (1+\mu\sum_{m=1}^{s}\delta_{2m-1})^2\delta_{2s}. \tag{2.11}$$

Thus we merely replace $A_s$ in (2.3) by

$$A_s' = A_s + \mu C_{s-1}$$

and use (2.4); note that the latter is an example of expression (8) in Shenton (1949).

Note that the c.f. derived from the addition of $\alpha^2+\beta^2$ to the mass at the origin, is the same as the two stage process, add $\alpha^2$ (or $\beta^2$) followed by $\beta^2$ (or $\alpha^2$).

For adding $\alpha^2$, produces the first modified c.f.

$$\frac{1}{\delta_1' n +}\;\frac{1}{\delta_2' +}\;\frac{1}{\delta_3' n +}\;\cdots,$$

where

$$\left\{\begin{aligned}\delta_{2s}' &= (1+\alpha^2\sum_{m=1}^{s}\delta_{2m-1})^2\delta_{2s},\\ \delta_{2s+1}' &= \frac{1}{\alpha^2}\left\{\frac{1}{1+\alpha^2\sum_{m=1}^{s}\delta_{2m-1}}-\frac{1}{1+\alpha^2\sum_{m=1}^{s+1}\delta_{2m-1}}\right\}.\end{aligned}\right.$$

Now add $\beta^2$ to the new "zero" term making it $\beta^2+(\alpha^2+e_0)$, with the modified form

$$\frac{1}{\delta_1'' n +}\;\frac{1}{\delta_2'' +}\;\frac{1}{\delta_3'' n +}\;\cdots$$

where

$$\begin{aligned}\delta_{2s}'' &= (1+\beta^2\sum_{m=1}^{s}\delta_{2m-1}')^2\delta_{2s}'\\ &= \left\{1+\frac{\beta^2\sum_{m=1}^{s+1}\delta_{2m-1}}{1+\alpha^2\sum_{m=1}^{s+1}\delta_{2m-1}}\right\}(1+\alpha^2\sum_{m=1}^{s+1}\delta_{2m-1})^2\delta_{2s}\\ &= \{1+(\alpha^2+\beta^2)\sum_{m=1}^{s+1}\delta_{2m-1}\}^2\delta_{2s}.\end{aligned}$$

Similarly

$$\delta_{2s+1}'' = \frac{\delta_{2s+1}'}{(1+\beta^2\sum_{m=1}^{s}\delta_{2m-1}')(1+\beta^2\sum_{m=1}^{s+1}\delta_{2m-1}')},$$

$$= \frac{\delta'_{2s+1}}{\left[1 + \dfrac{\beta^2 \sum_{m=1}^{s} \delta_{2m-1}}{1 + \alpha^2 \sum_{m=1}^{s} \delta_{2m-1}}\right]\left[1 + \dfrac{\beta^2 \sum_{m=1}^{s+1} \delta_{2m-1}}{1 + \alpha^2 \sum_{m=1}^{s+1} \delta_{2m-1}}\right]};$$

$$\text{i.e. } \delta''_{2s+1} = \frac{(1 + \alpha^2 \sum_{m=1}^{s} \delta_{2m-1})(1 + \alpha^2 \sum_{m=1}^{s+1} \delta_{2m-1})\delta'_{2s+1}}{\{1 + (\alpha^2 + \beta^2) \sum_{m=1}^{s} \delta_{2m-1}\}\{1 + (\alpha^2 + \beta^2) \sum_{m=1}^{s+1} \delta_{2m-1}\}}$$

$$= \frac{\delta_{2s+1}}{\{1 + (\alpha^2 + \beta^2) \sum_{m=1}^{s} \delta_{2m-1}\}\{1 + (\alpha^2 + \beta^2) \sum_{m=1}^{s+1} \delta_{2m-1}\}}$$

as expected.

**Example 2.2**

$$\ln(1 + \frac{1}{n}) = \frac{1}{n +}\ \frac{1}{2/1^2 +}\ \frac{1}{3n +}\ \frac{1}{4/2^2 +}\ \frac{1}{5n +}\ \frac{1}{6/3^2 +} \cdots$$
$$\sim \frac{1}{n} - \frac{1}{2n^2} + \frac{1}{3n^3} - \cdots .$$

What is the c.f. for

$$\frac{1 + \mu}{n} - \frac{1}{2n^2} + \frac{1}{3n^3} - \cdots \ ?$$

This series corresponds to a Stieltjes transform $\int_0^1 \frac{d\sigma(t)}{t + n}$, where the distribution (not standardized) is uniform on (0, 1) with the addition of a point mass $u$ at the origin. The c.f. using the result of Stieltjes is

$$\frac{\mu}{n} + \ln(1 + \frac{1}{n}) = \frac{1}{\delta_1 n +}\ \frac{1}{\delta_2 +}\ \frac{1}{\delta_3 n +}\ \frac{1}{\delta_4 +} \cdots , \qquad (\mu \geqslant 0)$$

where

$$\begin{cases} \delta_{2s+1} = (2s + 1)/\{(1 + \mu s^2)(1 + \mu(s + 1)^2\}, \\ \delta_{2s} = (2s / s^2)(1 + \mu s^2)^2 . \quad (s = 0, 1, \cdots) \end{cases}$$

For $n = 2$, $\mu = 1$, successive approximants to 0.905465 are:

| | | Convergents from the c.f. for $\ln(1+\frac{1}{n})$ | |
|---|---|---|---|
| $s$ | Convergent | $s$ | Convergent |
| 1 | $<1$ | 1 | $<1$ |
| 2 | $>0.888889$ | 2 | $>0.9$ |
| 3 | $<0.906250$ | 3 | $<0.906250$ |
| 4 | $>0.905325$ | 4 | $>0.905405$ |
| 5 | $<0.905473$ | 5 | $<0.905473$ |
| 6 | $>0.905463906$ | 6 | $>0.905464481$ |
| 7 | $<0.905465183$ | 7 | $<0.905465183$ |

It is surprising that odd approximants coincide? No, because an odd part of a c.f. for a Stieltjes transform is formed from the series with its first term truncated (see (1.25)).

**Example 2.3 Poisson Distribution Function Modified**

$$\sum_{x=0}^{\infty}\frac{e^{-\theta}\theta^{x}}{(x+n)x!} = \frac{1}{n+}\,\frac{\theta}{1+}\,\frac{1}{n+}\,\frac{\theta}{1+}\,\frac{2}{n+}\,\frac{\theta}{1+}\,\frac{3}{n+}\,\frac{\theta}{1+}\cdots, \qquad (n,\theta>0)$$

$$= \frac{1}{\frac{n}{0!}+}\,\frac{1}{\frac{0!}{\theta}+}\,\frac{1}{\frac{n\theta}{1!}+}\,\frac{1}{\frac{1!}{\theta^{2}}+}\,\frac{1}{\frac{n\theta^{2}}{2!}+}\,\frac{1}{\frac{2!}{\theta^{3}}+}\,\frac{1}{\frac{n\theta^{3}}{3!}+}\,\frac{1}{\frac{3!}{\theta^{4}}+}\cdots,$$

$$\sim \frac{1}{n}-\sum_{\lambda=1}^{1}\frac{S_1^{(\lambda)}\theta^{\lambda}}{n^{2}}+\sum_{\lambda=1}^{2}\frac{S_2^{(\lambda)}\theta^{\lambda}}{n^{3}}-\sum_{\lambda=1}^{3}\frac{S_3^{(\lambda)}\theta^{\lambda}}{n^{3}}+\cdots.$$

($S_k^{(\lambda)}$ Stirling number of the second kind.)

Now add mass $\mu$ ($\geqslant 0$) at the origin. Then

$$\frac{\mu}{n}+\sum_{x=0}^{\infty}\frac{e^{-\theta}\theta^{x}}{(x+n)x!} = \frac{1}{n\delta_1+}\,\frac{1}{\delta_2+}\,\frac{1}{n\delta_3+}\cdots$$

$$\sim \frac{1+\mu}{n}-\frac{\theta}{n^{2}}+\frac{\theta+\theta^{2}}{n^{3}}-\frac{\theta+3\theta^{2}+\theta^{3}}{n^{4}}+\cdots,$$

where

$$\delta_{2s} = (1 + \mu M_s)^2 (s-1)!/\theta^s ,$$
$$\delta_{2s+1} = (\theta^s/s!)/\{(1+\mu M_{s-1})(1+\mu M_s)\} ,$$
$$M_s = \sum_{k=0}^{s} \theta^k/k! . \qquad (M_{-1} = 0)$$

Numerical $n = 2, \mu = 3, \theta = 1$:

$$\frac{3}{2} + \sum \frac{e^{-1}}{x!(x+2)} = \frac{3}{2} + e^{-1}$$
$$= \frac{1}{\frac{1}{2}+} \; \frac{1}{\frac{16}{1}+} \; \frac{1}{\frac{2}{28}+} \; \frac{1}{\frac{49}{1}+} \; \frac{1}{\frac{2}{119}+} \; \frac{1}{\frac{289}{2}+}$$
$$\frac{1}{\frac{2}{459}+} \; \frac{1}{\frac{486}{1}+} \; \frac{1}{\frac{2}{1971}+} \; \cdots ,$$

with approximants 2, 1.777778, 1.875000, 1.861538, 1.868421, 1.867268, 1.867925, 1.867814, 1.867883.

**Example 2.4**

The factorial series

$$\frac{1}{n} - \frac{1!}{n^2} + \frac{2!}{n^3} - \cdots$$

sums to

$$\frac{1}{n+} \; \frac{1}{1+} \; \frac{1}{n+} \; \frac{1}{1/2+} \; \frac{1}{n+} \; \frac{1}{1/3+} \; \frac{1}{n+} \; \frac{1}{1/4+} \; \frac{1}{n+} \; \cdots . \qquad (n > 0)$$

The modified series

$$\frac{(1+\mu)}{n} - \frac{1!}{n^2} + \frac{2!}{n^3} - \cdots$$

sums to

$$f(n,\mu) = \frac{1+\mu}{n+} \; \frac{\frac{1}{1+\mu}}{1+} \; \frac{\frac{1+2\mu}{1+\mu}}{n+} \; \frac{\frac{2(1+\mu)}{1+2\mu}}{1+} \; \frac{\frac{2(1+3\mu)}{1+2\mu}}{n+}$$
$$\frac{\frac{3(1+2\mu)}{1+3\mu}}{1+} \; \frac{\frac{3(1+4\mu)}{1+3\mu}}{n+} \; \frac{\frac{4(1+3\mu)}{1+4\mu}}{1+} \; \frac{\frac{4(1+5\mu)}{1+4\mu}}{n+} \; \cdots ,$$

$$= \frac{1+\mu}{n+} \; \frac{1}{1+\mu+} \; \frac{1+2\mu}{n+} \; \frac{2(1+\mu)}{1+2\mu+} \; \frac{2(1+3\mu)}{n+}$$

$$\frac{3(1+2\mu)}{1+3\mu+} \; \frac{3(1+4\mu)}{n+} \; \frac{4(1+3\mu)}{1+4\mu+} \; \frac{4(1+5\mu)}{n+} \; \cdots .$$

It is clear that $f\,(n\,,0)$ reverts to the original and that it is not possible to substitute $\mu = -1$.

When $n = 5$, $\mu = 4$, we have

$$f\,(5,4) = \frac{5}{5+} \; \frac{\frac{1}{5}}{1+} \; \frac{\frac{9}{5}}{5+} \; \frac{\frac{10}{9}}{1+} \; \frac{\frac{26}{9}}{5+} \; \frac{\frac{27}{13}}{1+} \; \frac{\frac{51}{13}}{5+} \; \frac{\frac{52}{17}}{1+}$$

$$\frac{\frac{84}{17}}{5+} \; \frac{\frac{85}{21}}{1+} \; \frac{\frac{125}{21}}{5+} \; \frac{\frac{126}{25}}{1+} \; \frac{\frac{174}{25}}{5+} \; \cdots$$

with convergents

| $s$ | $C_s$ | $s$ | $C_s$ |
|---|---|---|---|
| 1 | 1 | 8 | 0.970418 |
| 2 | 0.9615 | 9 | 0.970423 |
| 3 | 0.9714 | 10 | 0.97042169 |
| 4 | 0.9700 | 11 | 0.97042232 |
| 5 | 0.97049 | 12 | 0.97042210 |
| 6 | 0.97039 | 13 | 0.97042220 |
| 7 | 0.970429 | | |

## 2.4 Modification of Series: Linear Transformation

Consider

$$f\,(n\,,k\,) = \sum_{s=0}^{\infty} (-1)^s\,(\mu'_{s+1} - k\,\mu'_s)/\,n^s\;, \tag{2.12a}$$

and its Stieltjes sum

$$f_s(n\,,k\,) = \int_0^{\infty} \frac{(t\,-k\,)d\,\sigma(t\,)}{n\,+t}. \qquad (k\,<0) \tag{2.12b}$$

The J-fraction form being

$$\frac{b_0}{n\,+a_1-} \; \frac{b_1}{n\,+a_2-} \; \frac{b_2}{n\,+a_3-} \cdots ,$$

relates to the orthogonal set $\{p_s(t)\}$ where

$$p_s(t) = (t - a_s)p_{s-1}(t) - b_{s-1}p_{s-2}(t), \qquad (s = 2, 3, \cdots) \qquad (2.12c)$$

with $p_0 = 1$, $p_1 = t - a_1$. The c.f. for $f_s(n, k)$ is readily constructed if we can set up the orthogonal polynomials $\{q_r(t)\}$ with respect to $(t - k)d\sigma(t)$, such that

$$\int_0^\infty q_r(t)q_s(t)(t - k)d\sigma(t) = \delta_{rs}\phi_s^*. \qquad (s = 0, 1, \cdots) \qquad (2.13)$$

But from Christoffel's theorem (Szegö 1939), at least formally,

$$q_s(t) = \begin{vmatrix} p_s(t) & p_{s+1}(t) \\ p_s(k) & p_{s+1}(k) \end{vmatrix} \div \{(k - t)p_s(k)\} \qquad (2.14a)$$

so that using the recurrence for $p_s(t)$ given in (2.12c) we find

$$\begin{aligned} q_s(t) = \; & p_s(t) + b_s p_{s-1}(t)p_{s-1}(k)/p_s(k) \\ & + b_s b_{s-1}p_{s-2}(t)p_{s-2}(k)/p_s(k) \\ & + \quad \cdots \quad + b_s b_{s-1} \cdots b_1/p_s(k). \end{aligned} \qquad (2.14b)$$

Evidently from (2.14a) and (2.14b)

$$\phi_s^* = -p_{s+1}(k)\phi_s/p_s(k). \qquad (2.15)$$

Let the recurrence for the set $\{q_r(\cdot)\}$ be

$$q_s(t) = (t - a_s^*)q_{s-1}(t) - b_{s-1}^* q_{s-2}(t), \qquad (2.16)$$

with $q_0(t) = 1$, $q_1(t) = t - a_1^*$. Multiplication by $q_{s-2}(t)$ and use of (2.13) gives

$$b_{s-1}^* \phi_{s-2}^* = \phi_{s-1}^*$$

so

$$b_{s-1}^* = \frac{p_{s-2}(k)p_s(k)}{p_{s-1}^2(k)} \cdot \frac{\phi_{s-1}}{\phi_{s-2}}.$$

But in the same way from (2.16)

$$\phi_s = b_s b_{s-1} \cdots b_1 b_0$$

giving

$$b_s^* = \frac{b_s p_{s-1}(k) p_{s+1}(k)}{p_s^2(k)} . \qquad (s = 1, 2, \cdots) \tag{2.17}$$

To determine $a_s^*$, equate coefficients $t^{s-1}$ in (2.16). If

$$q_s(t) = t^s + c_s^{(1)} t^{s-1} + \cdots$$

and

$$p_s(t) = t^s + \lambda_s^{(1)} t^{s-1} + \cdots$$

then

$$c_s^{(1)} = -a_s^* + c_{s-1}^{(1)} .$$

But from (2.14b)

$$c_s^{(1)} = \lambda_s^{(1)} + b_s p_{s-1}(k) / p_s(k)$$

so

$$\begin{aligned} a_s^* &= c_{s-1}^{(1)} - c_s^{(1)} \\ &= \lambda_{s-1}^{(1)} + b_{s-1} p_{s-2}(k) / p_{s-1}(k) - \lambda_s^{(1)} - b_s p_{s-1}(k) / p_s(k) \end{aligned}$$

where

$$\lambda_s^{(1)} = -a_s + \lambda_{s-1}^{(1)} .$$

Hence

$$a_s^* = a_s + b_{s-1} p_{s-2}(k) / p_{s-1}(k) - b_s p_{s-1}(k) / p_s(k) . \tag{2.18a}$$

An alternative, using (2.12c) is

$$a_s^* = a_{s+1} + \frac{p_{s+1}(k) p_{s-1}(k) - p_s^2(k)}{p_{s-1}(k) p_s(k)} . \tag{2.18b}$$

Hence there is the formal result that

$$f_s(n, k) = \frac{b_0(b_1 - k)}{n + a_1^* -} \; \frac{b_1^*}{n + a_2^* -} \; \frac{b_2^*}{n + a_3^* -} \cdots . \tag{2.19}$$

In summary then if the Stieltjes moment problem is determined and

$$\int_0^\infty \frac{d\sigma(t)}{n + t} = \frac{b_0}{n + a_1 -} \; \frac{b_1}{n + a_2 -} \cdots , \tag{2.20}$$

then

$$\int_0^\infty \frac{(t+k)d\sigma(t)}{n+t} = \frac{b_0\omega_1(k)}{n+a_2+D_2(k)-}\ \frac{\dfrac{b_1\omega_0(k)\omega_2(k)}{\omega_1^2(k)}}{n+a_3+D_3(k)-}\ \frac{\dfrac{b_2\omega_1(k)\omega_3(k)}{\omega_2^2(k)}}{n+a_4+D_4(k)-}\cdots \tag{2.21}$$

where with $k>0$,

(i) $$\frac{b_0}{k+a_1-}\ \frac{b_1}{k+a_2-}\ \cdots\ \frac{b_{s-1}}{k+a_s} = \frac{\chi_s(k)}{\omega_s(k)},$$

$(\omega_0 = 1,\quad \omega_1 = k+a_1,\ \text{etc.})$

(ii) $$D_s(k) = \frac{\omega_{s-1}(k)}{\omega_{s-2}(k)} - \frac{\omega_s(k)}{\omega_{s-1}(k)}.$$

### Example 2.5

As a simple case, take $\sigma(t) = \int_0^t e^{-x}\,dx\ ,\quad t>0.$

(a) $k = 1$

$$\int_0^\infty \frac{(t+1)e^{-t}\,dt}{n+t} = \frac{2}{n+1.5-}\ \frac{1.75}{n+A_2-}\ \frac{B_2}{n+A_3-}\ \cdots$$

where

| $s$ | $A_s$ | $B_s$ | $s$ | $A_s$ | $B_s$ |
|---|---|---|---|---|---|
| 2 | 0.3642856 01 | 0.5551020 01 | 13 | 0.2586182 02 | 0.1804456 03 |
| 3 | 0.5710084 01 | 0.1139014 02 | 14 | 0.2786683 02 | 0.2083775 03 |
| 4 | 0.7749930 01 | 0.1925377 02 | 15 | 0.2987133 02 | 0.2383118 03 |
| 5 | 0.9776819 02 | 0.2913397 02 | 16 | 0.3187541 02 | 0.2702482 03 |
| 6 | 0.1179649 02 | 0.4102606 02 | 17 | 0.3387911 02 | 0.3041867 03 |
| 7 | 0.1381169 02 | 0.5492712 02 | 18 | 0.3588251 02 | 0.3401269 03 |
| 8 | 0.1582390 02 | 0.7083524 02 | 19 | 0.3788563 02 | 0.3780687 03 |
| 9 | 0.1783398 02 | 0.8874907 02 | 20 | 0.3988851 02 | 0.4180121 03 |
| 10 | 0.1984250 02 | 0.1086677 03 | 21 | 0.4189119 02 | 0.4599569 03 |
| 11 | 0.2184982 02 | 0.1305903 03 | 22 | 0.4389368 02 | 0.5039030 03 |
| 12 | 0.2385620 02 | 0.1545164 03 | 23 | 0.4589601 02 | 0.5498503 03 |

(b) $k = n$

$$\int_0^\infty e^{-t}\,dt = \int_0^\infty \frac{(t+n)e^{-t}\,dt}{t+n} = 1, \qquad (n>0)$$

$$= \frac{\omega_1(n)}{n+3+D_2(n)-}\;\frac{1^2\frac{\omega_0\omega_2}{\omega_1^2}}{n+5+D_3(n)-}\;\frac{2^2\frac{\omega_1\omega_3}{\omega_2^2}}{n+7+D_4(n)-}\cdots .$$

In particular for $n = k = 0$,

$$\int_0^\infty e^{-x}\,dx = 1 = \frac{1}{2-}\;\frac{1\cdot 2}{4-}\;\frac{2\cdot 3}{6-}\cdots = \frac{1}{2-}\;\frac{1}{2-}\;\frac{1}{2-}\cdots .$$

Again if $n = k = 2$,

$$\int_0^\infty e^{-t}\,dt = 1,$$

$$= \frac{3}{10/3-}\;\frac{14/9}{116/21-}\;\frac{258/49}{2290/301-}\;\frac{20412/1849}{33640/3483-}\cdots ,$$

$$= \frac{3}{2+}\;\frac{4/3}{1+}\;\frac{7/6}{2+}\;\frac{33/14}{1+}\;\frac{172/77}{2+}\;\frac{1596/473}{1+}\;\frac{56133/17157}{2+}\;\frac{3409861/463239}{1+}\cdots$$

with approximants to unity, 1.5, 0.9, $1.0\dot{5}$, 0.983051, 1.010135, 0.996218, 1.002383, 0.998240. The basic c.f. is

$$\frac{1}{n+1-}\;\frac{1^2}{n+3-}\;\frac{2^2}{n+5-}\cdots .$$

**Example 2.6 An Irregular Case for Discussion**

For the Mills' ratio $R(t) = e^{t^2/2}\int_t^\infty e^{-x^2/2}dx$, we have

$$R(t) \sim \frac{1}{t} - \frac{1}{t^3} + \frac{1\cdot 3}{t^5} - \frac{1\cdot 3\cdot 5}{t^7} + \cdots ,$$

and

$$\frac{R(\sqrt{n})}{\sqrt{n}} \sim \frac{1}{n} - \frac{1}{n^2} + \frac{1\cdot 3}{n^3} - \frac{1\cdot 3\cdot 5}{n^4} + \cdots = \int_0^\infty \frac{d\sigma(t)}{t+n}, \qquad (n>0)$$

where

$$\sigma(t) = \int_0^t \frac{e^{-x/2}}{\sqrt{(2x\pi)}} dx .$$

From Chapter 1 (the even part of (1.51))

$$\frac{R(\sqrt{n})}{\sqrt{n}} = \frac{1}{n+1-}\ \frac{1\cdot 2}{n+5-}\ \frac{3\cdot 4}{n+9-}\ \frac{5\cdot 6}{n+13-}\ \cdots .$$

Then the series

$$\frac{1-k}{n} - \frac{1\cdot(3-k)}{n^2} + \frac{1\cdot 3(5-k)}{n^3} - \frac{1\cdot 3\cdot 5(7-k)}{n^4} + \cdots$$

formally sums to

$$\int_0^\infty \frac{(t-k)d\sigma(t)}{t+n} = 1 - \frac{(n+k)R(\sqrt{n})}{\sqrt{n}} .$$

Case (a) $k = 2$

*Series*: $-\frac{1}{n} - \frac{1}{n^2} + \frac{1\cdot 3^2}{n^3} - \frac{1\cdot 3\cdot 5^2}{n^4} + \frac{1\cdot 3\cdot 5\cdot 7^2}{n^5} - \cdots$

*Polynomials:*

| $s$ | $p_s(k)$ | $\omega_s(-k)$ | $s$ | $p_s(k)$ | $\omega_s(-k)$ |
|---|---|---|---|---|---|
| 0 | 1 | 1 | 4 | -103 | -103 |
| 1 | 1 | -1 | 5 | 257 | -257 |
| 2 | -5 | -5 | 6 | 4387 | 4387 |
| 3 | 23 | -23 | | | |

$$c.f.: \quad \frac{-1}{n-1+}\ \frac{10}{n+47/5-}\ \frac{276/25}{n+1509/(5\cdot 23)-}\ \cdots$$

continued as

| $s$ | $a^*_{s+1}$ | $b^*_s$ | $s$ | $a^*_{s+1}$ | $b^*_s$ |
|---|---|---|---|---|---|
| 3 | 18.983115239 | 29.206049149 | 9 | 40.987641769 | 302.645032163 |
| 4 | 40.565184542 | 31.201432746 | 10 | 45.211930033 | 380.242084965 |
| 5 | -22.802885959 | -615.716967706 | 11 | 49.950445987 | 456.955875215 |
| 6 | 38.654849644 | -237.652405575 | 12 | 55.867716057 | 524.673401899 |
| 7 | 34.464278340 | 124.823955344 | 13 | 66.532679193 | 547.854653824 |
| 8 | 37.215043780 | 223.327315458 | | | |

Approximants to the series

$$\frac{1}{n} + \frac{1}{n^2} - \frac{3^2}{n^3} + \frac{3\cdot 5^2}{n^4} - \cdots .$$

| | $n=1$ | $n=2$ | $n=3$ | $n=4$ | $n=10$ |
|---|---|---|---|---|---|
| 1 | $\infty$ | 1.00 | 0.5 | 0.33 | 0.111 |
| 2 | 1.04 | 0.53 | 0.356 | 0.267 | 0.105092 |
| 3 | 0.962 | 0.516 | 0.3504 | 0.2643 | 0.105082 |
| 4 | 0.9528 | 0.514 | 0.3499 | 0.2640 | 0.104942 |
| 5 | 0.9524 | 0.5144 | 0.34986 | 0.264041 | 0.1049418 |
| 10 | 0.967567 | 0.515766 | 0.350118 | 0.26410798 | 0.1049421735 |
| 11 | 0.967087 | 0.515749 | 0.3501162 | 0.26410776 | 0.10494217336 |
| 12 | 0.966925 | 0.515744 | 0.3501158 | 0.264107697 | 0.10494217334 |
| 13 | 0.966873 | 0.5157420 | 0.3501157 | 0.264107681 | 0.10494217334 |
| 14 | 0.966861 | 0.5157417 | 0.3501156 | 0.264107678 | 0.10494217334 |
| True | 0.968039 | 0.5157419 | 0.3501169 | 0.264107686 | 0.10494214 |

Comment: We can not prove convergence of the c.f. in this case. The anomolous signs for $a_5^*$, $a_6^*$, and $b_6^*$ should be noted; also observe the oscillatory tendency in successive convergents, making error analysis difficult. However we would opt for an approximation 0.967 for $n = 1$, 0.51574 for $n = 2$, and 0.104942173 for $n = 10$. Some caution in however needed, for loss of accuracy with later convergents can not be ruled out; computations were done using double-double precision on IBM System 360-75. We return to this example in the sequel, for clearly back-up assessments would be needed in practical situations where the precise value of the function might be out of reach.

Case (b). $k = -2$.

An appeal to the conditions for the existence of a unique solution to the Stieltjes moment problem shows in this case the Stieltjes c.f. exists and converges.

*Series*: $3/n - 1{\cdot}5/n^2 + 1{\cdot}3{\cdot}7/n^3 - 1{\cdot}3{\cdot}5{\cdot}9/n^4 + \cdots$

*Polynomials:*

| $s$ | $p_s(k)$ | $s$ | $p_s(k)$ | $s$ | $p_s(k)$ |
|---|---|---|---|---|---|
| 0 | 1 | 3 | -173 | 6 | 479851 |
| 1 | -3 | 4 | 2025 | 7 | -9156093 |
| 3 | 19 | 5 | -28787 | 8 | 196506001 |

*Continued Fractions:*

$$\int_0^\infty \frac{(t+2)d\sigma(t)}{t+n} = 1 - \frac{(n-2)R(\sqrt{n})}{\sqrt{n}} = \frac{3}{n+}\;\frac{p_1}{1+}\;\frac{q_1}{n+}\;\cdots,$$

$$\int_0^\infty \frac{d\sigma(t)}{t+n} = \frac{1}{n+}\;\frac{1}{1+}\;\frac{2}{n+}\;\frac{3}{1+}\;\frac{4}{n+}\;\cdots$$

$$= \frac{1}{n+1-}\;\frac{1{\cdot}2}{n+5-}\;\frac{3{\cdot}4}{n+9-}\;\cdots,$$

| $s$ | $p_s$ | $q_s$ |
|---|---|---|
| 1 | 1.666666667 | 2.533333333 |
| 2 | 3.694736842 | 4.669365722 |
| 3 | 5.730695124 | 6.729772337 |
| 4 | 7.759627503 | 8.764751535 |
| 5 | 9.782033665 | 11.98697123 |
| 6 | 10.60092919 | 14.25357457 |

(See Example 2.6).

Comment: The 12th and 13th convergents when $n = 9$, give the bounds 0.28928920, and 0.28928956, the true value being 0.289289303. Notice the regular magnitude pattern of the $p$'s and $q$'s.

## 2.5 Modification of Series: Change of $n$

It is readily seen that for the series $e_0/n - e_1/n^2 + e_2/n^3 - \cdots$ with Stieltjes sum

$$\int_0^\infty \frac{d\sigma(t)}{n+t} = \frac{b_0}{n+a_1-}\;\frac{b_1}{n+a_2-}\;\frac{b_2}{n+a_3-}\;\cdots,$$

that $\dfrac{e_0}{n+k} - \dfrac{e_1}{(n+k)^2} + \cdots$ sums to

$$\frac{b_0}{n+}\;\frac{\omega_1(k)}{1+}\;\frac{b_1/\omega_1(k)}{n+}\;\frac{\omega_2(k)/\omega_1(k)}{1+}\;\frac{b_2\omega_1(k)/\omega_2(k)}{n+}\;\frac{\omega_3(k)/\omega_2(k)}{1+}\;\cdots$$

and $\omega_1 = k + a_1$, $\omega_2 = (k + a_2)\omega_1 - b_1\omega_0$, etc.

Example 2.7

$$J_n(k) = \int_0^\infty \frac{e^{-t}\,dt}{n+k+t},$$

$$J_n(1) = \frac{1}{n+}\;\frac{2}{1+}\;\frac{1^2/2}{n+}\;\frac{7/2}{1+}\;\frac{2^2\cdot 2/7}{n+}\;\frac{34/7}{1+}\;\frac{3^2\cdot 7/34}{n+}\;\cdots$$

and when $n = 1$

$$J_1(1) = e^2E_1(2) = 0.361328619.$$

## 2.6 Modification of Series : Extending Series

Stieltjes (1918, pp544-5) considers the extended series

$$m/n - e_0/n^2 + e_1/n^3 - \cdots, \qquad (m > 0)$$

and its formal c.f. development

$$\frac{1}{a_1' n +}\ \frac{1}{a_2' +}\ \frac{1}{a_3' n +} \cdots,$$

where

$$\begin{cases} a_1' = 1/m, \\ a_{2s+1}' = \dfrac{1}{m - \sum_{r=1}^{s} a_{2r}} - \dfrac{1}{m - \sum_{r=1}^{s-1} a_{2r}}, \\ a_{2s}' = a_{2s-1}(m - \sum_{r=1}^{s-1} a_{2r})^2. \end{cases} \tag{2.22}$$

Clearly grave problems may arise in this case because of zero terms and the modified Stieltjes moment problem may not be determined.

**Example 2.8**

Consider Example 2.4, for which $a_{2s+1} = 1$, $a_{2s} = 1/s$, so that $\sum_{r=1}^{s} a_{2r} \sim \ln(s)$, $s \to \infty$. The c.f. for the series

$$3/n - 1/n^2 + 1/n^3 - 2!/n^4 + 3!/n^5 - \cdots$$

is

$$\frac{3}{n +}\ \frac{1/3}{1 +}\ \frac{2/3}{n +}\ \frac{3/2}{1 +}\ \frac{3/2}{n +}\ \frac{8/3}{1 +}\ \frac{3.857143}{n +}$$
$$\frac{3.142857}{1 +}\ \frac{5.090909}{n +}\ \frac{3.909090}{1 +}\ \frac{6.395352}{n +} \cdots.$$

| Convergents when $n = 5$ | | |
|---|---|---|
| $s$ | $C_{2s}$ | $C_{2s+1}$ |
| 0 | — | 0.6 |
| 1 | 0.5625 | 0.566667 |
| 2 | 0.565789 | 0.565957 |
| 3 | 0.565907 | 0.565919 |
| 4 | 0.565915 | 0.565916 |
| 5 | 0.5659155 | 0.56591563 |
| 6 | 0.56591555 | |

(The series may be summed to $3/n - (e^n/n)E_1(n)$ which equals 0.56591556 when $n = 5$.)

Comment: It is rather unrealistic to study this kind of case which is so artificial, for one immediately thinks of the source of the problem, and so its evaluation.

## 2.7 Further Transformations Involving Weight Function Modifications

### 2.7.1 Background

The modification studied in Section 2.4 is a weight function modification; if $k$ is zero it produces the odd-part of the J-fraction. Keep in mind also the possible change in sign involved in $(t - k)$, which is avoided if even powers of a function are considered. In this respect recall that central moments in statistics are defined by

$$\mu_s = \int_0^\infty (t - \mu_1')^s \, d\psi(t), \qquad (s = 0, 1, \cdots)$$

with $\mu_1' = 0$, $\mu_2$ the variance, $\mu_3 / \mu_2^{3/2}$ the skewness, and $\mu_4 / \mu_2^2$ the kurtosis ($\sqrt{\mu_2}$ is sometimes regarded as a scale factor).

The question then arises as to how a Stieltjes continued fraction $\frac{1}{\alpha_1 n +}\,\frac{1}{\alpha_2 +}\,\frac{1}{\alpha_3 n +}\cdots$ is modified when the corresponding distribution function becomes

$$\psi^*(t) = \int_0^t \{h(x)\}^2 d\psi(x). \qquad (h(x) \text{ real}) \tag{2.23}$$

There is particular interest in the form

$$S(z;\alpha,\psi) = \int_0^\infty \frac{\prod_{j=1}^{r} (t - \alpha_j)^2 d\psi(t)}{z + t}, \qquad (\alpha_j \text{ real}) \tag{2.24}$$

and more specifically for simplicity when the basic transform

$$F(z;\psi) = \int_0^\infty \frac{d\psi(t)}{z + t} = \frac{q_0}{z +}\,\frac{p_1}{1 +}\,\frac{q_1}{z +}\,\frac{p_2}{1 +}\,\frac{q_2}{z +}\cdots \tag{2.25}$$

is periodic. Even with these limitations the general problem has many ramifications. A point of especial interest is to study the relations of the modified transform (2.24) and the periodic c.f. (2.25) which produce c.f.s periodic in the limit.

For example, if

$$F(z) = \frac{1}{z+}\ \frac{1}{1+}\ \frac{1}{z+}\ \cdots = \frac{1}{2\pi}\int_0^4 \frac{\surd(4/t-1)dt}{z+t}, \tag{2.26}$$

then can we construct the c.f.s for transforms such as

$$H(z;a,b,c) = \frac{1}{2\pi}\int_0^4 \frac{(1-t)^{2a}(2-t)^{2b}(3-t)^{2c}\surd(4/t-1)dt}{z+t} \tag{2.27}$$

where $a$, $b$, $c$ are non-negative integers, and are the modified c.f.s periodic in some sense. It turns out that there is a periodicity, and that the $s$th partial numerators are rational fractions in $s$, the degrees depending on the form of the numerator of the integrand in (2.27). Note that c.f.s of this form arise as the solutions of differential equations (see Perron, (1957), pp283--293).

We mention that it is clear that the modified c.f.s are of Stieltjes form, and convergence questions present no particular problem.

Rather heavy algebra is often encountered in our examples, and use has, where expedient, been made of an algebraic computer language "REDUCE".

### 2.7.2 Basic Formulas

Consider the c.f. (2.25) with even-part

$$\frac{q_0}{z+p_1-}\ \frac{p_1q_1}{z+p_2+q_1-}\ \frac{p_2q_2}{z+p_3+q_2-}\ \cdots \tag{2.28}$$

and odd-part

$$\frac{q_0}{z} - \frac{q_0}{z}\left[\frac{p_1}{z+p_1+q_1-}\ \frac{q_1p_2}{z+p_2+q_2-}\ \frac{q_2p_3}{z+p_3+q_3-}\ \cdots\right]. \tag{2.29}$$

The fundamental orthogonal system $\{p_s(x)\}$ associated with $d\psi(t)$ is defined by

$$\begin{cases} p_s(s) = (x-a_s)p_{s-1} - b_{s-1}p_{s-2}(x), \quad (s=2,3,\cdots) \\ p_0(x) = 1, \quad p_1(x) = x - a_1, \end{cases} \tag{2.30}$$

where $a_s = p_s + q_{s-1}$, $b_s = p_s q_s$. If

$$\int_0^\infty p_s^2(t)d\psi(t) = \phi_s,$$

then from (2.30)

$$b_s = \phi_s / \phi_{s-1}, \qquad (s = 1, 2, 6 \cdots)$$

with $\phi_0 = q_0$.

Now let $\{q_s(x)\}$ be the orthogonal set for the weight function $\prod_{j=1}^{r}(x-\alpha_j)^2 d\psi(x)$. Then from a formula of Christoffel (Szegö, 1939, p28)

$$q_s(x) = \frac{\begin{vmatrix} p_s(x) & p_{s+1}(x) & & p_{s+2r}(x) \\ p_s(\alpha_1) & p_{s+1}(\alpha_1) & & p_{s+2r}(\alpha_1) \\ p_s^{(1)}(\alpha_1) & p_{s+1}^{(1)}(\alpha_1) & \cdots & p_{s+2r}^{(1)}(\alpha_1) \\ \cdots & \cdots & \cdots & \cdots \\ \cdots & \cdots & \cdots & \cdots \\ \cdots & \cdots & \cdots & \cdots \\ p_s(\alpha_r) & p_{s+1}(\alpha_r) & & p_{s+2r}(\alpha_r) \\ p_s^{(1)}(\alpha_r) & p_{s+1}^{(1)}(\alpha_r) & & p_{s+2r}^{(1)}(a_r) \end{vmatrix}}{\Delta_s(\underset{\sim}{\alpha})\prod_j^r (x-\alpha_j)^2} \tag{2.31}$$

where

(i) the determinant is of order $2r+1$,

(ii) $\Delta_s(\underset{\sim}{\alpha}) \equiv \Delta_s(\alpha_1, \alpha_2, \cdots, \alpha_r)$

$= | p_s(\alpha_1), p_{s+1}^{(1)}(\alpha_1), p_{s+2}(\alpha_2), p_{s+3}^{(1)}(\alpha_2), \cdots p_{s+2r-1}(\alpha_r), p_{s+2r}^{(1)}(\alpha_r) |$,

(in diagonal notation)

(iii) $p_s^{(1)}(x) = dp_s(x)/dx$ for $x = \alpha_j$, $j = 1, 2, \cdots, r$;

(iv) $\alpha_1, \alpha_2, \cdots$, are real in general.

Note that we have preferred to use orthogonal polynomials with highest coefficient unity; this applies to $\{p_s(x)\}$ and $\{q_s(x)\}$. Again multiple zeros are readily allowed for; thus if for example $\alpha_2 = \alpha_1$ then the rows in the determinant involving $\{p_s(\alpha_2)\}$, $\{p_s^{(1)}(\alpha_2)\}$ are replaced by $\{p_s^{(2)}(\alpha_1)\}$, $\{p_s^{(3)}(\alpha_1)\}$, superscripts referring to 2nd, 3rd derivatives.

We seek the partial numerators in the Stieltjes c.f.

$$\int_0^\infty \prod_{j=1}^{r} \frac{(t-\alpha_j)^2 d\psi(t)}{z+t} = \frac{Q_0}{z+} \; \frac{P_1}{1+} \; \frac{Q_1}{z+} \; \frac{P_2}{1+} \; \frac{Q_2}{z+} \cdots \tag{2.32}$$

with its associated even and odd parts. For example, the even part is

$$\frac{Q_0}{z+P_1-}\;\frac{P_1Q_1}{z+P_2+Q_1-}\cdots \tag{2.33}$$

which defines a corresponding set of orthogonal polynomials $\{q_s(x)\}$; Thus

$$\begin{cases} q_s(x) = (x-\alpha_s)q_{s-1}(x) - \beta_{s-1}q_{s-2}(x), \quad (s=2,3,\cdots) \\ q_0(x) = 1, \qquad q_1 = x-\alpha_1 = x - P_1. \end{cases} \tag{2.34}$$

From (2.4) for $Q_0$,

$$\int_0^\infty q_0^2(t)\prod_{j=1}^r (t-\alpha_j)^2 d\psi(t) = Q_0, \tag{2.35}$$

$$= \phi_0\Delta_1(\underset{\sim}{\alpha})/\,\Delta_0(\underset{\sim}{\alpha}). \qquad (\underset{\sim}{\alpha} = (\alpha_1,\alpha_2,\cdots,\alpha_r)$$

To determine $P_1$ and the general partial numerators in (2.30) we consider the odd-part of (2.5) and its associated orthogonal system $\{\hat{q}_s(x)\}$ where

$$\hat{q}_s(x) = -\frac{\begin{vmatrix} p_s(x) & p_{s+1}(x) & \cdots & p_{s+2r+1}(x) \\ p_s(0) & p_{s+1}(0) & \cdots & p_{s+2r+1}(0) \\ p_s(\alpha_1) & p_{s+1}(\alpha_1) & \cdots & p_{s+2r+1}(\alpha_1) \\ \cdots & \cdots & \cdots & \cdots \\ \\ \cdots & \cdots & \cdots & \cdots \\ p_s^{(1)}(\alpha_r) & p_{s+1}^{(1)}(\alpha_r) & \cdots & p_{s+2r+1}^{(1)}(\alpha_r) \end{vmatrix}_{(2r+2)}}{\tilde{\Delta}_s(0;\underset{\sim}{\alpha})x\prod_{j=1}^r (x-\alpha_j)^2} \tag{2.36}$$

where

$$\tilde{\Delta}_s(0;\underset{\sim}{\alpha}) = |p_s(0), p_{s+1}(\alpha_1), p_{s+2}^{(1)}(\alpha_2), \cdots p_{s+2r+1}^{(1)}(\alpha_r)|,$$

the negative sign appearing because of the order of the determinant. Clearly from (2.27) and its analogue for the modified case

$$Q_0P_1 = -\phi_0\tilde{\Delta}_1(0;\underset{\sim}{\alpha})/\,\tilde{\Delta}_0(0;\underset{\sim}{\alpha}). \tag{2.37}$$

Hence from (2.35)

$$P_1 = -\frac{\tilde{\Delta}_1(0;\underset{\sim}{\alpha})}{\tilde{\Delta}_0(0;\underset{\sim}{\alpha})}\frac{\Delta_0(\underset{\sim}{\alpha})}{\Delta_1(\underset{\sim}{\alpha})}. \tag{2.38}$$

### 2.7.3 Expressions for $P_s$, and $Q_s$

From (2.34) for the integral square

$$\Phi_s = \int_0^\infty q_s^2(t)\prod_{j=1}^r (t-\alpha_j)^2 d\,\psi(t),$$

we have

$$\beta_s = \Phi_s / \Phi_{s-1} = P_s Q_s. \qquad (s = 1, 2, \cdots) \tag{2.39}$$

But from (2.29)

$$\Phi_s = \phi_s \frac{\Delta_{s+1}(\underset{\sim}{\alpha})}{\Delta_s(\underset{\sim}{\alpha})}. \tag{2.40}$$

Again for the "odd" set of polynomials (2.36)

$$\hat{\Phi}_s = \int_0^\infty \hat{q}_s^2(t)t\prod_{j=1}^r (t-\alpha_j)^2 d\,\psi(t),$$

so that

$$\hat{\beta}_s = P_{s+1}Q_s = \hat{\Phi}_s / \hat{\Phi}_{s-1}. \tag{2.41}$$

But from (2.36)

$$\hat{\Phi}_s = -\phi_s \frac{\tilde{\Delta}_{s+1}(0;\underset{\sim}{\alpha})}{\tilde{\Delta}_s(0;\underset{\sim}{\alpha})}. \tag{2.42}$$

Hence from (2.39) - (2.42) we have

$$\frac{P_{s+1}}{P_s} = \frac{\hat{\Phi}_s}{\hat{\Phi}_{s-1}}\cdot\frac{\Phi_{s-1}}{\Phi_s},$$

and

$$P_{s+1} = \frac{\hat{\Phi}_s}{\hat{\Phi}_0}\cdot\frac{\Phi_0}{\Phi_s}P_1,$$

leading to

$$P_s = -\frac{\tilde{\Delta}_s(0;\underset{\sim}{\alpha})}{\tilde{\Delta}_{s-1}(0;\underset{\sim}{\alpha})}\frac{\Delta_{s-1}(\underset{\sim}{\alpha})}{\Delta_s(\underset{\sim}{\alpha})}, \tag{2.43}$$

and from (2.39)

$$Q_s = -\frac{\phi_s}{\phi_{s-1}}\frac{\tilde{\Delta}_{s-1}(0;\underset{\sim}{\alpha})\Delta_{s+1}(\underset{\sim}{\alpha})}{\tilde{\Delta}_s(0;\underset{\sim}{\alpha})\Delta_s(\underset{\sim}{\alpha})}. \qquad (s = 1, 2, \cdots) \tag{2.44}$$

### 2.7.4 Illustration:

From (2.28), the even part,

$$I_1(z) = \frac{1}{2\pi}\int_0^4 \frac{\sqrt{(4/t-1)}}{z+t}dt = \frac{1}{z+1-}\;\frac{1}{z+2-}\;\frac{1}{z+2-}\cdots.$$

What is the c.f. for

$$I_2(z) = \frac{1}{2\pi}\int_0^4 \frac{\sqrt{(4/t-1)}}{z+t}(1-t)^2 dt? \qquad (a=1,\ b=c=0 \text{ in } (2.29))$$

Here $p_s(x) = (x-2)p_{s-1}(x) - p_{s-2}(x)$, $(s = 2, 3, \cdots;\ p_0 = 1,\ p_1 = x-1)$, and $\phi_s = 1$. Moreover, by trial and error

$$\begin{cases} p_{3s}(1) = 1, & p_{3s+1}(1) = 0, & p_{3s+2}(1) = -1, \\ p_{3s}^{(1)}(1) = -s, & p_{3s+1}^{(1)}(1) = 2s+1, & p_{3s+2}^{(1)}(1) = -(s+1). \end{cases} \tag{2.45}$$

From (2.31)

$$\Delta_{3s}(\alpha_1) = \begin{vmatrix} 1 & 0 \\ -s & 2s+1 \end{vmatrix} = 2s+1,$$

$$\Delta_{3s+1}(\alpha_1) = \begin{vmatrix} 0 & -1 \\ 2s+1 & -(s+1) \end{vmatrix} = 2s+1.$$

Also, for the odd-part, since $p_s(0) = (-1)^s$,

$$\tilde{\Delta}_{3s} = (-1)^{3s}\begin{vmatrix} 1 & -1 & 1 \\ 1 & 0 & -1 \\ -s & 2s+1 & -(s+1) \end{vmatrix} = (-1)^{3s}(2s+1),$$

$$\tilde{\Delta}_{3s+1} = (-1)^{3s+1}\begin{vmatrix} 1 & -1 & 1 \\ 0 & -1 & 1 \\ 2s+1 & -(s+1) & -(s+1) \end{vmatrix} = (-1)^{3s+1}(2s+2).$$

Hence from (2.35), (2.43), and (2.44),

$$Q_0 = 1/1 = 1, \qquad P_1 = -\frac{(-1)2}{1}\cdot\frac{1}{1} = 2.$$

To complete the determination of the partial numerators in the c.f. for $I_2(z)$, we need

$$\Delta_{3s+2}(\alpha_1) = \begin{vmatrix} -1 & 1 \\ -(s+1) & -(s+1) \end{vmatrix} = 2(s+1),$$

$$\tilde{\Delta}_{3s+2}(\alpha_1) = (-1)^{3s+2}\begin{vmatrix} 1 & -1 & 1 \\ -1 & 1 & 0 \\ -(s+1) & -(s+1) & 2s+3 \end{vmatrix} = (-1)^{3s+2}2(s+1).$$

Thus,

$$P_{3s} = 1, \qquad Q_{3s} = \frac{2s}{2s+1};$$

$$P_{3s+1} = \frac{2s+2}{2s+1}, \qquad Q_{3s+1} = 1; \qquad (P_{3s+1} + Q_{3s} = 2)$$

$$P_{3s+2} = \frac{2s+1}{2s+2}, \qquad Q_{3s+2} = \frac{2s+3}{2s+2}. \qquad (P_{3s+2} + Q_{3s+2} = 2)$$

Hence,

$$I_2(z) = \frac{1}{z+}\;\frac{2}{1+}\;\frac{1}{z+}\;\frac{\frac{1}{2}}{1+}\;\frac{\frac{3}{2}}{1+}\;\frac{1}{1+}\;\frac{\frac{2}{3}}{z+}\;\frac{\frac{4}{3}}{1+}\;\frac{1}{z+}\;\frac{\frac{3}{4}}{1+}\;\frac{\frac{5}{4}}{z+}\cdots,$$

$$= -(z+1) + (z+1)^2\left\{-\frac{1}{2} + \frac{1}{2}\sqrt{\left(\frac{4}{z} + 1\right)}\right\}. \qquad (z > 0)$$

The partial numerators ultimately tend to unity, and consist of ratios of first degree polynomials. Moreover, $I_2(z)$ may be regarded as the fractional part (or residue) of $(z+1)^2 I_1(z)$.

For the series developments we have

$$I_1(z) \sim \sum_{s=0}^{\infty} \frac{(-1)^s (2s)!}{z^{s+1} s!(s+1)!},$$

$$I_2(z) \sim \sum_{s=0}^{\infty} \frac{(-1)^s 3(2s)!(3s^2+3s+2)}{z^{s+1} s!(s+3)!},$$

**Numerical:** $0.749938846 < I_2(\frac{1}{2}) < 0.750025750$
(14 terms of c.f.) (15 terms of c.f.)

### 2.7.5 Further Illustrations

For a particular case of the modified Stieltjes integral (2.29) we decide on the modification factors in the integrals; it may be a single factor $(t_0 - t)^{2r}$ ($t_0$ real, $r = 1, 2, \cdots$). In the basic c.f. (2.28) we must now set up $\{p_s^{(k)}(t_0)\}$ for $k = 0, 1, \cdots, 2r-1$. The problems here are those of pattern recognition if we merely generate the derivatives for $s = 0, 1, 2, \cdots$, to a certain point. Trial and error sometimes gives a quick answer, but if there are periodicities of size eight or more, success starts to fade. We shall return briefly to this aspect in the sequel.

There remains the problem of evaluating the required sets of the $\Delta$'s and $\tilde{\Delta}$'s ((2.31) and (2.36)). An appeal to an algebraic language is necessary when determinants of order six or more are involved.

Keep in mind also the fact that although the modified integral can be expanded in descending powers of $z$ and this series converted to a c.f. by an algorithm we still will lack proof of any guessed formulas for the $P$'s and $Q$'s. Moreover there will be cases where we decide on the c.f. form in (2.27) but do not know its integral formula; quite a serious problem in actual practice. A series development would then be necessary based on the c.f. (Stieltjes considered this problem in letter 194 to Hermite, April 15, 1889; *Correspondance d'Hermite et de Stieltjes*, Gauthier-Villars, 1905).

One further point concerns whether the zeros of the modified integrand are contained in the interval of orthogonality; it appears at this stage that if they are there is enhanced interest.

### 2.7.6 $H(z\,; a, b, c)$ (see (2.29); note $\phi_s = 1$)

Derivatives (see (2.45)) are

$$\left\{\begin{array}{lll} p_{3s}(3) = (-1)^s, & p_{3s+1}(3) = 2(-1)^s, & p_{3s+2}(3) = (-1)^s, \\ p_{3s}^{(1)}(3) = 3s(-1)^{s-1}, & p_{3s+1}^{(1)}(3) = (-1)^s, & p_{3s+2}^{(1)}(3) = 3(s+1)(-1)^s; \end{array}\right. \tag{2.46}$$

$$\left\{\begin{array}{ll} p_{2s}(2) = (-1)^s, & p_{2s+1}(2) = (-1)^s, \\ p_{2s}^{(1)}(2) = (-1)^{s-1}s, & p_{2s+1}^{(1)}(2) = (-1)^s(s+1). \end{array}\right.$$

**Table 2.1 Series for $H(z;a,b,c)$**

| | $a=1$ $b=0$ $c=1$ | $a=3$ $b=c=0$ | $a=b=c=1$ |
|---|---:|---:|---:|
| 1 | 3 | 15 | 10 |
| 2 | -1 | -51 | -2 |
| 3 | 2 | 178 | 4 |
| 4 | -6 | -628 | -14 |
| 5 | 20 | 2236 | 52 |
| 6 | -70 | -8025 | -195 |
| 7 | 252 | 29004 | 734 |
| 8 | -923 | -105477 | -2770 |
| 9 | 3418 | 385698 | 10476 |
| 10 | -12750 | -1417341 | -39695 |
| 11 | 47804 | 5231460 | 150670 |
| 12 | -179911 | -19386859 | -572812 |
| 13 | 679098 | 72105322 | 2180952 |
| 14 | -2569560 | -269068375 | -8315535 |
| 15 | 9742800 | 1007096400 | 31747590 |
| 16 | -37008495 | -3779939025 | -121360650 |
| 17 | 140810370 | 14223573450 | 464477340 |
| 18 | -536568150 | -53648589150 | -1779696360 |
| 19 | 2047508460 | 202795445484 | 6826516944 |
| 20 | -7823457600 | -768140995428 | -26212175364 |
| 21 | 29930313504 | 2915049573200 | 100748304920 |
| 22 | -114639598980 | -11082010648322 | -387600766430 |
| 23 | 439585446824 | 42199701399176 | 1492540856812 |
| 24 | -1687384157806 | -160943539197714 | -5752360510260 |
| 25 | 6483735700740 | 614709981940308 | 22188539129528 |
| 26 | -24937766681996 | -2351052333145005 | -85656319563745 |
| 27 | 96004663054552 | 9003587163875472 | 330921007386354 |
| 28 | -369923905053969 | -34522230786695841 | -1279411877176328 |
| 29 | 1426597975384022 | 132520604230683606 | 4949998991781088 |
| 30 | -5506108158496540 | -509263764444100547 | -19164467819750519 |
| 31 | 21268039724231160 | 1959081583840199000 | 74246190461887990 |
| 32 | -82212194082688999 | -7543797793034509433 | -287823832563270458 |
| 33 | 318022807895719346 | 29075980643686575354 | 1116463260856395900 |
| 34 | -1231066287815391846 | -112167246461176035066 | -4333289198236572532 |
| 35 | 4768637180535085900 | 433078874069915705772 | 16828189078334349800 |
| 36 | -18483572769115348748 | -1673479218237519174720 | -65387528864703509940 |

(a) $a=1,\ b=0,\ c=1 \qquad H(z;1,0,1)$

The Δ's;

$$\Delta_{3s} = 16(s+1)(3s+1), \qquad \tilde{\Delta}_{3s} = 144(s+1)^2;$$

$$\Delta_{3s+1} = 48(s+1)^2, \qquad \tilde{\Delta}_{3s+1} = -144(s+1)^2;$$

$$\Delta_{3s+2} = 16(s+1)(3s+5), \qquad \tilde{\Delta}_{3s+2} = 144(s+1)(s+2).$$

$P$'s and $Q$'s:

$$P_{3s} = \frac{3s+2}{3s+1}, \qquad Q_{3s} = \frac{3s}{3s+1};$$

$$P_{3s+1} = \frac{3s+1}{3s+3}, \qquad Q_{3s+1} = \frac{3s+5}{3s+3};$$

$$P_{3s-1} = \frac{3(s+1)}{3s+2}, \qquad Q_{3s-1} = \frac{3s+1}{3s+2};$$

$$H(z;1,0,1) = \frac{3}{z+}\ \frac{\frac{1}{3}}{1+}\ \frac{\frac{5}{3}}{z+}\ \frac{\frac{6}{5}}{1+}\ \frac{\frac{4}{5}}{z+}\ \frac{\frac{5}{4}}{1+}\ \frac{\frac{3}{4}}{z+}\ \frac{\frac{2}{3}}{1+}\ \frac{\frac{4}{3}}{z+}\ \frac{\frac{9}{8}}{1+}\ \frac{\frac{7}{8}}{z+}\ \cdots . \tag{2.47}$$

(Note that (i) $P_s + Q_s = 2$, so that the denominator parameters in the J-form of the odd-part are 2; also that (ii) $P$'s and $Q$'s $\to 1$ as $s \to \infty$; (iii) that numerical work in integer arithmetic was carried through to $P_{18}$).

(b) $a = 3,\ \ b = c = 0;\ \ H(z;3,0,0)$

**Further Derivatives**

We need derivatives of $p_s(x)$ up to the 5th order and evaluated at $x=1$. We examine each series and see that they have a period of 3. We have the results:

| | $j=3s$ | $j=3s+1$ | $j=3s+2$ |
|---|---|---|---|
| $p_j(1)$ | $1$ | $0$ | $-1$ |
| $p_j^{(1)}(1)$ | $-s$ | $2s+1$ | $-(s+1)$ |
| $p_j^{(2)}(1)$ | $-s(3s+1)$ | $0$ | $(s+1)(3s+2)$ |
| $p_j^{(3)}(1)$ | $3s^2(s+1)$ | $-3s(2s+1)(s+1)$ | $3s(s+1)^2$ |
| $p_j^{(4)}(1)$ | $s(9s+2)(s+1)(s-1)$ | $4s(2s+1)(s+1)$ | $-s(s+1)(9s+7)(s+2)$ |
| $p_j^{(5)}(1)$ | $-s(s+1)(s-1)(9s^2+30s+4)$ | $9s(s+1)(s-1)(2s+1)(s+2)$ | $-s(s+1)(s+2)(9s^2-12s-17)$ |

For the computer algebra implementation we set up a matrix of $7\times 9$ as below: Every derivative has a period 3. Note that $p_j^{(k)} + p_{j+1}^{(k)} + p_{j+2}^{(k)} = kp_{j+1}^{(k-1)}$. We also need derivatives at $x=0$.

Scheme for the Δ's and $\tilde{\Delta}$'s.

| | 1 | 2 | 3 |
|---|---|---|---|
| $p_j(0)$ | $(-1)^{(s+1)}$ | $(-1)^s$ | $(-1)^{(s+1)}$ |
| $p_j(1)$ | $1$ | $0$ | $-1$ |
| $p_j^{(1)}(1)$ | $-s$ | $2s+1$ | $-(s+1)$ |
| $p_j^{(2)}(1)$ | $-s(3s+1)$ | $0$ | $(s+1)(3s+2)$ |
| $p_j^{(3)}(1)$ | $3s^2(s+1)$ | $-3s(2s+1)(s+1)$ | $3s(s+1)^2$ |
| $p_j^{(4)}(1)$ | $s(9s+2)(s+1)(s-1)$ | $4s(2s+1)(s+1)$ | $-s(s+1)(9s+7)(s+2)$ |
| $p_j^{(5)}(1)$ | $-s(s+1)(s-1)(9s^2+30s+4)$ | $9s(s+1)(s-1)(2s+1)(s+2)$ | $-s(s+1)(s+2)(9s^2-12s-17)$ |

| | 4 5 6 | 7 8 9 |
|---|---|---|
| $p_j(0)$ | | |
| $p_j(1)$ | 1st 3 columns | 1st 3 columns |
| $p_j^{(1)}(1)$ | $s$ replaced | $s$ replaced |
| $p_j^{(2)}(1)$ | by $s+1$ | by $s+2$ |
| $p_j^{(3)}(1)$ | | |
| $p_j^{(4)}(1)$ | | |
| $p_j^{(5)}(1)$ | | |

We compute determinants of orders six and seven.

Computation of determinants.

$\Delta_{3s}$ columns 1 to 6 rows 2 to 7.

$$= 384(s+1)^3(2s+1)^2(2s+3)^2(9s^2+18s+10)$$

$\tilde{\Delta}_{3s}$ columns 1 to 7 rows 1 to 7.

$$= 384(s+1)^2(s+2)(2s+1)(2s+3)^2(18s^3+45s^2+32s+5)$$

$\Delta_{3s+1}$ columns 2 to 7 rows 2 to 7.

$$= 192(s+1)^2(s+2)(2s+1)(2s+3)^3(18s^2+63s+50)$$

$\tilde{\Delta}_{3s+1}$ columns 2 to 8 rows 1 to 7.

$$= -192(s+1)^2(s+2)^2(2s+3)^2(36s^2+108s+85)$$

$\Delta_{3s+2}$ columns 3 to 8 rows 2 to 7.

$$= 192(s+1)(s+2)^2(2s+3)^3(2s+5)(18s^2+45s+23)$$

$\tilde{\Delta}_{3s+2}$ columns 3 to 9 rows 1 to 7.

$$= 384(s+2)^3(2s+3)^2(2s+5)(18s^3+99s^2+167s+86).$$

The $P$'s and $Q$'s,

$$P_j = -\frac{\Delta_{j-1}}{\Delta_j}\frac{\tilde{\Delta}_j}{\tilde{\Delta}_{j-1}}, \qquad Q_j = -\frac{\Delta_{j+1}}{\Delta_j}\frac{\tilde{\Delta}_{j-1}}{\tilde{\Delta}_j};$$

$$P_{3s+1} = \frac{(9s^2+18s+10)(36s^2+108s+85)}{(18s^2+63s+50)(18s^2+27s+5)},$$

$$P_{3s+2} = \frac{2(2s+1)(18s^2+63s+50)(18s^2+81s+86)}{(2s+3)(18s^2+45s+23)(36s^2+108s+85)},$$

$$P_{3s+3} = \frac{(s+3)(18s^2+63s+50)(18s^2+45s+23)}{2(s+2)(18s^2+81s+86)(9s^2+36s+37)};$$

$$Q_{3s+1} = \frac{2(2s+5)(18s^2+45s+23)(18s^2+27s+5)}{(2s+3)(18s^2+63s+50)(36s^2+108s+85)},$$

$$Q_{3s+2} = \frac{(9s^2+36s+37)(36s^2+108s+85)}{(18s^2+45s+23)(18s^2+81s+86)},$$

$$Q_{3s+3} = \frac{(18s^2+99s+131)(18s^2+81s+86)}{2(9s^2+36s+37)(18s^2+63s+50)}.$$

$$H(z;3,0,0) = \frac{15}{z+}\,\frac{P_1}{1+}\,\frac{Q_1}{z+}\cdots \tag{2.48}$$

where

| | |
|---|---|
| $P_1$=17/5 | $Q_1$=23/255 |
| $P_2$=1720/1173 | $Q_2$=3145/1978 |
| $P_3$=1725/6364 | $Q_3$=5633/3700 |
| $P_4$=8473/6550 | $Q_4$=12040/29999 |
| $P_5$=14541/9847 | $Q_5$=9389/7955 |
| $P_6$=11266/22755 | $Q_6$=22940/16113 |
| $P_7$=18245/16244 | $Q_7$=43623/77252 |
| $P_8$=31744/23051 | $Q_8$=2581/2368 |
| $P_9$=1147/1856 | $Q_9$=1203/899 |
| $P_{10}$=106285/99448 | $Q_{10}$=1745920/2645397 |
| $P_{11}$=1378237/1055520 | $Q_{11}$=82829/78560 |
| $P_{12}$=38496/55483 | $Q_{12}$=57938/45313 |
| $P_{13}$=123509/118295 | $Q_{13}$=2559583/3546785 |
| $P_{14}$=7412760/5903293 | $Q_{14}$=355225/342718 |
| $P_{15}$=202783/272220 | $Q_{15}$=56887/46020 |
| $P_{16}$=19825/19234 | $Q_{16}$=494184/646295 |
| $P_{17}$=1687213/1383785 | $Q_{17}$=337025/328409 |
| $P_{18}$=1137740/1455727 | $Q_{18}$=1518774/1260805 |
| $P_{19}$=448409/438470 | $Q_{19}$=2607511/3274806 |

$P_{20}$=6826144/5727867 $Q_{20}$=1170733/1148020
$P_{21}$=2278161/2815760 $Q_{21}$=2931355/2483408
$P_{22}$=1503085/1477348 $Q_{22}$=9976672/12160661
$P_{23}$=1264533/1080554 $Q_{23}$=38033/37454
$P_{24}$=167506/201699 $Q_{24}$=1047484/902061 .

(The values of $P$'s and $Q$'s were identical with the values calculated from the series development of

$$H(z\,;3,0,0) = -z^5-5z^4-11z^3-12z^2-9z+2 + \frac{1}{2}(z+1)^6\{-1+\surd(\frac{4}{z}+1)\};$$

as an alternative one can use the difference interpretation of the integral as an operator on the series for $H(z\,;0,0,0)$ .)

(c) $H(z\,;1,1,1)$ $a=b=c=1$

For the case $(1-t)^2(2-t)^2(3-t)^2$, we notice $p_s(0)$ has period 2, $p_s(1)$ and $p_s^{(1)}(1)$ have period 3, $p_s(2)$ and $p_s^{(1)}(2)$ have period 2 or 4; and $p_s(3)$ and $p_s^{(1)}(3)$ have period 6. Before we proceed to construct determinants, we must see that the joint periodicity is 6 resulting in the same periodicity for the $P$'s and $Q$'s. We construct a $7\times 15$ matrix as follows:

| | 1 | 2 | 3 | 4 | 5 | 6 |
|---|---|---|---|---|---|---|
| $p_j(0)$ | 1 | -1 | 1 | -1 | 1 | -1 |
| $p_j(1)$ | 1 | 0 | -1 | 1 | 0 | -1 |
| $p_j^{(1)}(1)$ | $-(2s-1)$ | $4s-1$ | $-2s$ | $-2s$ | $4s+1$ | $-(2s+1)$ |
| $p_j(2)$ | $(-1)^s$ | $(-1)^{s+1}$ | $(-1)^{s+1}$ | $(-1)^s$ | $(-1)^s$ | $(-1)^{s+1}$ |
| $p_j^{(1)}(2)$ | $(-1)^s(3s-1)$ | $(-1)^s(3s-1)$ | $(-1)^{s+1}3s$ | $(-1)^{s+1}3s$ | $(-1)^s(3s+1)$ | $(-1)^s(3s+1)$ |
| $p_j(3)$ | -1 | -2 | -1 | 1 | 2 | 1 |
| $p_j^{(1)}(3)$ | $3(2s-1)$ | -1 | $-6s$ | $-6s$ | 1 | $3(2s+1)$ |

| | 7-12 | 13-15 |
|---|---|---|
| $p_j(0)$<br>$p_j(1)$<br>$p_j^{(1)}(1)$<br>$p_j(2)$<br>$p_j^{(1)}(2)$<br>$p_j(3)$<br>$p_j^{(1)}(3)$ | 1st 6 columns $s$ replaced by $s+1$ | 1st 3 columns $s$ replaced by $s+2$ |

The $P$'s and $Q$'s.

We evaluated the determinants as before, but now there are 6 $\Delta$s and 6 $\tilde{\Delta}$s which we omit.

$$P_{6s-2} = (6s+5)/\{3(2s+1)\},$$
$$P_{6s-1} = 9(4s^2+4s+1)/(36s^2+36s+5),$$
$$P_{6s} = 2(36s^3+48s^2+13s+1)/(72s^3+108s^2+38s+1),$$
$$P_{6s+1} = (36s^2+36s+1)/(36s^2+36s+5),$$
$$P_{6s+2} = (6s+5)/\{3(2s+1)\},$$
$$P_{6s+3} = (2s+1)/\{2(s+1)\};$$

$$Q_{6s-2} = (6s+1)/\{3(2s+1)\},$$
$$Q_{6s-1} = (36s^2+36s+1)/(36s^2+36s+5),$$
$$Q_{6s} = 2s(36s^2+60s+25)/(72s^3+108s^2+38s+1),$$
$$Q_{6s+1} = 9(4s^2+4s+1)/(36s^2+36s+5),$$
$$Q_{6s+2} = (6s+1)/\{3(2s+1)\},$$
$$Q_{6s+3} = (2s+3)/\{2(s+1)\}.$$

$$H(z;1,1,1) = \frac{10}{z+}\;\frac{P_1}{1+}\;\frac{Q_1}{z+}\cdots \tag{2.49}$$

where

| | | | |
|---|---|---|---|
| $P_1=1/5$ | $Q_1=9/5$ | $P_{12}=1014/1085$ | $Q_{12}=1156/1085$ |
| $P_2=5/3$ | $Q_2=1/3$ | $P_{13}=217/221$ | $Q_{13}=225/221$ |
| $P_3=1/2$ | $Q_3=3/2$ | $P_{14}=17/15$ | $Q_{14}=13/15$ |
| $P_4=11/9$ | $Q_4=7/9$ | $P_{15}=5/6$ | $Q_{15}=7/6$ |
| $P_5=81/77$ | $Q_5=73/77$ | $P_{16}=23/21$ | $Q_{16}=19/21$ |
| $P_6=196/219$ | $Q_6=242/219$ | $P_{17}=441/437$ | $Q_{17}=433/437$ |
| $P_7=73/77$ | $Q_7=81/77$ | $P_{18}=2888/3031$ | $Q_{18}=3174/3031$ |
| $P_8=11/9$ | $Q_8=7/9$ | $P_{19}=433/437$ | $Q_{19}=441/437$ |
| $P_9=3/4$ | $Q_9=5/4$ | $P_{20}=23/21$ | $Q_{20}=19/21$ |
| $P_{10}=17/15$ | $Q_{10}=13/15$ | $P_{21}=7/8$ | $Q_{21}=9/8$ |
| $P_{11}=225/221$ | $Q_{11}=217/221$ | $P_{22}=29/27$ . | |

**Remarks**

(i) Notice that $P_s + Q_s = 2$, and $P_s$ and $Q_s \to 1$ as $s \to \infty$.

(ii) For the odd-part we have

$$H(z;1,1,1) = \frac{10}{z} - \frac{10}{z}\left(\frac{P_1}{z+2-}\;\frac{Q_1P_2}{z+2-}\cdots\right) \tag{2.50}$$

where

$$Q_{6s+1}P_{6s+2} = \frac{(6s+3)(6s+5)}{36s^2+36s+5},$$

$$Q_{6s+2}P_{6s+3} = \frac{6s+1}{6s+6},$$

$$Q_{6s+3}P_{6s+4} = \frac{6s+11}{6s+6},$$

$$Q_{6s+4}P_{6s+5} = \frac{(6s+7)(6s+9)}{36s^2+108s+77},$$

$$Q_{6s+5}P_{6s+6} = \frac{(36s^2+108s+73)(72s^3+312s^2+434s+196)}{(36s^2+108s+77)(72s^3+324s^2+470s+219)},$$

$$Q_{6s+6}P_{6s+7} = \frac{(36s^2+108s+73)(72s^3+336s^2+506s+242)}{(36s^2+108s+77)(72s^3+324s^2+470s+219)}.$$

(iii) $$H(z;1,1,1) = -z^5-11z^4-48z^3-105z^2-119z-63 + \{(z+1)(z+2)(z+3)\}^2\left\{-\frac{1}{2}+\frac{1}{2}\sqrt{\left(\frac{4}{z}+1\right)}\right\}.$$

(iv) Numerical by using the c.f. for $z = 1$

| | Increasing | Decreasing |
|---|---|---|
| 1 | 8.333333333333333 | 9.333333333333333 |
| 2 | 8.933333333333333 | 9.000000000000000 |
| 3 | 8.980392156862745 | 8.990476190476191 |
| 4 | 8.986719787516600 | 8.987951807228916 |
| 5 | 8.987441562013017 | 8.987623762376238 |
| 6 | 8.987557215762614 | 8.987584650112867 |
| 7 | 8.987574833389811 | 8.987578791249537 |
| 8 | 8.987577121894280 | 8.987577639751553 |
| 9 | 8.987577463187219 | 8.987577542940865 |
| 10 | 8.987577512375064 | 8.987577523021988 |
| 11 | 8.987577518775506 | 8.987577520351704 |
| 12 | 8.987577519768278 | 8.987577520002571 |
| 13 | 8.987577519915856 | 8.987577519949737 |
| 14 | 8.987577519935898 | 8.987577519940588 |
| 15 | 8.987577519938922 | 8.987577519939637 |
| 16 | 8.987577519939362 | 8.987577519939461 |
| 17 | 8.987577519939422 | 8.987577519939436 |

The true value when $n = 1$, is 8.987577519939435.

### 2.7.7 Miscellaneous Results

We now give further cases, omitting details.

**2.7.7.1** $H(z;0,0,1) = \frac{5}{z+}\ \frac{P_1}{1+}\ \frac{Q_1}{z+} \cdots$

$$P_{3s+1} = \frac{(2s+2)(6s+1)}{(2s+1)(6s+5)}, \qquad Q_{3s+1} = \frac{6s+3}{6s+5}; \qquad (2.51)$$

$$P_{3s+2} = \frac{6s+5}{6s+6}, \qquad Q_{3s+2} = \frac{6s+7}{6s+6};$$

$$P_{3s+3} = \frac{6s+9}{6s+7}, \qquad Q_{3s+3} = \frac{(2s+2)(6s+11)}{(2s+3)(6s+7)}.$$

**2.7.7.2** $J(z,p) = \frac{1}{2\pi}\int_{\alpha}^{\beta} \frac{\sqrt{\{(t-\alpha)(\beta-t)\}}}{t(z+t)}(t-p)^2 dt$ (See Stieltjes, 1918, pp.509-510)

$$\alpha = (\sqrt{p} - \sqrt{q})^2, \quad \beta = (\sqrt{p} + \sqrt{q})^2, \quad (q, p > 0;\ q = 3p)$$

$$J(z,p) = \frac{7p^3}{z+}\ \frac{P_1}{1+}\ \frac{Q_1}{z+} \cdots \qquad (2.52)$$

| $r$ | $P_r$ | $Q_r$ |
|---|---|---|
| $6s$ | $\frac{3p(12s+1)}{12s+3}$ | $\frac{p\,12s(12s+7)}{(12s+1)(12s+3)}$ |
| $6s+1$ | $\frac{3p(12s+3)(12s+4)}{(12s+1)(12s+7)}$ | $\frac{p(12s+1)(12s+10)}{(12s+4)(12s+7)}$ |
| $6s+2$ | $\frac{3p(12s+7)(12s+8)}{(12s+4)(12s+10)}$ | $\frac{p(12s+4)(12s+11)}{(12s+8)(12s+10)}$ |
| $6s+3$ | $\frac{3p(12s+10)}{12s+8}$ | $\frac{p(12s+8)}{12s+11}$ |
| $6s+4$ | $\frac{3p(12s+12)}{12s+11}$ | $p$ |
| $6s+5$ | $\frac{3p(12s+11)}{12s+12}$ | $\frac{p(12s+15)}{12s+12}$ |

$$J(z,p) = \frac{\sqrt{\{z^2+2(p+q)z+(p-q)^2\}}-z+6p-q}{2z}(z+p)^2 - zp + p^2.$$

Basic c.f.: $\dfrac{p}{z+}\,\dfrac{q}{1+}\,\dfrac{p}{z+}\,\dfrac{q}{1+}\cdots$.

Note that we may write

$$\frac{(x+1)^2}{2x}\{\sqrt{(x^2+8x+4)}-x+1\} = \frac{7}{x+}\,\frac{P_1/p}{1+}\,\frac{Q_1/p}{x+}\cdots,$$

the partial numerators now being independent of $p$.

**2.7.7.3** If

$$\int_0^\infty \frac{d\psi(t)}{z+t} = \frac{1}{z+}\,\frac{1}{1+}\,\frac{2}{z+}\,\frac{1}{1+}\,\frac{2}{z+}\cdots,$$

$$= \{1-z+\sqrt{(z^2+6z+1)}\}/(4z), \qquad (z>0)$$

then

$$\int_0^\infty \frac{(1-t)^2 d\psi(t)}{z+t} = \frac{2}{z+}\,\frac{P_1}{1+}\,\frac{Q_1}{z+}\cdots \tag{2.53}$$

where

| $r$ | $P_r$ | $Q_r$ |
|---|---|---|
| $4s-3$ | $\dfrac{4s-1}{4s-3}$ | $\dfrac{4(2s-1)}{4s-1}$ |
| $4s-2$ | $\dfrac{2s(4s-3)}{(2s-1)(4s-1)}$ | $\dfrac{4s-1}{2s-1}$ |
| $4s-1$ | $\dfrac{2s-1}{2s}$ | $\dfrac{4s+1}{2s}$ |
| $4s$ | $1$ | $\dfrac{8s}{4s+1}$ |

Note:

(a) $P_{4s}(1) = (-1)^s 4^s$, $\qquad P_{4s+1}(1) = 0$,

$P_{4s+2}(1) = (-1)^s 2\cdot 4^s$, $\qquad P_{4s+3}(1) = (-1)^s 4^{s+1}$;

(b) $P_{4s}^{(1)}(1)=(-1)^{s+1}s\,2^{2s+1}$, $\qquad P_{4s+1}^{(1)}(1)=(-1)^s(4s+1)4^s$,

$P_{4s+2}^{(1)}(1)=(-1)^{s+1}(4s+2)4^s$, $\quad P_{4s+3}^{(1)}(1)=0$;

(c) $P_s(0)=(-1)^s$;

(d) $P_{4s-1}+Q_{4s-1}=P_{4s+1}+Q_{4s}=3$.

**2.7.7.4** If

$$f(z) = \frac{1}{z+}\,\frac{2}{1+}\,\frac{2}{z+}\,\frac{1}{1+}\,\frac{1}{z+}\,\frac{2}{1+}\,\frac{2}{z+}\cdots, \tag{2.54}$$

we may show that the c.f. for the Stieltjes modified form (the factor being $(t-2)^2$ in the weight function) in the limit is periodic with ultimate period 1, 1, 2, 2.

## 2.8 Series Arising from Complicated Structures

### 2.8.1 General

The examples so far considered have all arisen from the expansion of simple definite integrals, for the most part Stieltjes' transforms. There are many situations where the genesis of the series is a multiple integral or partial differential equation, complicated enough to inhibit straightforward analysis. Many examples occur in perturbation series in fluid mechanics and have been described by Van Dyke (1974, 1975). One example (due to Reddall, 1972) quoted by Van Dyke indicates the flavor of the studies;

$$\begin{aligned} x_s &= t - \frac{2}{3}t^2 + \frac{151}{189}t^3 - \frac{3611}{3591}t^4 + \frac{40138286}{34312005}t^5 \\ &\quad + \cdots - 30.3507t^{14} - 1.2253t^{15}\cdots. \end{aligned}$$

Van Dyke remarks that Reddall spent three weeks calculating the fifth term by hand, and then computed 15 terms in one minute on IBM 360/67 machine. The series is noteworthy in having the irregular sign pattern $+-+-+-++-++-+--$.

The lack of a simple function which generates the series creates a completely new situation. If the series, for which only a finite set of terms can be obtained, looks convergent, assessments can be based on it; but divergent series can arise and error analysis of assessments now takes on a more complicated form. The problem is quite distinct from that of classical asymptotic series as treated in Hardy (1949), Erdelyi (1956), and de Bruijn (1961).

Van Dyke (1974) remarks on p. 436 "Authenticated examples of divergent power series are rare". He quotes an example of Bender and Wu

(1969) who studied the ground-state energy of an anharmonic oscillator, taking the series in powers of $\beta$ for an eigenvalue as far as the term in $\beta^{75}$. However we shall show that divergent series occur frequently in mathematical statistics. Some examples will be described here.

### 2.8.2 Series for the Standard Deviation in Exponential Sampling

For a random sample $(x_1, x_2, \cdots, x_n)$ the standard deviation may be taken as $\sqrt{m_2}$, where

$$m_2 = \sum_{j=1}^{n} (x_j - \bar{x})^2/n . \qquad (\bar{x} = \sum x_j/n)$$

Whereas the moments of $m_2$ present little difficulty, for example, $E(m_2) = (1 - 1/n)\mu_2$, the odd ones for $\sqrt{m_2}$ produce infinite series in powers of $n^{-1}$. Exact answers are known in a few cases; for the normal the result for the mean involves gamma functions (see Bowman and Shenton, 1988, pp.37-39).

The problems can be highlighted by considering $E(\sqrt{m_2})$; higher moments introduce only minor modifications. Now if the density of $m_2$ was known, or the joint density of $m_1'$ and $m_2'$ $(m_s' = \sum x_j^s/n)$, then

$$\begin{aligned} E(\sqrt{m_2}) &= \underbrace{\int\int \cdots \int}_{n-\text{fold}} \sqrt{\frac{\sum(x_j - \bar{x})^2}{n}}\, f(x_j)dx_j \\ &= \int\int \sqrt{m_2}\,\psi(m_1', m_2')dm_1'dm_2' \qquad (2.55) \\ &= \int \sqrt{m_2}\,\phi(m_2)dm_2 , \end{aligned}$$

at least in the continuous case. For $n = 2, 3, 4$, these integrals may be evaluated exactly (or at worst fairly accurately by quadrature). Thus in exponential sampling exact answers are given by Lam (1978, 1980), the case $n = 4$ being rather troublesome. However in general $\psi(m_2)$ exists but is unknown. We do know however, that if moments exist then $0 < E(\sqrt{m_2}) < \sigma\sqrt{(1 - 1/n)}$.

One approach to the general problem is through simulation studies. Another is to develop computer extended power series in $n^{-1}$ for the moment concerned (we refer to this approach as COETS; for details see Bowman, 1986, pp 156-168; Dusenberry and Bowman, 1977; Shenton and Bowman, 1975).

In the case of an exponential population (density $f(x) = \exp(-x)$, $x > 0$) the series (Table 2.2) shows a regular sign pattern of period two, and modulus increase about that of the double factorial series. Higher moments can be set up using COETS.

**Table 2.2** $E(\sqrt{m_2}) \sim \alpha_0 + \alpha_1/n + \cdots$ **from an Exponential Population**

| $s$ | $\alpha_s$ | $r_s$ | $s$ | $\alpha_s$ | $r_s$ |
|---|---|---|---|---|---|
| 0 | 1.000000000 00 | | 15 | -1.369547748 32 | 912.285 |
| 1 | -1.500000000 00 | 1.500 | 16 | 1.418623118 38 | 1035.833 |
| 2 | 6.125000000 00 | 4.083 | 17 | -1.656091271 38 | 1167.393 |
| 3 | -1.513125000 02 | 24.704 | 18 | 2.164488706 41 | 1306.986 |
| 4 | 8.642210938 03 | 57.115 | 19 | -3.148511720 44 | 1454.621 |
| 5 | -8.505438320 05 | 98.417 | 20 | 5.070050970 47 | 1610.301 |
| 6 | 1.251222284 08 | 147.109 | 21 | -8.994383654 50 | 1774.022 |
| 7 | -2.538401906 10 | 202.874 | 22 | 1.750110840 54 | 1945.782 |
| 8 | 6.743174153 12 | 265.646 | 23 | -3.719990761 57 | 2125.574 |
| 9 | -2.262406892 15 | 335.511 | 24 | 8.605808037 60 | 2313.395 |
| 10 | 9.335091184 17 | 412.618 | 25 | -2.159403093 64 | 2509.239 |
| 11 | -4.640837914 20 | 497.139 | 26 | 5.858683167 67 | 2713.103 |
| 12 | 2.734561982 23 | 589.239 | 27 | -1.713655047 71 | 2924.983 |
| 13 | -1.884275123 26 | 689.059 | 28 | 5.389234680 74 | 3144.877 |
| 14 | 1.501227222 29 | 796.713 | | | |

(Some loss of accuracy is unavoidable; the last entry may have 2 or so correct first digits. A plot of the ratio $r_s = |\alpha_s / \alpha_{s-1}|$ against $s$ shows a remarkable smoothness).

For the J-fraction for $E(\sqrt{m_2})$,

$$E(\sqrt{m_2}) \sim \frac{nb_0}{n+a_1-} \; \frac{b_1}{n+a_2-} \; \cdots$$

we have

| $s$ | $b_s$ | $a_s$ | $s$ | $b_s$ | $a_s$ |
|---|---|---|---|---|---|
| 0 | 1 | - | 8 | -1.968 05 | -2.148 03 |
| 1 | 3.875 | 1.5 | 9 | 4.224 05 | 1.720 03 |
| 2 | 8.753 02 | 3.518 01 | 10 | 8.132 05 | 1.831 03 |
| 3 | 6.058 03 | 1.228 02 | 11 | 1.274 06 | 2.170 03 |
| 4 | 1.985 04 | 2.550 02 | 12 | 1.861 06 | 2.581 03 |
| 5 | 4.388 04 | 4.438 02 | 13 | 2.607 06 | 3.040 03 |
| 6 | 3.428 04 | 7.564 02 | 14 | — | 3.544 03 |
| 7 | -5.892 06 | 2.944 03 | | | |

showing some anomolous signs. Nonetheless the last few convergents, using an extended version of the coefficients for $n = 2$, gave 0.5958, 0.5962, and 0.5966 for the true value 0.5. Similarly the last convergent led to 0.688 for $n = 3$, 0.746 for $n = 4$, and 0.785 for $n = 5$; these may be compared with 0.6506, and 0.7271 for $n = 3$, and 4 respectively given by Lam (1980), and 0.775 for $n = 5$ resulting from 200,000 cycle simulations. Approximants would be expected to improve for larger values of $n$.

Next consider the possibility of a Stieltjes c.f. form

$$E(\surd m_2) = \frac{n}{n+}\ \frac{p_1}{1+}\ \frac{q_1}{n+}\ \cdots \tag{2.56}$$

where the $p$'s and $q$'s are expected to be positive. For $n = 1$ this leads to a contradiction since $E(\surd m_2)$ is now zero and the c.f. asserts

$$1/(1+p_1) < E(\surd m_2) < 1 .$$

This leads to the possibility

$$E(\surd m_2) = 1 - \frac{1.5}{n+}\ \frac{p_1}{1+}\ \frac{q_1}{n+}\ \cdots , \tag{2.57}$$

for which the coefficients turn out to be

| $s$ | $p_s$ | $q_s$ | $s$ | $p_s$ | $q_s$ |
|---|---|---|---|---|---|
| 1 | 4.083333 | 20.620748 | 8 | 644.021915 | 744.298724 |
| 2 | 38.828929 | 66.365495 | 9 | 812.513397 | 926.700271 |
| 3 | 92.719150 | 130.635053 | 10 | 1000.516932 | 1129.027319 |
| 4 | 164.676968 | 213.837495 | 11 | 1208.041097 | 1351.278971 |
| 5 | 255.455363 | 316.628897 | 12 | 1435.093097 | 1593.453091 |
| 6 | 365.523157 | 439.220062 | 13 | 1681.681057 | 1855.542420 |
| 7 | 495.031129 | 581.822712 | 14 | 1947.823516 | |

with some possibility of loss of accuracy in the last 2 or so coefficients. The pronounced monotonicity is noteworthy, suggesting the unlikelihood of a turn-around and some sign changes. There are the numerically oriented approximations

$$p_s \sim \pi^2 s^2 , \qquad q_s \sim \pi^2 s(s+1/2) \tag{2.58}$$

which may be compared with the approximate values $p_s \sim s(s-1)\pi^2$, and $q_s \sim s^2\pi^2$ for the double factorial series. For the Stieltjes form

$$\frac{1}{k_1 n+}\ \frac{1}{k_2+}\ \cdots$$

the approximations (2.58) lead to

$$k_{2s+1} = \prod_{m=1}^{s} (p_m / q_m) \sim 1/\{2\surd(2s)\} ,$$

$$k_{2s} \sim 2\surd 2/(\pi^2 s^{3/2}) ,$$

suggesting the divergence of $\sum k_s$.

Some assessments are as follows:

| | $n=2$ | | $n=3$ | | $n=4$ | | $n=10$ | |
|---|---|---|---|---|---|---|---|---|
| $s$ | $C_{2s+1}$ | $C_{2s}$ | $C_{2s+1}$ | $C_{2s}$ | $C_{2s+1}$ | $C_{2s}$ | $C_{2s+1}$ | $C_{2s}$ |
| 11 | 0.4877 | 0.5674 | 0.6429 | 0.6733 | 0.7227 | 0.7369 | 0.876117 | 0.876843 |
| 12 | 0.4899 | 0.5640 | 0.6439 | 0.6717 | 0.7233 | 0.7361 | 0.876162 | 0.876780 |
| 13 | 0.4918 | 0.5611 | 0.6448 | 0.6704 | 0.7238 | 0.7354 | 0.876197 | 0.876730 |
| 14 | -- | 0.5585 | -- | 0.6693 | -- | 0.7348 | -- | 0.876690 |
| True | 0.5 | | 0.6506 | | 0.7271 | | 0.8758* | |

(* Simulation of 20,000 cycles. For $n=50$, $C_{24}=C_{25}=0.971794$ to this accuracy).

There is no inconsistency in evidence here and one has some confidence in the conjecture that

$$E(\sqrt{m_2}) = 1 - \int_0^\infty \frac{d\sigma(t)}{t+n}, \qquad (n=1,2,\cdots) \tag{2.59}$$

where $\sigma(t)$ is a distribution function. In particular

$$\int_0^\infty \frac{d\sigma(t)}{t+1} = 1. \tag{2.60}$$

As further support for the assessments, the Levin $t$-algorithm (Bowman and Shenton, 1988, pp.222-228) was used. The best stopping member of the sequences (signaled by comparison with exact values for $n=2,3,4$) turns out to be $r=16$ with sequence values shown in Table 2.3. For $n=10$, the Levin value is 0.876399 in excellent agreement with the c.f. Again for $n=25$, the c.f. yields the value 0.945976 against 0.945970 for Levin.

### 2.8.3 $E(\sqrt{m_2})$ in Normal Sampling

#### 2.8.3.1 J-Fractions for Even and Odd Parts

The series divides naturally into even and old parts (Table 1.8, Bowman and Shenton, 1988) and an analysis of the magnitudes is given in Table 2.4.

There are the interesting approximations

$$\sqrt{(|e_{2s}/e_{2s-2}|)} \sim (2s-5/2)\pi,$$
$$\sqrt{(|e_{2s+1}/e_{2s-1}|)} \sim (2s-½)/\pi. \tag{2.61}$$

Of course, the exact value of the expectation is

$$y(n) = E(\sqrt{m_2}) = \frac{(n-1)\Gamma(\tfrac{1}{2}n)}{\sqrt{(2n)}\Gamma(\tfrac{1}{2}n + \tfrac{1}{2})}, \qquad (n = 2, 3, \cdots) \quad (2.62)$$

$$= \frac{(2m-1)}{2\sqrt{(m\pi)}} \int_0^\infty e^{-mx}(1-e^{-x})^{-1/2}dx . \qquad (m = \tfrac{1}{2}n)$$

Setting up a recursion for the coefficients in the power series expansion of $(1-e^{-x})^{-1/2}$ we find

$$2se_s + \frac{(2s-3)e_{s-1}}{2!} + \frac{(2s-5)e_{s-2}}{3!} + \cdots + \frac{e_1}{s!} = \frac{e_0}{(s+1)!} \qquad (2.63)$$

for $s = 2, 3, \cdots;\ e_0 = 1,\ e_1 = 1/4$. The coefficients may be generated from this, the only limit being storage and other computing problems.

**Table 2.3 Levin's t-Algorithm $\{\alpha_r\}$ for $E(\sqrt{m_2})$ Exponential Population**

| | Sample size $n$ | | | | | |
|---|---|---|---|---|---|---|
| $r$ | 1 | 2 | 3 | 4 | 5 | 10 |
| 2 | 0.583 | 0.702 | 0.761035 | 0.798047 | 0.824162 | 0.891098 |
| 3 | 0.473 | 0.618 | 0.699072 | 0.751617 | 0.788421 | 0.878088 |
| 4 | 0.307 | 0.547 | 0.662392 | 0.730372 | 0.775181 | 0.875928 |
| 5 | 0.209 | 0.522 | 0.654464 | 0.727659 | 0.774413 | 0.876352 |
| 6 | 0.179 | 0.522 | 0.656291 | 0.729336 | 0.775663 | 0.876580 |
| 7 | 0.181 | 0.526 | 0.658161 | 0.730197 | 0.776051 | 0.876548 |
| 8 | 0.186 | 0.527 | 0.658058 | 0.729908 | 0.775781 | 0.876483 |
| 9 | 0.184 | 0.525 | 0.656952 | 0.729299 | 0.775438 | 0.876453 |
| 10 | 0.173 | 0.521 | 0.655715 | 0.728770 | 0.775185 | 0.876436 |
| 11 | 0.155 | 0.518 | 0.654749 | 0.728407 | 0.775025 | 0.876425 |
| 12 | 0.136 | 0.515 | 0.654016 | 0.728142 | 0.774908 | 0.876417 |
| 13 | 0.114 | 0.512 | 0.653230 | 0.727853 | 0.774779 | 0.876409 |
| 14 | 0.096 | 0.510 | 0.652833 | 0.727728 | 0.774732 | 0.876408 |
| 15 | 0.078 | 0.509 | 0.652461 | 0.727615 | 0.774690 | 0.876407 |
| 16 | 0.023 | 0.503 | 0.651103 | 0.727175 | 0.774516 | 0.876399 |
| 17 | 0.061 | 0.508 | 0.654488 | 0.727577 | 0.774680 | 0.876406 |
| 18 | -0.125 | 0.513 | 0.653393 | 0.727894 | 0.774793 | 0.876410 |
| Exact | 0.000 | 0.500 | 0.650620 | 0.727087 | -- | 0.8760* |
| % Error | | 0.53 | 0.07 | 0.01 | - | |

(* Simulation of 200,000 cycles.)

**Table 2.4 Analysis of Coefficients**

Population: Normal Statistic: Standard Deviation
Moment: $E(\sqrt{m_2})$ Even and Odd Sequences

| $s$ | $\frac{e_{2s}}{e_{2s-2}}$ | $\left\|\frac{e_{2s}}{e_{2s-2}}\right\|^{1/2}$ | $r_s$ | $\frac{e_{2s+1}}{e_{2s-1}}$ | $\left\|\frac{e_{2s+1}}{e_{2s-1}}\right\|^{1/2}$ | $r_s'$ |
|---|---|---|---|---|---|---|
| 1 | 0.2188 | 0.4677 | - | 0.0938 | 0.3062 | 0.4775 |
| 2 | 0.1317 | 0.3629 | 0.4665 | 0.8385 | 0.9157 | 0.1141 |
| 3 | 1.2304 | 1.1092 | 1.1141 | 2.7692 | 1.6641 | 1.7507 |
| 4 | 3.1073 | 1.7627 | 1.7507 | 5.4860 | 2.3422 | 2.3873 |
| 5 | 5.7514 | 2.3982 | 2.3873 | 8.9826 | 2.9971 | 3.0239 |
| 6 | 9.1869 | 3.0310 | 3.0239 | 13.2698 | 3.6428 | 3.6606 |
| 7 | 13.4289 | 3.6645 | 3.6606 | 18.3562 | 4.2844 | 4.2972 |
| 8 | 18.4836 | 4.2993 | 4.2972 | 24.2466 | 4.9241 | 4.9338 |
| 9 | 24.3519 | 4.9348 | 4.9338 | 30.9437 | 5.5627 | 5.5704 |
| 10 | 31.0332 | 5.5707 | 5.5704 | 38.4489 | 6.2007 | 6.2070 |
| 11 | 38.5267 | 6.2070 | 6.2070 | 46.7630 | 6.8384 | 6.8437 |
| 12 | 46.8320 | 6.8434 | 6.8437 | 55.8866 | 7.4757 | 7.4803 |
| 13 | 55.9485 | 7.4799 | 7.4803 | 65.8199 | 8.1129 | 8.1169 |
| 14 | 65.8761 | 8.1164 | 8.1169 | 76.5631 | 8.7500 | 8.7535 |
| 15 | 76.6146 | 8.7530 | 8.7535 | 88.1164 | 9.3870 | 9.3901 |
| 16 | 86.1640 | 9.3896 | 9.3901 | 100.4799 | 10.0240 | 10.0268 |
| 17 | 100.5241 | 10.0262 | 10.0268 | 113.6537 | 10.6608 | 10.6634 |
| 18 | 113.6949 | 10.6628 | 10.6634 | 127.6378 | 11.2977 | 11.3000 |
| 19 | 127.6764 | 11.2994 | 11.3000 | 142.4322 | 11.9345 | 11.9366 |
| 20 | 142.4686 | 11.9360 | 11.9366 | 158.0371 | 12.5713 | 12.5732 |
| 21 | 158.0715 | 12.5726 | 12.5732 | 174.4524 | 13.2080 | 13.2099 |
| 22 | 174.4849 | 13.2093 | 13.2099 | 191.6781 | 13.8448 | 13.8465 |
| 23 | 191.7090 | 13.8459 | 13.8465 | 209.7143 | 14.4815 | 14.4831 |
| 24 | 209.7438 | 14.4825 | 14.4831 | 228.5609 | 15.1182 | 15.1197 |
| 25 | 228.5891 | 15.1192 | 15.1197 | 248.2181 | 15.7549 | 15.7563 |
| 26 | 248.2450 | 15.7558 | 15.7563 | 268.6857 | 16.3916 | 16.3930 |
| 27 | 268.7116 | 16.3924 | 16.3930 | 289.9639 | 17.0283 | 17.0296 |
| 28 | 289.9887 | 17.0291 | 17.0296 | 312.0526 | 17.6650 | 17.6662 |
| 29 | 312.0764 | 17.6657 | 17.6662 | 334.9518 | 18.3017 | 18.3028 |

(Here $r_s = (2s - 5/2)/\pi$, and $r_s' = (2s - ½)/\pi$.)

From numerical analysis of Table 2.4 we have the approximation

$$\begin{cases} |e_{2s}| \sim 0.2116(2/\pi)^{2s}\Gamma^2(s - 1/4), \\ |e_{2s+1}| \sim 0.1143(2/\pi)^{2s}\Gamma^2(s + 3/4), \end{cases} \tag{2.64}$$

so that the series is relatively benign for the first dozen or so terms, which in fact give little warning of the ultimate divergency.

The first point about this case concerns the J-fraction for $y(n)$. Suppose we are searching for a summation algorithm, how do the various forms perform? If

$$y(n) = \frac{nb_0}{n+a_1-}\;\frac{b_1}{n+a_2-}\;\cdots$$

it turns out that the $b$'s are negative (excepting $b_0$). $a_1>0$, $a_2$ to $a_9$ are negative and alternate in sign thereafter. This is not very encouraging.

Consider the even and odd parts defined by

$$y_e(n) = \{y(n)+y(-n)-2\}/2, \tag{2.65a}$$

$$y_o(n) = \{y(n)-y(-n)-2e_1/n\}/2. \tag{2.65b}$$

We now have

$$ny_o(n) = \frac{-p_0^*}{n^2+}\;\frac{p_1^*}{1+}\;\frac{q_1^*}{n^2+}\;\frac{p_2^*}{1+}\;\frac{q_2^*}{n^2+}\;\cdots,\quad (p_0^* = 0.0703125) \tag{2.66a}$$

$$y_e(n) = \frac{-p_0}{n^2+}\;\frac{p_1}{1+}\;\frac{q_1}{n^2+}\;\frac{p_2}{1+}\;\frac{q_2}{n^2+}\;\cdots,\quad (p_0 = 0.21875) \tag{2.66b}$$

for which the first few partial numerators are:

| $s$ | $y_o(n)$ $p_s^*$ | $q_s^*$ | $y_e(n)$ $p_s$ | $q_s$ |
|---|---|---|---|---|
| 1 | 0.838541667 | 1.93067417 | 0.131696429 | 1.09870611 |
| 2 | 3.89678156 | 5.88056825 | 2.10181904 | 4.15252976 |
| 3 | 8.94478582 | 11.8352963 | 6.05008488 | 9.20293048 |
| 4 | 15.9890141 | 19.7925819 | 12.0013848 | 16.2506931 |
| 5 | 25.0310506 | 29.7515640 | 19.9548759 | 25.2965089 |
| 6 | 36.0715727 | 41.7117990 | 29.9100214 | 36.3408246 |
| 7 | 49.1109503 | 55.6730188 | 41.8664726 | 49.3839383 |
| 8 | 64.1494145 | 71.6350449 | 55.8239900 | 64.4260590 |
| 9 | 81.1871230 | 89.5977503 | 71.7824019 | 81.4673394 |
| 10 | 100.224190 | 109.561041 | 89.7415795 | 100.507895 |
| 11 | 121.260700 | 131.524843 | 109.701424 | 121.547817 |
| 12 | 144.296720 | 155.489103 | 131.661856 | 144.587177 |
| 13 | 169.332298 | 181.453783 | 155.622813 | 169.626033 |
| 14 | 196.367456 | 181.369003 | 181.584244 | 196.134692 |

(Note: $p_s^* \sim s^2$, $q_s^* \sim s(s+1)$; $p_s \sim s(s-1)$, $q_s \sim s^2$. The entry for $q_{14}^*$ is clearly suspect.)

They are positive as far as the calculations go. Moreover, since

$$y(n) = 1 - (3/4)/n + y_o(n) + y_e(n)$$

it follows that using the Stieltjes odd (even) approximants from (2.66a) and (2.66b), we shall find approximants to $y(n)$ less (greater) than the true value. For example, the 28th and 29th approximants to $y(2) = 0.5641895835$ are

$$y_{28}(2) = 1 - 0.75/2 - 0.053234993 - 0.007575335 = 0.56418967,$$

$$y_{29}(2) = 1 - 0.75/2 - 0.053235032 - 0.007575461 = 0.56418951.$$

### 2.8.3.2 Reciprocal Relations and Stieltjes c.f.s

Stieltjes (1905) proved the equivalent of the expression

$$y(n) = \left(1 - \frac{1}{n}\right)\sqrt{\left(1 + \frac{2}{4n - 1 +}\;\frac{1\cdot 3}{4n +}\;\frac{3\cdot 5}{4n +}\;\frac{5\cdot 7}{4n +}\cdots\right)} \tag{2.67}$$

which incidentally provides monotonic sequences of bounding approximants to $y(n)$ when $4n > 1$. By elementary manipulation

$$\frac{n^2y^2(n) - (n-1)^2}{n^2y^2(n) + (n-1)^2} = \frac{n}{4n^2 +}\;\frac{1}{4/3 +}\;\frac{1}{4n^2/5 +}\;\frac{1}{4/7 +}\;\frac{1}{4n^2/9 +}\cdots \tag{2.68}$$

so that the left side is an odd function in $n$, leading to

$$y(-n) = (1 - 1/n^2)/y(n)$$

which defines $y(n)$ for negative $n$ (complex values not being of especial interest). It is worth mentioning that in letter 153 to Hermite (written around 1888; see Stieltjes, 1905), Stieltjes uses the expressions

$$\frac{\Gamma(a)\Gamma(n)}{\Gamma(a+n)} = \int_0^\infty \left(\frac{1 - e^{-y}}{y}\right)^{a-1} y^{a-1}e^{-ny}\,dy, \tag{2.69a}$$

$$\frac{\Gamma(a)\Gamma(n-a+1)}{\Gamma(n+1)} = \int_0^\infty \left(\frac{e^{y} - 1}{y}\right)^{a-1} y^{a-1}e^{-ny}\,dy, \tag{2.69b}$$

along with power series for the first factors in the integrand, to obtain the series

$$\frac{\Gamma(n-a+1)}{\Gamma(n+1)} = \frac{1}{n^a}\left\{1 + \frac{a}{n}c_1 + \frac{a(a+1)}{n^2}c_2 + \cdots\right\}, \tag{2.70a}$$

$$\frac{\Gamma(n)}{\Gamma(n+a)} = \frac{1}{n^a}\left\{1 - \frac{a}{n}c_1 + \frac{a(a+1)}{n^2}c_2 - \cdots\right\}, \tag{2.70b}$$

where $c_1, c_2, \cdots$ are polynomials in $a$. He points out that if $n$ is replaced by $-n$ in (2.70b) then $(-n)^a$ has to be replaced by $n^a \sin(n\pi)/(\sin(n-a)\pi)$. The case $a = \frac{1}{2}$, $n \to \frac{1}{2}n$ is clearly related to our $y(n)$ in (2.62). The unanswered problem is whether there is a single expression for $y(n)$ valid over a domain to at least include positive and negative $n$. Again, Stieltjes' discovery of (2.67) was after his discussion of the form in (2.70), but he apparently did not associate the former with the reciprocity problem.

### 2.8.3.3 Integral for Reciprocal Function

The relation for $y(n)$, suggests studying the function

$$f(n) = y(n)/(1 - 1/n) \tag{2.71}$$

for which, at least formally,

$$f(n)f(-n) = 1. \tag{2.72}$$

How will summation algorithms work for $f(n)$? First of all from (2.62)

$$f(n) = \sqrt{(m/\pi)}\int_0^\infty e^{-mx}(1-e^{-x})^{-\frac{1}{2}}dx \tag{2.73}$$

so that if

$$f(n) \sim \sum b_s / n^s,$$

then expanding the integrand,

$$2sb_s + \frac{(2s-1)(2s-3)}{2!}b_{s-1} + \frac{(2s-1)(2s-3)(2s-5)}{3!}b_{s-2} \tag{2.74}$$

$$+ \cdots + \frac{(2s-1)(2s-3)\cdots 1}{s!}b_1 = \frac{(2s-1)(2s-3)\cdots 1}{(s+1)!}.$$

$$(s = 1, 2, \cdots; b_0 = 1)$$

For example,

$$f(n) \sim 1 + 1/(4n) + 1/(32n^2) - 5/(128n^3) - 21/(2048n^4) + \cdots$$

and an extended tabulation brings out the fact that, apart from the first term, there is a periodicity of four in the sign pattern, and that the coefficients are integers apart from a power of 2 in the denominators. For example:

$b_s = N_s / 2^{\phi(s)}$

| $s$ | $N_s$ | $\phi_s$ | $s$ | $N_s$ | $\phi_s$ |
|---|---|---|---|---|---|
| 0 | 1 | 0 | 7 | -39325 | 18 |
| 1 | 1 | 2 | 8 | -334477 | 23 |
| 2 | 1 | 5 | 9 | 28717403 | 25 |
| 3 | -5 | 7 | 10 | 59697183 | 28 |
| 4 | -21 | 11 | 11 | -8400372435 | 30 |
| 5 | 399 | 13 | 12 | -341429291905 | 34 |

The Stieltjes c.f. has been computed for a few terms, and

$$n^{-1}f(n) = \frac{1}{n-}\ \frac{1/4}{1+}\ \frac{1/8}{n-}\ \frac{p_2}{1+}\ \frac{p_2}{n-}\ \frac{p_3}{1+}\ \frac{p_3}{n-}\ \frac{p_4}{1+}\ \frac{p_4}{n-}\ \cdots \tag{2.75}$$

where

| $s$ | $p_s$ | $s$ | $p_s$ | $s$ | $p_s$ |
|---|---|---|---|---|---|
| 2 | 1.37500000 | 12 | 8.95466981 | 22 | 16.2947783 |
| 3 | 0.732438017 | 13 | 4.03118846 | 23 | 7.43616714 |
| 4 | 2.93279411 | 14 | 10.4338010 | 24 | 17.7495811 |
| 5 | 1.37362135 | 15 | 4.70713644 | 25 | 8.12324957 |
| 6 | 4.46116414 | 16 | 11.9064989 | 26 | 19.2011091 |
| 7 | 2.02809281 | 17 | 5.38598397 | 27 | 8.81186645 |
| 8 | 5.97095265 | 18 | 13.3737577 | 28 | 20.6496781 |
| 9 | 2.69058071 | 19 | 6.06730750 | 29 | 9.50187496 |
| 10 | 7.46772663 | 20 | 14.8363244 | 30 | 22.0955700 |
| 11 | 3.35872173 | 21 | 6.75078574 | | |

The interesting form of (2.75) can not escape notice (see McCabe, 1983), and the fact the even and odd partial numerators belong to different classes. Numerical studies ($n = 1(1)5$) bring out a periodicity of four with respect to increasing and decreasing sequences (in this connection compare the c.f. for $\overline{R}(t)$ in (1.52)); for example, the 3(4)57 approximants exceed the true value whereas the 5(4)55 approximants are deficient. To pursue this further, a contraction process applied to (2.75) yields

$$f(n) = 1 + \frac{1}{4n - \frac{1}{2} +}\ \frac{4^2 p_1 p_2}{4n +}\ \frac{4^2 p_2 p_3}{4n +}\ \cdots \tag{2.76}$$

with $p_1 = 1/8$, and this clearly leads to bounding sequences since the $p$ 's (as far as calculated) are positive. The resemblance to the Stieltjes form

$$f^2(n) = 1 + \frac{2}{4n-1+}\ \frac{1\cdot 3}{4n +}\ \frac{3\cdot 5}{4n +}\ \cdots$$

is remarkable. Can one conclude that Stieltjes integral transforms $\psi_1(\cdot)$ and $\psi_2(\cdot)$ exist such that

$$\begin{cases} f(n) = 1 + \dfrac{1}{4n - \tfrac{1}{2} + n\psi_1(n^2)}, \\ f^2(n) = 1 + \dfrac{2}{4n - 1 + n\psi_2(n^2)}, \end{cases} \tag{2.77}$$

where

$$\psi_i(n^2) = \int_0^\infty \frac{d\,\sigma_i(t)}{t + n^2}. \qquad (i = 1, 2)$$

#### 2.8.3.4 Numerical Case

The 55th and 57th convergents of (2.75) yield $1.253198 < f(1) < 1.253417$ for $\sqrt{(\pi/2)} = 1.253314$.

Similarly, $1.12837910 < f(2) < 1.12837922$ for $2/\sqrt{\pi} = 1.128379167$.

Again for the even and odd parts of $f(n)$,

$$\begin{cases} f_e(n) = \dfrac{n^2}{n^2-}\;\dfrac{p_1}{1+}\;\dfrac{q_1}{n^2+}\;\cdots, \\ f_o(n) = \dfrac{0.25n}{n^2+}\;\dfrac{p_1^*}{1+}\;\dfrac{q_1^*}{n^2+}\;\cdots, \end{cases} \tag{2.78}$$

where apart from $p_1$, the $p$'s and $q$'s are positive and increase with $s$, and where

$$f_e(n) = \{f(n) + f(-n)\}/2 = \sqrt{(2n/\pi)}\int_0^\infty e^{-2nx}\sqrt{(\coth x)}dx,$$

$$f_o(n) = \{f(n) - f(-n)\}/2 = \sqrt{(2n/\pi)}\int_0^\infty e^{-2nx}\sqrt{(\tanh x)}dx.$$

(The first few partial numerators in (2.78) are given in Bowman and Shenton, 1988, p39).

#### 2.8.3.5 $E(\sqrt{m_2})$: Conjectured Bounds

Lastly, a deduction from (2.76) gives the tentative inequalities ($m > \tfrac{1}{2}$)

$$h_1 < h_2 < \frac{\sqrt{m}\,\Gamma(m)}{\Gamma(m + \tfrac{1}{2})} < H_2 < H_1, \tag{2.79}$$

where, with $M = 16m$,

$$H_1 = (M+1)/(M-1),$$
$$h_1 = (M^2+M+11)/(M^2-M+11),$$
$$H_2 = \frac{11M^3+11M^2+830M+709}{11M^3-11M^2+830M-709},$$
$$h_2 = \frac{709M^4+709M^3+150969M^2+143170M+1072189}{709M^4-709M^3+150969M^2-143170M+1072189}.$$

**References:** Further details are given in Bowman and Shenton (1978). The ratio $y(n)$, or one closely related to it, has attracted the attention of mathematicians from time to time over the last century or longer (see for example, Perron, 1957, pp.31-6; also Mitrinović, 1970, pp.286-8). Recently, inequalities for $y^2(n)$ have been derived from purely statistical concepts (see for example, Gurland (1956), Gokhale (1962), and Uppuluri (1966); also Gautschi (1964) has derived results for $n$ not restricted to integers, and from a mathematical viewpoint). Remarks on related c.f.s are given by McCabe (1983).

### 2.8.4 Moments of the Standard Deviation in Logistic Sampling

The density is

$$f(x) = e^{-x}/(1+e^{-x})^2, \qquad (-\infty < x < \infty) \qquad (2.80)$$

which is symmetric about $x = 0$. The central moments are

$$\mu_{2s} = (2\pi)^{2s}\,|B_{2s}|\,(1-2^{1-2s}) \qquad (2.81)$$

in terms of Bernoulli numbers. In particular

$$\mu_2 = \sigma^2 = \pi^2/3.$$

It is convenient to consider the standardized variate $y = x/\sigma$, for which $E(y) = 0$, $Var(y) = 1$. Series for the mean and variance of $\sqrt{m_2}$ are given in Table 2.5. The sign pattern has periodicity two, and the magnitude is between $s!$ and $(2s)!$, the latter being a drastic reduction factor. Approximants, Table 2.6 using the 2 component Borel algorithm (see Chapter 5) referred to as 2cB with $a = 1$ are in good agreement for $n \geqslant 10$ with Monte-Carlo simulations (MC), but deteriorate alarmingly for the smaller values of $n$, and especially for higher moments. This phenomenon is doubtless due to over-correcting in the choice of 2cB (quite apart from the choice of $a$).

**Table 2.5 Series for $E(\sqrt{m_2})$ and $Var(\sqrt{m_2})$ from the Logistic Density**

| | $E(\sqrt{m_2})$ | | | $Var(\sqrt{m_2})$ | | |
|---|---|---|---|---|---|---|
| $s$ | $e_s$ | $r$ | $e_s/(2s)!$ | $e_s$ | $r$ | $e_s/(2s)!$ |
| 0 | 1.00000000 00 | - | 1.000000 | - | - | - |
| 1 | -0.90000000 00 | 0.95 | -0.450000 | 0.80000000 00 | - | 0.400000 |
| 2 | 0.45357143 00 | 0.71 | 0.018899 | -1.71714286 00 | 1.47 | -0.07158 |
| 3 | -4.76035714 00 | 3.24 | -0.006612 | 1.03371429 01 | 2.45 | 0.14357 |
| 4 | 8.15741778 01 | 4.14 | 0.002023 | -1.71922725 02 | 4.08 | -0.004264 |
| 5 | -2.46240478 03 | 5.49 | -0.000679 | 5.07596141 03 | 5.43 | 0.001399 |
| 6 | 1.11019595 05 | 6.72 | 0.000232 | -2.26568179 05 | 6.68 | -0.000473 |
| 7 | -6.89667790 06 | 7.88 | -0.000079 | 1.39962015 07 | 7.86 | 0.000161 |
| 8 | 5.60483303 08 | 9.02 | 0.000026 | -1.13351144 09 | 9.00 | -0.000054 |
| 9 | -5.72802419 10 | 10.13 | -0.000009 | 1.15977069 11 | 10.12 | 0.000018 |
| 10 | -7.24409764 12 | 11.23 | 0.000003 | -1.45922580 13 | 11.22 | 0.000006 |
| 11 | -1.09922817 15 | 12.32 | -0.000001 | 2.21155486 15 | 12.31 | 0.000002 |
| 12 | 1.97583837 17 | 13.41 | 0.000000 | -3.97153542 17 | 13.40 | -0.000001 |
| 13 | -4.15109617 19 | 14.50 | -0.000000 | 8.33786540 19 | 14.49 | 0.000000 |
| 14 | 1.00794121 22 | 15.58 | 0.000000 | -2.02337352 22 | 15.58 | -0.000000 |
| 15 | -2.80143602 24 | 16.67 | -0.000000 | 5.62105474 24 | 16.67 | 0.000000 |

(Here $r = \sqrt{|e_s/e_{s-1}|}$.)

**Table 2.6 Approximants to the Mean and Variance of $\sqrt{m_2}$ from the Logistic Density**

| $n$ | | $E(\sqrt{m_2})$ | $\mu_2(\sqrt{m_2})$ | $\mu_3(\sqrt{m_2})$ | $\mu_4(\sqrt{m_2})$ |
|---|---|---|---|---|---|
| 2 | S | 0.6655 | 0.2191 | 0.0737 | 0.1236 |
| | T | 0.5513 | 0.1960 | 0.1125 | 0.2020 |
| 3 | S | 0.7532 | 0.1700 | 0.0487 | 0.0798 |
| | MC | 0.7063 | 0.1668 | 0.0734 | 0.1413 |
| 4 | S | 0.8046 | 0.1393 | 0.0350 | 0.0562 |
| | MC | 0.7811 | 0.1390 | 0.0451 | 0.0811 |
| 5 | S | 0.8385 | 0.1181 | 0.0265 | 0.0418 |
| | MC | 0.8342 | 0.1208 | 0.0334 | 0.0581 |
| | MC | 0.8267 | 0.1192 | 0.0346 | 0.0614 |
| 10 | S | 0.9140 | 0.0675 | 0.0099 | 0.0147 |
| | MC | 0.9132 | 0.0676 | 0.0119 | 0.0179 |
| | MC | 0.9111 | 0.0663 | 0.0105 | 0.0164 |
| 20 | S | 0.9559 | 0.0365 | 0.00319 | 0.00473 |
| | MC | 0.9547 | 0.0368 | 0.00347 | 0.00472 |
| 50 | S | 0.9822 | 0.01538 | 0.00061 | 0.00076 |
| | MC | 0.9823 | 0.01554 | 0.00065 | 0.00079 |

(Note: S refers to 16 terms of 2cB with $a = 1$, T refers to the exact theoretical value; MC refers to 20,000 Monte Carlo simulations.)

A closer inspection of sequences (not given here) shows that 2cB with $a = 1$ is a case of over-dilution. A less drastic modification is to use $3^s e_s / (2s)!$, for which the coefficients in $E(\sqrt{m_2})$ become 1, -1.35, 0.1701, -0.1785, 0.1649, -0.1730, 0.1758, -0.1767, 0.1758. Applying 2cB to the series (omitting $e_0$, and $e_1$) with $a = 5$, $k = 1/3$, now yields, using six terms altogether, $F_4(2) = 0.5747$, $F_4(3) = 0.7145$, $F_4(4) = 0.7848$, and $F_4(5) = 0.8271$, in much better agreement with MC. As a further check on the assessments, Levin approximants are given in Table 2.7.

**Table 2.7 Levin's t-Algorithm for $E(\sqrt{m_2})$ under Logistic Sampling**

| $s$ | $n = 1$ | $n = 2$ | $n = 3$ | $n = 5$ | $n = 10$ |
|---|---|---|---|---|---|
| 10 | 0.05455 | 0.5557 | 0.7083 | 0.82574 | 0.912335 |
| 11 | 0.03357 | 0.5537 | 0.7079 | 0.82570 | 0.912334 |
| 12 | 0.01560 | 0.5523 | 0.7076 | 0.82567 | 0.912333 |
| 13 | -0.004812 | 0.5508 | 0.7073 | 0.82555 | 0.912333 |
| 14 | -0.05680 | 0.5469 | 0.7065 | 0.82559 | 0.912331 |
| 15 | -0.2980 | 0.5309 | 0.7040 | 0.82537 | 0.912327 |
| 16 | -1.500 | 0.4528 | 0.6938 | 0.82456 | 0.912315 |
| Exact | 0.0000 | 0.5513 | | | |
| MC | - | - | 0.7063 | 0.83442 | 0.9132 |
| | | | | 0.8267 | 0.9111 |
| 2cB* | | | 0.6945 | 0.8240 | 0.9168 |

(*Uses 2cB with $a = 1$ and 3 terms truncated, and 10 terms of the series altogether.)

### 2.8.5 $E(\sqrt{m_2})$ from a Uniform Distribution $U(-\sqrt{3}, \sqrt{3})$, $\sigma = 1$

The coefficients (Table 2.8) show an indication of a periodicity of ten (5 negative terms followed by 5 positive terms) with within half-period increasing moduli.

**Table 2.8 Series for $E(\sqrt{m_2})$ from Uniform Distribution $\mu_1' = 0$, $\sigma = 1$**

| $s$ | $e_s$ | $e_s / e_{s-1}$ | $s$ | $e_s$ | $e_s / e_{s-1}$ |
|---|---|---|---|---|---|
| 0 | 1.000000000 00 | - | 8 | 0.464829620 03 | 7.6115 |
| 1 | -0.600000000 00 | -0.6000 | 9 | 0.281710837 04 | 6.0605 |
| 2 | -0.371428571 00 | 0.6190 | 10 | 0.996487311 04 | 3.5373 |
| 3 | -0.462857143 00 | 1.2462 | 11 | -0.768692340 05 | -7.7140 |
| 4 | -0.724259740 00 | 1.5648 | 12 | -0.245242757 07 | 31.9039 |
| 5 | -0.646120280 00 | 0.8921 | 13 | -0.390525965 08 | 15.9241 |
| 6 | 0.529261819 01 | -8.1914 | 14 | -0.470781601 09 | 12.0551 |
| 7 | 0.610693024 02 | 11.5386 | 15 | -0.395478160 10 | 8.4005 |

Continued Fraction forms:

**2.8.5.1 One term truncated**

$$E(\sqrt{m_2}) = 1 - \left\{\frac{0.6}{n+}\ \frac{-0.6190}{1+}\ \frac{-0.6271}{n+}\ \frac{-0.6331}{1+}\ \frac{3.6181}{n+}\right.$$

$$\frac{-13.4021}{1+}\ \frac{1.4641}{n+}\ \frac{-2.4601}{1+}\ \frac{-117.7363}{n+}$$

$$\left.\frac{140.4516}{1+}\ \frac{-0.4416}{n+}\ \frac{-14.1029}{1+}\ \cdots\right\}.$$

Approximants to $1 - E(\sqrt{m_2})$:

| | $n = 1$ | | $n = 2$ | | $n = 3$ | |
|---|---|---|---|---|---|---|
| $s$ | $C_{2s-1}$ | $C_{2s}$ | $C_{2s-1}$ | $C_{2s}$ | $C_{2a-1}$ | $C_{2s}$ |
| 1 | 0.6000 | 1.575 | 0.300 | 0.434 | 0.200 | 0.252 |
| 2 | -0.909 | 0.320 | 0.546 | 0.701 | 0.271 | 0.278 |
| 3 | -0.474 | 0.533 | 0.575 | 1.177 | 0.273 | 0.285 |
| 4 | 1.181 | 0.320 | 0.238 | 0.621 | 0.319 | 0.270 |
| 5 | 1.159 | 0.505 | 0.227 | 0.989 | 0.316 | 0.283 |
| 6 | 0.434 | 0.510 | 0.859 | 1.015 | 0.282 | 0.283 |

Approximants to $1 - E(\sqrt{m_2})$:

| | $n = 4$ | | $n = 5$ | | $n = 10$ | |
|---|---|---|---|---|---|---|
| $s$ | $C_{2s-1}$ | $C_{2s}$ | $C_{2s-1}$ | $C_{2s}$ | $C_{2s-1}$ | $C_{2s}$ |
| 1 | 0.150 | 0.177 | 0.120 | 0.137 | 0.06006 | 0.06396 |
| 2 | 0.184 | 0.185 | 0.140 | 0.140 | 0.06424 | 0.06426 |
| 3 | 0.184 | 0.186 | 0.1401 | 0.1408 | 0.06426 | 0.06410 |
| 4 | 0.190 | 0.181 | 0.1421 | 0.1375 | 0.06423 | 0.06423 |
| 5 | 0.190 | 0.186 | 0.1421 | 0.1406 | 0.064227 | 0.064263 |
| 6 | 0.186 | 0.186 | 0.1405 | 0.1406 | 0.064257 | 0.064293 |
| | $1-0.186=0.814$ | | $1-0.1405=0.8595$ | | $1-0.06427=0.93573$ | |

(True values are 0, 0.577350, 0.744628; 0.8187 (Levin), 0.8624 (MC), 0.9365 (MC).)

**2.8.5.2 Three terms truncated**

$$E(\sqrt{m_2}) \sim 1 - \frac{0.6}{n} - \frac{0.371429}{n^2} - \frac{1}{n^2}\left\{\frac{0.462857}{n+}\ \frac{p_1}{1+}\ \frac{q_1}{n+}\ \cdots\right\}. \tag{2.83}$$

Approximants to $1 - E(\sqrt{m_2})$:

| | $n = 2$ | | $n = 3$ | | $n = 4$ | | $n = 5$ | | $n = 10$ | |
|---|---|---|---|---|---|---|---|---|---|---|
| $s$ | $C_{2s-1}$ | $C_{2s}$ | $C_{2s-1}$ | $C_{2s}$ | $C_{2s-1}$ | $C_{2s}$ | $C_{2s-1}$ | $C_{2s}$ | $C_{2s-1}$ | $C_{2s}$ |
| 1 | 0.231 | 1.063 | 0.154 | 0.322 | 0.116 | 0.190 | 0.093 | 0.135 | 0.046 | 0.055 |
| 2 | 0.588 | 1.433 | 0.269 | 0.353 | 0.174 | 0.201 | 0.128 | 0.142 | 0.054 | 0.060 |
| 3 | -3.389 | 0.526 | 0.505 | 0.159 | 0.246 | 0.015 | 0.168 | -0.081 | 0.0510 | 0.0519 |
| 4 | 0.762 | -0.470 | 0.414 | 0.661 | 0.217 | 0.261 | 0.147 | 0.171 | 0.0531 | 0.0485 |
| 5 | -1.373 | -0.395 | 0.543 | 0.700 | 0.248 | 0.265 | 0.165 | 0.173 | 0.0458 | 0.0495 |
| 6 | -1.420 | -0.476 | 0.540 | 0.658 | 0.247 | 0.260 | 0.1650 | 0.1705 | 0.0456 | 0.0484 |
| | Unreliable | | 0.698 to 0.686 | | 0.811 to 0.811 | | 0.8585 to 0.8583 | | 0.9358 to 0.9358 | |

For $n = 10$, the series gives the preferred sum

$$\begin{aligned} E(\sqrt{m_2}) &\sim 1 - 0.06 - 0.003714 - 0.000463 - 0.000072 - 0.000006 \\ &\quad + 0.000005 + 0.000006 + 0.000005 + 0.00003 + 0.000001\,, \\ &= 0.935765 \end{aligned} \tag{2.84}$$

in good agreement with the c.f. assessments.

Comment: The algorithm is fairly successful for $n \geqslant 4$.

### 2.8.5.3 Levin's $t$-Algorithm

H. L. Rietz (1931) has given exact values for $n = 2, 3$ (with notational differences). Using these, the stopping rule applied to the sequences indicated $\alpha_{11}$ to be optimal.

| $E(\sqrt{m_2})$ from Uniform ($\sigma = 1$): Levin's $t$ Algorithm | | | | | |
|---|---|---|---|---|---|
| | 1 | 2 | 3 | 5 | 10 |
| $\alpha_{10}$ | -1.41 | 0.4951 | 0.7315 | 0.86023 | 0.935754 |
| $\alpha_{11}$ | -0.50 | 0.5514 | 0.7435 | 0.86156 | 0.935763 |
| $\alpha_{12}$ | 0.56 | 0.9738 | 0.8145 | 0.86373 | 0.935764 |
| Exact | 0.00 | 0.5774 | 0.7446 | — | — |

(Note that for $n = 10$, $\alpha_{12}$ thru $\alpha_{25}$ have the value 0.969008997 with only a single unit changing in the 9th decimal place; the c.f. (version (a)) leads to 0.9690090.)

### 2.8.5.4 Levin's U-Algorithm

We can use this version which Levin introduced to cope with less extreme cases; basically it is defined as

$$\beta_r = \sum_{j=0}^{r}(-1)^r \binom{r}{j}\left(\frac{j+1}{r+1}\right)^{r-2}\frac{A_{j+1}}{a_j} \Big/ \sum_{j=0}^{r}(-1)^r \binom{r}{j}\left(\frac{j+1}{r+1}\right)^{r-2}\frac{1}{a_j}.$$

$$(r \geqslant 2)$$

The stopping rule signals $\beta_{11,}$ the assessments being

| $n$ | 2 | 3 | 4 | 5 | 10 |
|---|---|---|---|---|---|
| $\beta_{11}$ | 0.524907 | 0.737946 | 0.818702 | 0.861124 | 0.935762 |

in good agreement with previous assessments for $n \geqslant 4$.

### 2.8.6 Student's *t*

We define this statistic as

$$t = (m_1' - \mu_1')/ s_x \tag{2.85a}$$

where $m_1'$, $s_x$ refer to the sample mean and standard deviation and $E(m_1') = \mu_1'$, $E(s_x^2) = \mu_2$.

$$t^* = (m_1' - \mu_1')/ \sqrt{m_2} = t / \sqrt{(1 - 1/n)}, \tag{2.85b}$$

which we sometimes find more convenient to use and more consistent with the usual notation for the skewness ($\sqrt{b_1} = m_3/ m_2^{3/2}$) and kurtosis ($b_2 = m_4/ m_2^2$) statistics.

#### 2.8.6.1 Sampling from the Exponential

The mean value of $t^*$ (Table 2.9) has alternating signs for $s \geqslant 2$, and the coefficients increase about as fast as $(2s)!$. Note that

$$½\sqrt{|e_s / e_{s-1}|} \sim s\,,$$

and also the smoothness of the plot of $|e_s / e_{s-1}|$ ( Figure 2.1a) and the effect of the modifications $|e_s / (2s - 1)!|$, and $|e_s / \Gamma(2s - ½)|$ ( Figure 2.1b).

Some comparisons for assessments of $E(t^*)$ are shown in Table 2.10. There is good agreement for $n \geqslant 10$, reasonable agreement for $5 \leqslant n < 10$, and thereafter considerable deterioration. It is possible that $E(t^*)$ is infinite for $n = 2$, involving a double integral with $|x - y|$ in the denominator. If this is correct then we should expect a certain wild behavior in the algorithms for $n = 2+$. The usual approach to singularities (Baker, 1975, pp.274-279; Gaunt and Guttmann, 1973, pp.206-210) is to sum a logarithmic derivative series, of one kind or another. For example, if

$$nE(t^*) \sim e_1 + e_2 x + \cdots, \qquad (x = 1/n)$$
$$= (1 - \lambda x)^k \phi(x) \qquad (2.86)$$

where $\phi(x)$ is regular in the vicinity of $x = 1/\lambda$, then

**Table 2.9 Series for $E(t^*)$ for the Exponential Density**

| $s$ | $e_s$ | $r_1$ | $r_2$ | $r_3$ | $r_4$ |
|---|---|---|---|---|---|
| 0 | 0.0 | - | - | - | - |
| 1 | -1.000000000 00 | - | - | 1.000 | 1.128 |
| 2 | -3.500000000 00 | 3.500 | 0.935 | 0.583 | 1.053 |
| 3 | 1.237500000 01 | 3.536 | 0.940 | 0.103 | 0.236 |
| 4 | -7.288125000 02 | 58.894 | 3.837 | 0.145 | 0.389 |
| 5 | 5.902222656 04 | 80.984 | 4.500 | 0.163 | 0.495 |
| 6 | -7.536102285 06 | 127.682 | 5.650 | 0.189 | 0.633 |
| 7 | 1.361212199 09 | 180.625 | 6.720 | 0.219 | 0.796 |
| 8 | -3.272621808 11 | 240.420 | 7.753 | 0.250 | 0.977 |
| 9 | 1.004917106 14 | 307.068 | 8.762 | 0.283 | 1.173 |
| 10 | -3.825960311 16 | 380.724 | 9.756 | 0.315 | 1.380 |
| 11 | 1.766005263 19 | 461.585 | 10.742 | 0.346 | 1.593 |
| 12 | -9.710400939 21 | 549.851 | 11.724 | 0.376 | 1.811 |
| 13 | 6.270053964 24 | 645.705 | 12.705 | 0.404 | 2.031 |
| 14 | -4.698137921 27 | 749.298 | 13.687 | 0.431 | 2.252 |
| 15 | 4.043910913 30 | 860.748 | 14.669 | 0.457 | 2.474 |
| 16 | -3.963595921 33 | 980.139 | 15.654 | 0.482 | 2.695 |
| 17 | 4.390900409 36 | 1107.530 | 16.640 | 0.506 | 2.915 |
| 18 | -5.456317608 39 | 1242.953 | 17.628 | 0.528 | 3.135 |
| 19 | 7.564793424 42 | 1386.428 | 18.617 | 0.550 | 3.355 |
| 20 | -1.163435829 46 | 1537.961 | 19.608 | 0.570 | 3.573 |
| 21 | 1.974992247 49 | 1697.552 | 20.601 | 0.590 | 3.792 |
| 22 | -3.683746145 52 | 1865.195 | 21.594 | 0.610 | 4.010 |
| 23 | 7.518107136 55 | 2040.886 | 22.588 | 0.628 | 4.228 |
| 24 | -1.672491851 59 | 2224.618 | 23.583 | 0.647 | 4.445 |
| 25 | 4.041384040 62 | 2416.385 | 24.578 | 0.664 | 4.663 |
| 26 | -1.057299071 66 | 2616.181 | 25.574 | 0.682 | 4.880 |
| 27 | 2.985813411 69 | 2824.001 | 26.571 | 0.698 | 5.097 |
| 28 | -9.076399059 72 | 3039.841 | 27.567 | 0.715 | 5.314 |

(Here $r_1 = |e_s/e_{s-1}|$, $r_2 = \frac{1}{2}\sqrt{|e_s/e_{s-1}|}$, $r_3 = |e_s/(2s-1)!|$ and $r_4 = |e_s/\Gamma(2s-\frac{1}{2})|$.)

Figure 2.1a $E(t^*)$ Statistic from Exponential Distribution

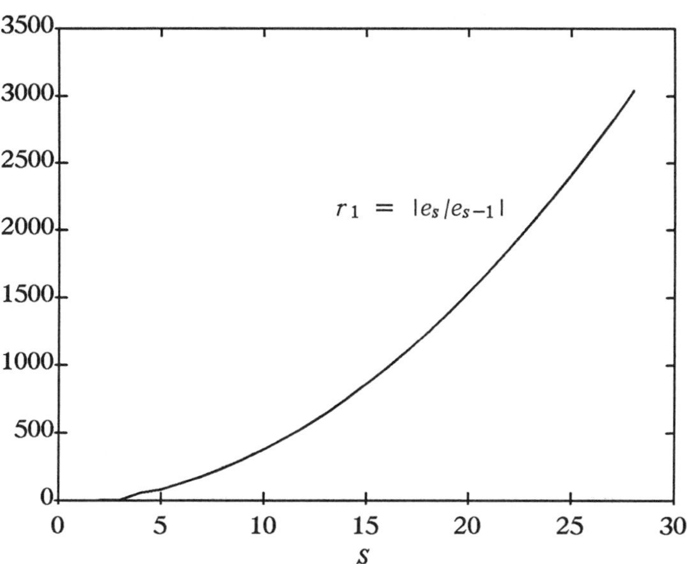


Figure 2.1b $E(t^*)$ Statistics from Exponential Distribution

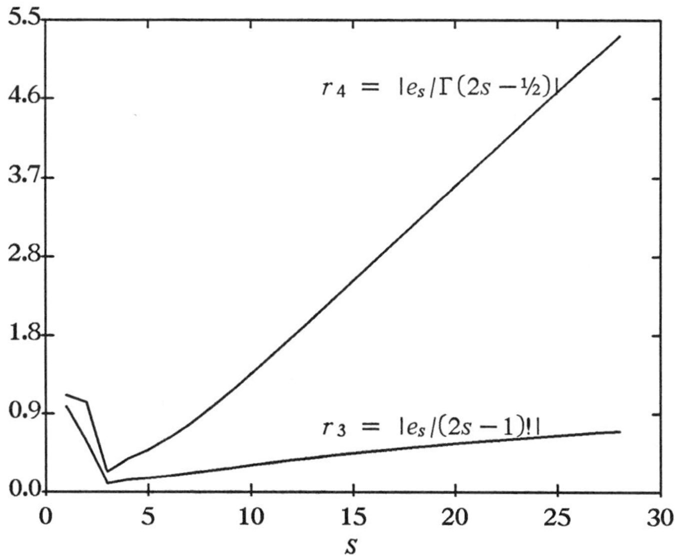

$$\frac{d}{dx}\ln\{nE(t^*)\} = -\frac{k\lambda}{1-\lambda x} + \frac{\phi'(x)}{\phi(x)}, \tag{2.87}$$

so that $(1-\lambda x)\frac{d}{dx}\ln\{nE(t^*)\}$ approaches $-k\lambda$ as $x \to 1/\lambda$. Assuming $\lambda$ known (or guessed) we sum this series by an algorithm when $n = \lambda$. The

**Table 2.10** $E(t^*)$ **in Exponential Sampling**

| | | $n = 4$ | $n = 5$ | $n = 10$ |
|---|---|---|---|---|
| (a) | $F_{10}$ | - | -0.3540 | -0.134855 |
| | $F_{11}$ | - | -0.3545 | -0.134855 |
| | $F_{12}$ | - | -0.3549 | -0.134910 |
| (b) | $F_{10}$ | - | -0.3596 | -0.135108 |
| | $F_{11}$ | - | -0.3598 | -0.135114 |
| | $F_{12}$ | - | -0.3600 | -0.135119 |
| (c) | Padé | -0.6339 [4\|5] | -0.3746 [11\|12] | -0.13548 [11\|12] |
| | | -0.6137 [5\|6] | -0.3736 [12\|13] | -0.13544 [12\|13] |
| | | -0.6001 [6\|7] | -0.3727 [13\|42] | -0.13541 [13\|14] |
| (d) | MC | -0.5496 | -0.3687 | -0.1423* |
| | | | -0.3637 | -0.1353 |
| | | | -0.3627 | |
| (e) | $F_5$ | -0.5766 | -0.3782 | -0.13587 |
| | $F_6$ | -0.5750 | -0.3776 | -0.13581 |
| (f) | $F_3$ | -0.4083 | -0.2965 | -0.1249 |
| | $F_3^*$ | -0.4572 | -0.3249 | -0.1303 |
| | $F_4$ | -0.4330 | -0.3116 | -0.1282 |
| | $F_4^*$ | -0.4690 | -0.3318 | -0.1316 |
| | $F_5$ | -0.4498 | -0.3215 | -0.1324 |
| | $F_5^*$ | -0.4779 | -0.3268 | -0.1324 |

( (a) is 2cB with $a = 8$, starting at $n^{-4}$ term;
(b) is 2cB with $a = 7.5$, starting at $n^{-4}$ term;
(c) Padé J-fraction $P[s \mid s+1]$;
(d) Monte Carlo simulation of 20,000 runs; * refers to 50,000 runs;
(e) 2cB with $(1-2/n)^{-1}$ factored out, $a = 1$;
(f) $F$ is 2cB with $\surd(1-2/n)^{-1}$ factored out, $a = 1$;
$F^*$ is 2cB with $\surd(1-2/n)^{-1}$ factored out, one term truncated, $a = 4$.
Note that for $n = 50$ the assessments in (a), (b), (c) indicate a value of -0.02135396 with 3 or so units different in the last digit. a simulation pointed to the value -0.02185. )

12th, 13th, and 14th 2cB approximants ($a = 1$) gave for $-nk$

| | | | |
|---|---|---|---|
| $n = 2$ | 1.4148 | 1.4041 | 1.3959 |
| $n = 3$ | 0.9883 | 0.9274 | 0.9022 |
| $n = 4$ | 0.6865 | 0.6539 | 0.6252. |

suggesting $(1 - 2/n)^{-1/2}$ as a singularity. The corresponding last three Padé approximants for $n = 2$ were 1.0581, 1.1101, and 1.1493 in fair agreement. A few approximants are given under (f) in Table 2.10 and show some improvement.

The series (Table 2.11) for $\mu_2(t^*)$, $\mu_3(t^*)$, and $\mu_4(t^*)$ are similar in structure to the series for $E(t^*)$. For the variance, the sign pattern is established from the $n^{-2}$ term onwards; for $\mu_3(t^*)$ for the $n^{-5}$ term onwards, and similarly for $\mu_4(t^*)$. Using fourteen coefficients for the variance, Padé approximants to the possible singularity derived from the sum $(1 - \lambda x)\frac{d}{dx}\{n\,\mu_2(t^*)\}$ with $x = 1/n$, were for the last three convergents

| | | | | |
|---|---|---|---|---|
| $n = 2$ | 3.6170 | 3.8615 | 4.0551 | $(k = -2)$ |
| $n = 3$ | 3.6381 | 3.7507 | 3.8259 | $(k = -1)$ |
| $n = 4$ | 3.6240 | 3.6702 | 3.6937 | |
| $n = 5$ | 3.6053 | 3.6177 | 3.6200. | |

These are anything but conclusive, and one would accept the possibility of singularities at $n = 2$, and $n = 3$, but probably not at $n \geqslant 4$. In fact in normal sampling (an extreme case)

$$\begin{cases} E(t^*) = 0, \\ Var\,(t^*) = (n-1)/\{n(n-3)\}, \\ \mu_3(t^*) = 0, \\ \mu_4(t^*) = 3(n-1)^2/\{(n-3)(n-5)\}. \end{cases}$$

We try $\mu_2(t^*) \sim (1-2/n)^3\Phi(n)$ and $\mu_2(t^*) \sim (1-2/n)^{-2}(1-3/n)\Phi^*(n)$ where $\Phi(\cdot)$ and $\Phi^*(\cdot)$ are regular functions for $|n| > 3$. Some results are shown in Table 2.12. We should point out that these, compared to the MC are acceptable for $n \geqslant 10$, and not too discrepant at $n = 5$, whereas 2cB truncating the $e_1$, $e_2$, $e_3$ terms and using $a = 9$ on the rest gives the first few approximants

| | | | |
|---|---|---|---|
| $n = 5$ | 0.869 | 0.886 | 0.901 |
| $n = 10$ | 0.230 | 0.231 | 0.232. |

these being too small. At $n = 4$ the approximants are very discrepant.

The 2cB and Padé algorithmic approaches are not satisfactory for the third and fourth moments especially for $10 \leqslant n < 20$ or so, but become acceptable for $n \geqslant 25$, and the Padé method seems an improvement over 2cB.

### 2.8.6.2 Stieltjes Continued Fractions

Converting the series to a Stieltjes continued fraction form we find the remarkable fact that, as far as calculations go, the partial numerators are positive (Table 2.13); this applies to $E(t)$ and $E(t^*)$. Moreover the coefficients are increasing and it seems unlikely they will go through stationary values.

Numerically, for $E(t^*)$ and $n = 3$,

| Odd | -0.13 | -0.25 | -0.34 | -0.42 | -0.49 | -0.54 | -0.59 | -0.63 | -0.66 | -0.68 | -0.71 | -0.73 | -0.75 |
|---|---|---|---|---|---|---|---|---|---|---|---|---|---|
| Even | -6.6 | -2.5 | -1.82 | -1.57 | -1.43 | -1.34 | -1.28 | -1.24 | -1.21 | -1.18— | 1.16 | -1.15 | -1.13 |

Table 2.11 Higher Moments of $t^*$ in Exponential Sampling

| $s$ | $\mu_2$ | $\mu_3$ | $\mu_4$ |
|---|---|---|---|
| 1 | 1.000000000000000 00 | 0.0 | 0.0 |
| 2 | 1.000000000000000 01 | -4.000000000000000 00 | 3.000000000000000 00 |
| 3 | -2.000000000000000 00 | -8.000000000000000 01 | 1.020000000000000 02 |
| 4 | 2.287500000000000 03 | -2.235000000000000 02 | 1.200000000000000 03 |
| 5 | -2.071580000000000 05 | -2.737650000000000 06 | 3.554100000000000 04 |
| 6 | 2.981018462500000 07 | 2.723414531250000 06 | -2.599818000000000 06 |
| 7 | -5.915759421875000 09 | -4.480830699375000 08 | 4.866640717500000 08 |
| 8 | 1.538585459524687 12 | 9.874858351953515 10 | -1.151781254190000 11 |
| 9 | -5.054959987010634 14 | -2.804264583383120 13 | 3.477802246395111 13 |
| 10 | 2.042252385495834 17 | 9.936134257423676 15 | -1.299782623032332 16 |
| 11 | -9.939241934046327 19 | -4.288867884066757 18 | 5.882743369464925 18 |
| 12 | 5.732620184824879 22 | 2.213789720086532 21 | -3.169104027566526 21 |
| 13 | -3.866387677176922 25 | -1.346261300735221 24 | 2.003815117070274 24 |
| 14 | 3.015383132838937 28 | 9.527468695999074 26 | -1.469869241557990 27 |
| 15 | -2.693379996589244 31 | -7.765332908957068 29 | 1.238480466545676 30 |
| 16 | 2.732400588300134 34 | 7.224002820878996 32 | -1.188373771959265 33 |
| 17 | -3.125252721025359 37 | -7.610542303698070 35 | 1.288797204449500 36 |
| 18 | 4.003791188155508 40 | 9.016640984060900 38 | -1.569134982146965 39 |
| 19 | -5.711432970333974 43 | -1.193863282234463 42 | 2.131845492738941 42 |
| 20 | 9.023895184011915 46 | 1.756711636752818 45 | -3.214387534922715 45 |
| 21 | -1.571511769258019 50 | -2.858018489776168 48 | 5.352184809470102 48 |
| 22 | 3.003301742881138 53 | 5.117141308441820 51 | -9.796819265524511 51 |
| 23 | -6.273099770174026 56 | -1.004011216210859 55 | 1.963159084056317 55 |
| 24 | 1.426768791623564 60 | 2.150308849948739 58 | -4.290253648525459 58 |

**Table 2.12 $\mu_2(t^*)$ in Exponential Sampling**

| | | $n = 4$ | $n = 5$ | $n = 10$ |
|---|---|---|---|---|
| (a) | $F_6$ | 2.7912 | 1.2546 | 0.2416 |
| | $F_7$ | 2.8357 | 1.2703 | 0.2428 |
| | $F_8$ | 2.8682 | 1.2815 | 0.2436 |
| (b) | $F_{13}$ | 4.4848 | 1.5676 | 0.2520 |
| | $F_{14}$ | 4.4428 | 1.5579 | 0.2524 |
| | $F_{15}$ | 4.4024 | 1.5486 | 0.2520 |
| (c) | MC | 3.3086 | 1.3575<br>1.3426 | 0.2384 |
| (d) | Padé | ? | ? | 0.2706 [4\|5] |
| | | ? | ? | 0.2629 [5\|6] |
| | | ? | ? | 0.2582 [6\|7] |

( (a) 2cB with $(1 - 2/n)^{-3}$ factored out, $a = 1$;
(b) 2cB with $(1 - 2/n)^{-2}(1 - 3/n)^{-1}$ factored out, $a = 1$;
(c) Monte-Carlo, 200,000 runs;
(d) Padé J-fraction. Note that for $n = 50$ the Padé assessment is 0.02415, compared to the simulation value 0.02439. )

**Table 2.13 Stieltjes c.f.s for $E(t^*)$ and $E(t)$ in Exponential Sampling**

| $s$ | $p_s^*$ | $q_s^*$ | $p_s$ | $q_s$ |
|---|---|---|---|---|
| 0 | - | 2.46250000 01 | - | 2.32500000 01 |
| 1 | 3.48553299 01 | 4.07370661 01 | 3.64301075 01 | 3.96751537 01 |
| 2 | 8.84812073 01 | 9.89888264 01 | 9.10604639 01 | 9.69166934 01 |
| 3 | 1.59171861 02 | 1.75220760 02 | 1.62843858 02 | 1.72072837 02 |
| 4 | 2.48394659 02 | 2.70410956 02 | 2.53251496 02 | 2.66084113 02 |
| 5 | 3.56998604 02 | 3.85025389 02 | 3.63153444 02 | 3.79402132 02 |
| 6 | 4.85279253 02 | 5.19193395 02 | 4.92852002 02 | 5.12149724 02 |
| 7 | 6.33319766 02 | 6.72943806 02 | 6.42436025 02 | 6.64350868 02 |
| 8 | 8.01152182 02 | 8.46283770 02 | 8.11943026 02 | 8.36007968 02 |
| 9 | 9.88795750 02 | 1.03921519 03 | 1.00139771 03 | 1.02711867 03 |
| 10 | 1.19626231 03 | 1.25174760 03 | 1.21081739 03 | 1.23768538 03 |
| 11 | 1.42354312 03 | 1.48392815 03 | 1.44019836 03 | 1.46775292 03 |
| 12 | 1.67052185 03 | 1.73605317 03 | 1.68942649 03 | 1.71761648 03 |
| 13 | 1.93649175 03 | - | 1.95778058 03 | - |

(Here $E(t^*) = \dfrac{-1}{n - 3.5 + S^*(n)}$ $\qquad E(t) = \dfrac{-1}{n - 3 + S(n)}$

$$S^*(n) = \frac{q_0^*}{n+}\ \frac{p_1^*}{1+}\ \frac{q_1^*}{n+}\ \frac{p_2^*}{1+}\cdots \qquad S(n) = \frac{q_0}{n+}\ \frac{p_1}{1+}\ \frac{q_1}{n+}\ \frac{p_2}{1+}\cdots$$

Note $t^* = (m_1' - \mu_1')/\sqrt{m_2}$, $t = (m_1' - \mu_1')/s_x$. It may be conjectured, that, for example, $E(t^*) = -1/\{n - 3{\cdot}5 + \int_0^\infty \frac{d\,\phi(t)}{t+n}\}$, $\phi(t)$ being a distribution function.)

For $E(t)$ and $n = 3$ the last six convergents are

| Odd | -0.62 | -0.63 | -0.64 |
|---|---|---|---|
| Even | -0.85 | -0.85 | -0.84 |

showing a wide spread.

**Further cases** (last three odd and even convergents)

| $E(t^*)$ | | | | | |
|---|---|---|---|---|---|
| $n = 4$ | | $n = 5$ | | $n = 10$ | |
| -0.4748 | -0.5693 | -0.3466 | -0.3746 | -0.1347 | -0.1358 |
| -0.4847 | -0.5638 | -0.3504 | -0.3727 | -0.1348 | -0.1354 |
| -0.4897 | -0.5638 | -0.3504 | -0.3727 | -0.1348 | -0.1354 |
| $E(t)$ | | | | | |
| $n = 4$ | | $n = 5$ | | $n = 10$ | |
| -0.4230 | -0.4785 | -0.3124 | -0.3310 | -0.1278 | -0.1284 |
| -0.4273 | -0.4770 | -0.3140 | -0.3304 | -0.1279 | -0.1284 |
| -0.4310 | -0.4757 | -0.3153 | -0.3300 | -0.1279 | -0.1284 |

If there is a singularity at $n = 2$, then

$$S^*(2) = 1.5, \qquad S(2) = 1,$$

but there are not enough terms to provide convincing evidence.

Since $E(t^*)$ and $E(t)$ are dependent, the question arises as to whether the bounds from the Stieltjes fractions are consistent. Define $y^*(n) = -E(t^*)$, $y(n) = -E(t)$, so that

$$y^*(n) = \sqrt{n}\,y(n)/\sqrt{(n-1)}.$$

It turns out that the bounds from $y(n)$ converted to $y^*(n)$ are sharper. For example, converted bounds are

$$0.4977 < y^*(4) < 0.5496 \quad \text{compared to} \quad 0.4887 < y^*(4) < 0.5638,$$
$$0.3525 < y^*(5) < 0.3690 \quad \text{compared to} \quad 0.3504 < y^*(5) < 0.3727,$$
$$0.1348 < y^*(10) < 0.1353 \quad \text{compared to} \quad 0.1348 < y^*(10) < 0.1354;$$

we also have the sharper bounds which give $y^*(20) = 0.05844$ to 5 decimal places, and $y^*(50) = 0.0213539$ to 7 decimal places.

We can now return to the tabulation of $E(t^*)$ in Table 2.10. For $n = 4$, only the MC passes. For $n = 5$, the 2cB algorithms in (a) and (b), and MC pass; the rest fail. For $n = 10$, the failures are Padé, the first MC run, (e), and (f). For $n = 50$, (a), (b) and (c) are acceptable, MC borderline (however $\sigma$ for this MC run of 50,000 is about 0.0007, and a $3\sigma$ interval would be 0.01975 to 0.02395).

### 2.8.6.3 Further Comments on Higher Moments

Stieltjes fractions for $E(t)$ and $E(t^*)$ exist also for the 2nd, 3rd, and 4th central moments. Thus

$$\mu_2(t) = \frac{1}{n-9+} \; \frac{93}{n+a_2-} \; \frac{b_2}{n+a_3-} \; \cdots, \tag{2.89a}$$

$$\mu_3(t) = \frac{-4}{n-18.5+} \; \frac{316}{n+\hat{a}_2-} \; \frac{\hat{b}_2}{n+\hat{a}_3-} \; \cdots, \tag{2.89b}$$

$$\mu_4(t) = \frac{3}{n-32+} \; \frac{691}{n+a_2^*-} \; \frac{b_2^*}{n+a_3^*-} \; \cdots, \tag{2.89c}$$

where the $a$'s and $b$'s are positive, and where omitting the first convergent the remaining c.f. has a Stieltjes form (as far as our calculations go; i.e. to $n^{-14}$). As in Section 2.8.6.2 bounds can be set up, but these may have wide spreads for small $n$. In the even part of the c.f. this will appear as negative approximants for $\mu_2$ and $\mu_4$. For example:

| $n$ | $\mu_2(t)$ | | | | | | |
|---|---|---|---|---|---|---|---|
| 5 | -0.25 | -0.602 | -1.332 | -3.699 | 4.92 | 4.72 | 2.89 |
| 10 | 1.0 | 0.325 | 0.268 | 0.248 | 0.239 | 0.233 | 0.230 |

| $n$ | $\mu_4(t)$ | | | | | |
|---|---|---|---|---|---|---|
| 10 | -0.01 | -0.05 | -0.18 | -1.14 | 0.89 | 0.49 |
| 20 | -0.01 | 0.13 | 0.04 | 0.0355 | 0.0336 | 0.0328 |

Thus the expected pattern of decreasing positive approximants is established, later for small $n$, sooner for large $n$.

There are strong indications that mathematical properties of these moments await discovery.

### 2.8.6.4 Student's $t$ from Convoluted Exponentials

The moment series for $t$ in sampling from the density $f(x) = x\exp(-x)$, $x \geqslant 0$, are given in Table 2.14. They all show a sign pattern of period 2 (excepting the first 1 to 3 terms) and increase as fast as $(2s)!$ approximately. Moreover, as in Section 2.8.6.2, there are Stieltjes fractions. For example,

$$\mu_2(t) = \frac{1}{n-5.5+}\;\frac{32.125}{n+A_2-}\;\frac{B_2}{n+A_3-}\;\cdots, \tag{2.90a}$$

$$\mu_3(t) = \frac{-2.828}{n-11.5+}\;\frac{113.32}{n+\hat{A}_2-}\;\frac{\hat{B}_2}{n+\hat{A}_3-}\;\cdots, \tag{2.90b}$$

$$\mu_4(t) = \frac{3}{n-19+}\;\frac{236}{n+A_2^*-}\;\frac{B_2^*}{n+A_3^*}\;\cdots, \tag{2.90c}$$

and note the decreasing values of the first few coefficients.

| $n$ | $\mu_2(t)$ | | | | | | |
|---|---|---|---|---|---|---|---|
| 5 | -2.0 | 1.32 | 0.86 | 0.75 | 0.696 | 0.670 | 0.654 |
| 10 | 0.22 | 0.18 | 0.172 | 0.1686 | 0.1681 | 0.1677 | |

| $n$ | $\mu_3(t)$ | | | | | |
|---|---|---|---|---|---|---|
| 10 | 0.19 | -0.14 | -0.10 | -0.0899 | -0.0863 | -0.0846 |
| 20 | -0.016 | -0.013 | -0.012 | -0.0121 | -0.01205 | -0.01203 |

**Table 2.14 Series for the First Four Moments of $t$ from Convoluted Exponential Density**

| $s$ | $\mu_1'$ | $\mu_2$ | $\mu_3$ | $\mu_4$ |
|---|---|---|---|---|
| 1 | -0.707107 | 1.00000 00 | - | - |
| 2 | -1.32583 00 | 5.50000 00 | -2.82843 00 | 3.00000 00 |
| 3 | 3.08807 00 | -1.87500 00 | -3.25269 01 | 5.70000 01 |
| 4 | -1.07152 02 | 4.76609 02 | -5.35412 01 | 3.75000 02 |
| 5 | 4.87321 03 | -2.40473 04 | -4.02842 03 | 7.25653 03 |
| 6 | -3.51263 05 | 1.95348 06 | 2.20423 05 | -2.97694 05 |
| 7 | 3.54916 07 | -2.16710 08 | -2.05432 07 | 3.16310 07 |
| 8 | -4.73707 09 | 3.12836 10 | 2.53191 09 | -4.17382 09 |
| 9 | 8.01811 11 | -5.66544 12 | -3.99306 11 | 6.98966 11 |
| 10 | -1.67165 14 | 1.25356 15 | 7.80268 13 | -1.43937 14 |
| 11 | 4.19955 16 | -3.32077 17 | -1.84532 16 | 3.56725 16 |
| 12 | -1.24970 19 | 1.03675 20 | 5.18706 18 | -1.09613 19 |
| 13 | 4.34459 21 | -3.76546 22 | -1.70814 21 | 3.58110 21 |
| 14 | -1.74446 24 | 1.57399 25 | 6.51218 23 | -1.41493 24 |

The changes in the structure of the c.f.s for higher convolutions is interesting, because the density $f(x) = e^{-x}x^{\rho-1}/\Gamma(\rho)$ itself approaches normality as $\rho \to \infty$. Moreover, we should therefore expect $E(t)$ and $\mu_3(t)$ to approach zero, and $Var(t) \to (n-1)/\{n(n-3)\}$, $\mu_4(t) \to 3(n-1)^2/\{(n-3)(n-5)\}$. These indicate that the Stieltjes continued fraction form will break down for the moment series as $\rho$ increases. However there will still be J-fractions. For example, for $\rho = 25$, and $n = 10$ the last three approximants (using coefficients to $n^{-14}$) are:

| | | | |
|---|---|---|---|
| $\mu_1'(t)$: | -0.0220225 | -0.0220222 | -0.0220221 |
| $\mu_2(t)$: | 0.131231 | 0.131223 | 0.131221 |
| $\mu_3(t)$: | -0.015085 | -0.015066 | -0.015057 |
| $\mu_4(t)$: | 0.7564 | 0.7546 | 0.7539 |

(The normal sampling values are 0, 0.1286, 0, 0.6943.)

### 2.8.7 Skewness Statistic in Gamma Sampling

We have studied various summation algorithms for the first few moments of $\sqrt{b_1} = m_3/m_2^{3/2}$ ($m_2$ and $m_3$ are the 2nd and 3rd central sample moments) in gamma sampling, the density being

$$f(x;\rho) = e^{-x}x^{\rho-1}/\Gamma(\rho), \qquad (x>0, \rho>0)$$

(since this subject is treated in Bowman and Shenton (1988), only a brief outline is given here). We take two values of the shape parameter $\rho$ and use COETS to derive the moment series coefficients given in Table 2.15.

**Table 2.15 Moments of $\sqrt{b_1}$ in Gamma Sampling**

| $s$ | $\mu_1'$ | $r_s$ | $\mu_2$ | $r_s$ | $\mu_3$ | $r_s$ | $\mu_4$ | $r_s$ |
|---|---|---|---|---|---|---|---|---|
| | | | | $\rho = 8/3$ | | | | |
| 0 | 1.2247449 00 | | | | | | | |
| 1 | -1.1367163 01 | 9 | 2.3718750 01 | | | | | |
| 2 | 1.7252696 02 | 15 | -1.5376904 03 | 65 | 2.9803281 03 | | 1.6877373 03 | |
| 3 | -5.9228858 03 | 34 | 1.2304471 05 | 80 | -7.1704288 05 | 241 | 1.0725074 06 | 635 |
| 4 | 3.2756993 05 | 55 | -1.2379146 07 | 101 | 1.4515393 08 | 202 | -6.6334206 08 | 618 |
| 5 | -2.5738957 07 | 79 | 1.5473153 09 | 125 | -3.0474567 10 | 210 | 2.6691551 11 | 402 |
| 6 | 2.6847187 09 | 104 | -2.3593240 11 | 152 | 7.0319358 12 | 231 | -9.9375632 13 | 372 |
| 7 | -3.5568411 11 | 132 | 4.3140724 13 | 183 | -1.8179355 14 | 259 | 3.7730973 16 | 380 |
| 8 | 5.8038090 13 | 163 | -9.3209217 15 | 216 | 5.2941862 17 | 291 | -1.5190023 19 | 403 |
| 9 | -1.1401341 16 | 196 | 2.3498401 18 | 252 | -1.7363690 20 | 328 | 6.5998086 21 | 434 |
| 10 | 2.6494802 18 | 232 | -6.8384883 20 | 291 | 6.3966370 22 | 368 | -3.1198751 24 | 473 |
| 11 | -7.1817740 20 | 271 | 2.2761526 23 | 333 | -2.6370435 25 | 412 | 1.6099601 27 | 516 |
| 12 | 2.2447173 23 | 313 | -8.5951512 25 | 378 | 1.2115856 28 | 459 | -9.0758360 29 | 564 |
| | | | | $\rho = 20$ | | | | |
| $s$ | $\mu_1'$ | $r_s$ | $\mu_2$ | $r_s$ | $\mu_3$ | $r_s$ | $\mu_4$ | $r_s$ |
| 0 | 4.4721360-01 | | | | | | | |
| 1 | -3.1696264 00 | 7 | 7.8750000 00 | | | | | |
| 2 | 1.6702904 01 | 5 | -1.2045206 02 | 15 | 1.6292936 02 | | 1.8604687 02 | |
| 3 | -1.5256124 02 | 9 | 2.5780729 03 | 21 | -1.1346409 04 | 70 | 8.3254414 03 | 45 |
| 4 | 2.4482075 03 | 16 | -7.1881854 04 | 28 | 6.1678183 05 | 54 | -1.8518429 06 | 222 |
| 5 | -5.4332131 04 | 22 | 2.4260968 06 | 34 | -3.4104868 07 | 55 | 2.0124463 08 | 109 |
| 6 | 1.5522191 06 | 29 | -9.8102511 07 | 40 | 2.0357743 09 | 60 | -1.9242776 10 | 96 |
| 7 | -5.5018657 07 | 35 | 4.6631189 09 | 48 | -1.3354406 11 | 66 | 1.8253225 12 | 95 |
| 8 | 2.3470773 09 | 43 | -2.5654775 11 | 55 | 9.6767574 12 | 72 | -1.7956280 14 | 98 |
| 9 | -1.1786045 11 | 50 | 1.6133908 13 | 63 | -7.7442779 14 | 80 | 1.8682483 16 | 104 |
| 10 | 6.8506831 12 | 58 | -1.1478338 15 | 71 | 6.8291397 16 | 88 | -2.0749421 18 | 111 |
| 11 | -4.5483129 14 | 66 | 9.1577973 16 | 80 | -6.6144987 18 | 97 | 2.4703351 20 | 119 |
| 12 | 3.4122422 16 | 75 | -8.1329070 18 | 89 | 7.0123735 20 | 106 | -3.1576583 22 | 128 |

($r_s$ gives the absolute value of the ratio of successive terms.)

The sign patterns are for the most part alternating with period two, but the moduli show an unusual pattern. This is particularly true for small $\rho$ (large skewness in the population); thus for $\rho = 8/3$ ($\beta_1 = 1.5$) and the third and fourth central moments there is a surge in value in the first two coefficients which tapers off but becomes renewed later in the series. In addition divergence is quite strong but in the region of the double factorial. These factors strain summation algorithms especially for small to moderate sample sizes. Rational fraction approaches are shown in Table 2.16.

The continued fractions are

$$E(\sqrt{b_1}) \sim \frac{n\, q_0^{(1)}}{n+} \; \frac{p_1^{(1)}}{n+} \; \cdots,$$

$$Var(\sqrt{b_1}) \sim \frac{q_0^{(2)}}{n+} \; \frac{p_1^{(2)}}{1+} \; \frac{q_1^{(2)}}{n+} \; \cdots,$$

$$\mu_3(\sqrt{b_1}) \sim \frac{n^{-1} q_0^{(3)}}{n+} \; \frac{p_1^{(3)}}{1+} \; \frac{q_1^{(3)}}{n+} \; \cdots,$$

$$\mu_4(\sqrt{b_1}) \sim \frac{n^{-1} q_0^{(4)}}{n+} \; \frac{p_1^{(4)}}{1+} \; \frac{q_1^{(4)}}{n+} \; \cdots,$$

and the approximants are

$$C_{2s}^{(1)} = \frac{n\, q_0^{(1)}}{n+} \cdots \frac{p_s^{(1)}}{n}, \qquad C_{2s+1}^{(1)} = \frac{n\, q_0^{(1)}}{n+} \cdots \frac{q_s^{(1)}}{n}, \quad \text{etc.}$$

**Table 2.16 Padé sequences for the Moments of the Skewness ($\sqrt{b_1}$)**

| Partial Numerators of c.f. $E(\sqrt{b_1})$ | | | | |
|---|---|---|---|---|
| | $\rho = 20,\ \beta_1 = 0.2$ | | $\rho = 8/3,\ \beta_1 = 1.5$ | |
| $s$ | $p_s^{(1)}$ | $q_s^{(1)}$ | $p_s^{(1)}$ | $q_s^{(1)}$ |
| 0 | | 4.47213595-01 | | 1.22474487 00 |
| 1 | 7.08750000 00 | -1.81782407 00 | 9.28124991 00 | 5.89641198 00 |
| 2 | -1.12017242 01 | 3.14077101 01 | 4.92996108 01 | 7.45088548 00 |
| 3 | 2.51608519 00 | 4.62966657 01 | 2.25723133 02 | -1.14058171 02 |
| 4 | -6.84192022 00 | -6.45469559 01 | -4.48164727 01 | 2.24997452 02 |
| 5 | 1.19662812 02 | 7.48105514 00 | 5.27072380 01 | 2.10335898 02 |
| 6 | 6.77338242 01 | | 1.12050355 02 | |

| Approximants to $E(\sqrt{b_1})$ | | | | | | | |
|---|---|---|---|---|---|---|---|
| | | $\rho = 20$ | | | $\rho = 8/3$ | | |
| $n$ | $s$ | $C_{2s}^{(1)}$ | $C_{2s+1}^{(1)}$ | Levin's $\alpha_{12}$ | $C_{2s}^{(1)}$ | $C_{2s+1}^{(1)}$ | Levin's $\alpha_{12}$ |
| 5 | 1 | 0.18499011 | 0.13857422 | | 0.42879471 | 0.66139104 | |
| | 2 | 0.21334598 | 0.11222246 | | 0.45794720 | 0.48996475 | |
| | 3 | 0.09042211 | 0.11044364 | | 0.45876732 | 0.41491384 | |
| | 4 | 0.11296012 | 0.11025767 | | 0.46089998 | 0.37621783 | |
| | 5 | 0.11647620 | 0.10368890 | 0.1153 | 0.47165966 | 0.35425212 | 0.3221 |
| 10 | 1 | 0.26171973 | 0.23963706 | | 0.63519994 | 0.77326676 | |
| | 2 | 0.34882169 | 0.23004965 | | 0.66411202 | 0.67884439 | |
| | 3 | 0.22736627 | 0.22958169 | | 0.66482977 | 0.64730247 | |
| | 4 | 0.22988184 | 0.22953427 | | 0.66675869 | 0.63412208 | |
| | 5 | 0.23027209 | 0.23414432 | 0.2303 | 0.67945483 | 0.62781043 | 0.6199 |
| 20 | 1 | 0.33019924 | 0.32178158 | | 0.83653866 | 0.90160903 | |
| | 2 | 0.30913523 | 0.31921623 | | 0.85786059 | 0.86262010 | |
| | 3 | 0.31896536 | 0.31913921 | | 0.85829099 | 0.85368096 | |
| | 4 | 0.31915961 | 0.31913138 | | 0.85956200 | 0.85088267 | |
| | 5 | 0.31918139 | 0.31920169 | 0.31919 | 0.89373986 | 0.84982221 | 0.848895 |
| 50 | 1 | 0.39169135 | 0.38986522 | | 1.03299515 | 1.05034213 | |
| | 2 | 0.38931542 | 0.38956268 | | 1.04214057 | 1.04273798 | |
| | 3 | 0.38955601 | 0.38955917 | | 1.04225578 | 1.04187706 | |
| | 4 | 0.38955942 | 0.38955855 | | 1.04288666 | 1.04172164 | |
| | 5 | 0.38955959 | 0.38955962 | 0.3895596 | 1.04149597 | 1.04168417 | 1.041667 |
| 100 | 1 | 0.41761512 | 0.41710401 | | 1.12072736 | 1.12605244 | |
| | 2 | 0.41703828 | 0.41705548 | | 1.12435540 | 1.12443739 | |
| | 3 | 0.41705517 | 0.41705526 | | 1.12438144 | 1.12434966 | |
| | 4 | 0.41705527 | 0.41705528 | | 1.12427451 | 1.12434089 | |
| | 5 | 0.41705527 | 0.41705527 | 0.41705527 | 1.12433704 | 1.12433960 | 1.124339 |
| 200 | 1 | 0.43190786 | 0.43177232 | | 1.17042962 | 1.17191800 | |
| | 2 | 0.43176420 | 0.43176536 | | 1.17163017 | 1.17163857 | |
| | 3 | 0.43176534 | 0.43176535 | | 1.17163418 | 1.17163254 | |
| | 4 | 0.43176535 | 0.43176535 | | 1.17163181 | 1.17163224 | |
| | 5 | 0.43176535 | 0.43176535 | - | 1.17163222 | 1.17163224 | - |

| Preferred Assessments of $E(\sqrt{b_1})$ | | |
|---|---|---|
| $n$ | $\rho = 20$ | $\rho = 8/3$ |
| 5 | 0.12 | 0.3 (?) |
| 10 | 0.230 | 0.62 |
| 20 | 0.3192 | 0.849 |
| 50 | 0.389560 | 1.0417 |
| 100 | 0.4170553 | 1.124338 |
| 200 | 0.43176535 | 1.171632 |
| 250 | 0.43479315 (Levin) | 1.18172245 (Levin) |

| Partial Numerators of c.f. $Var(\sqrt{b_1})$ | | | | |
|---|---|---|---|---|
| | $\rho = 20,\ \beta_1 = 0.2$ | | $\rho = 8/3,\ \beta_1 = 1.5$ | |
| $s$ | $p_s^{(2)}$ | $q_s^{(2)}$ | $p_s^{(2)}$ | $q_s^{(2)}$ |
| 0 | | 7.87500000 00 | | 2.37187497 01 |
| 1 | 1.52955000 01 | 6.18252867 00 | 6.48301624 01 | 1.51890093 01 |
| 2 | 2.10103876 01 | 1.06801191 01 | 1.08460806 02 | 3.12990229 01 |
| 3 | 4.30705751 01 | -8.72187794-02 | 1.75478762 02 | 3.90797891 01 |
| 4 | -1.12005781 04 | 1.12602244 04 | 3.25522745 02 | -1.88945118 01 |
| 5 | 1.33496223-01 | 8.43938239 01 | -1.30019892 03 | 1.72046183 03 |
| 6 | 2.49054858 01 | | 2.40654180 01 | |

| Approximants to $Var(\sqrt{b_1})$ | | | | | | | |
|---|---|---|---|---|---|---|---|
| | | $\rho = 20$ | | | $\rho = 8/3$ | | |
| $n$ | $s$ | $C_{2s}^{(2)}$ | $C_{2s+1}^{(2)}$ | Levin's $\alpha_{12}$ | $C_{2s}^{(2)}$ | $C_{2s+1}^{(2)}$ | Levin's $\alpha_{12}$ |
| 5 | 1 | 0.38801705 | 0.66517349 | | 0.33966339 | 1.12647077 | |
| | 2 | 0.45160141 | 0.52184035 | | 0.38147754 | 0.56744839 | |
| | 3 | 0.46195941 | 0.46180497 | | 0.38828097 | 0.42811970 | |
| | 4 | 0.46195948 | 0.41330856 | | 0.38905262 | 0.38596729 | |
| | 5 | 0.46855130 | 0.37038290 | 0.3748 | 0.38906365 | 0.37254938 | 0.4916(??) |
| 10 | 1 | 0.31132020 | 0.40484560 | | 0.31696777 | 0.66369396 | |
| | 2 | 0.34519760 | 0.36183629 | | 0.35158496 | 0.42589479 | |
| | 3 | 0.34925027 | 0.34922347 | | 0.35675617 | 0.37153863 | |
| | 4 | 0.34925029 | 0.34265705 | | 0.35731206 | 0.35622391 | |
| | 5 | 0.35215655 | 0.33916462 | 0.338318 | 0.35731989 | 0.35162004 | 0.3652 |
| 20 | 1 | 0.22311626 | 0.24855016 | | 0.27960279 | 0.41723966 | |
| | 2 | 0.23627926 | 0.23893624 | | 0.30418303 | 0.32887833 | |
| | 3 | 0.23726801 | 0.23726530 | | 0.30732596 | 0.31161822 | |
| | 4 | 0.23726801 | 0.23677015 | | 0.30763179 | 0.30734408 | |
| | 5 | 0.23956625 | 0.23661884 | 0.236582 | 0.30763598 | 0.30618633 | 0.3072 |
| 50 | 1 | 0.12060555 | 0.12379676 | | 0.20655505 | 0.23784212 | |
| | 2 | 0.12288449 | 0.12300523 | | 0.21728521 | 0.22091847 | |
| | 3 | 0.12295359 | 0.12295354 | | 0.21821938 | 0.21867103 | |
| | 4 | 0.12295359 | 0.12294765 | | 0.21828972 | 0.21826588 | |
| | 5 | 0.12294508 | 0.12294691 | 0.1229468 | 0.21829062 | 0.21818119 | 0.218181 |
| 100 | 1 | 0.06830275 | 0.06883446 | | 0.14389812 | 0.15176935 | |
| | 2 | 0.06874292 | 0.06875036 | | 0.14784737 | 0.15176935 | |
| | 3 | 0.06874825 | 0.06874825 | | 0.14805888 | 0.14810200 | |
| | 4 | 0.06874825 | 0.06874815 | | 0.14807018 | 0.14806852 | |
| | 5 | 0.06874814 | 0.06874815 | 0.06876815 | 0.14807031 | 0.14806351 | 0.1480629 |
| 200 | 1 | 0.03657763 | 0.03665572 | | 0.08956212 | 0.09113687 | |
| | 2 | 0.03664821 | 0.03664855 | | 0.09060301 | 0.09065306 | |
| | 3 | 0.03664849 | 0.03664849 | | 0.09063089 | 0.09063306 | |
| | 4 | 0.03664849 | 0.03664849 | | 0.09063177 | 0.09063172 | |
| | 5 | 0.03664849 | 0.03664849 | - | 0.09063178 | 0.09063160 | - |

| Preferred Assessments of $Var(\sqrt{b_1})$ | | |
|---|---|---|
| $n$ | $\rho = 20$ | $\rho = 8/3$ |
| 5 | 0.4 (?) | 0.4 (?) |
| 10 | 0.34 | 0.36 |
| 20 | 0.240 | 0.307 |
| 50 | 0.12295 | 0.2182 |
| 100 | 0.0687482 | 0.148063 |
| 200 | 0.03664849 | 0.0906317 |
| 250 | 0.02974208 (Levin) | 0.07598750 (Levin) |

| Partial Numerators of c.f. $\mu_3(\sqrt{b_1})$ | | | | |
|---|---|---|---|---|
| | $\rho = 20, \beta_1 = 0.2$ | | $\rho = 8/3, \beta_1 = 1.5$ | |
| $s$ | $p_s^{(3)}$ | $q_s^{(3)}$ | $p_s^{(3)}$ | $q_s^{(3)}$ |
| 0 | | 1.29293600 02 | | 2.98032806 03 |
| 1 | 6.96400533 01 | -1.52808451 01 | 2.40591930 02 | -3.81578336 01 |
| 2 | -3.32845433 00 | 2.64099672 02 | -3.98549344 01 | 6.28685056 02 |
| 3 | -1.72413266 02 | -3.70811755 00 | -2.33360125 02 | -4.82154023 01 |
| 4 | 1.00069434 02 | 2.01595638 01 | 4.62305409 02 | 7.39708043 01 |
| 5 | 6.61768565 01 | 1.70244664-01 | 4.13371692 02 | -5.41145573 03 |

| Approximants to $\mu_3(\sqrt{b_1})$ | | | | | | | |
|---|---|---|---|---|---|---|---|
| | | $\rho = 20$ | | | $\rho = 8/3$ | | |
| $n$ | $s$ | $C_{2s}^{(3)}$ | $C_{2s+1}^{(3)}$ | Levin's $\alpha_{10}$ | $C_{2s}^{(3)}$ | $C_{2s+1}^{(3)}$ | Levin's $\alpha_{10}$ |
| 5 | 1 | 0.43657354 | -1.12875597 | | 2.42705700 | -19.05592643 | |
| | 2 | -9.17019418 | -1.15348611 | | 4.97325999 | -20.78432363 | |
| | 3 | -0.20249938 | 4.42739038 | | -0.93840773 | -37.14264355 | |
| | 4 | -0.16341266 | -0.01229274 | 0.3126 | -0.45704879 | 8.52585449 | - |
| 10 | 1 | 0.20458218 | -0.13368790 | | 1.18931526 | -3.95038194 | |
| | 2 | -0.37058296 | -0.13884842 | | 2.57790633 | -4.35877841 | |
| | 3 | 0.00053986 | 0.20434985 | | -0.24205202 | -8.64662264 | |
| | 4 | 0.00848912 | 0.02276682 | 0.0329 | -0.06380850 | 1.53316306 | - |
| 20 | 1 | 0.09087978 | 0.02585049 | | 0.57183813 | -0.60822848 | |
| | 2 | 0.00952986 | 0.02491890 | | 1.45605086 | -0.69975705 | |
| | 3 | 0.03982758 | 0.04839804 | | 0.03527964 | -1.94811590 | |
| | 4 | 0.04079275 | 0.04156805 | 0.04161 | 0.08712213 | 0.30493182 | -0.02 (??) |
| 50 | 1 | 0.02723659 | 0.02168192 | | 0.20512119 | 0.05592516 | |
| | 2 | 0.02122298 | 0.02161361 | | -1.60524210 | 0.04510821 | |
| | 3 | 0.02202551 | 0.02209955 | | 0.09683642 | 9.50717893 | |
| | 4 | 0.02204754 | 0.02205311 | 0.022052 | 0.10239714 | 0.11061171 | 0.1028 |
| 100 | 1 | 0.00960442 | 0.00894228 | | 0.08750437 | 0.06094218 | |
| | 2 | 0.00891714 | 0.00893556 | | 0.03931166 | 0.05919869 | |
| | 3 | 0.00895131 | 0.00895247 | | 0.06380597 | 0.07054263 | |
| | 4 | 0.00895187 | 0.00895192 | 0.0089519 | 0.06435000 | 0.06469908 | 0.06444 |
| 200 | 1 | 0.00302124 | 0.00295804 | | 0.03382186 | 0.02996408 | |
| | 2 | 0.00295690 | 0.00295755 | | 0.02888757 | 0.02974966 | |
| | 3 | 0.00295793 | 0.00295794 | | 0.03001068 | 0.03013019 | |
| | 4 | 0.00295794 | 0.00295794 | - | 0.03003803 | 0.03004522 | - |

| Preferred Assessments of $\mu_3(\sqrt{b_1})$ | | |
|---|---|---|
| $n$ | $\rho = 20$ | $\rho = 8/3$ |
| 5 | -- | ?? |
| 10 | -- | ?? |
| 20 | 0.0412 | ?? |
| 50 | 0.02205 | 0.10 |
| 100 | 0.0089519 | 0.0644 |
| 200 | 0.00295794 | 0.03004 |
| 250 | 0.002010306 (Levin) | 0.0023285 (Levin) |

| Partial Numerators of c.f. $\mu_4(\sqrt{b_1})$ | | | | |
|---|---|---|---|---|
| | $\rho = 20, \beta_1 = 0.2$ | | $\rho = 8/3, \beta_1 = 1.5$ | |
| $s$ | $p_s^{(4)}$ | $q_s^{(4)}$ | $p_s^{(4)}$ | $q_s^{(4)}$ |
| 0 | | 1.86046875 02 | | 1.68773727 03 |
| 1 | -4.47491600 01 | 2.67180961 02 | -6.35470614 02 | 1.25396708 03 |
| 2 | -9.47060694 01 | -6.58299387 00 | -1.06595566 02 | -5.35371223 01 |
| 3 | 3.77863905 01 | 6.64982418 01 | 1.85400186 02 | 2.95462482 02 |
| 4 | 4.31011422 01 | 3.51937353 01 | 1.55776949 02 | 2.19180953 02 |
| 5 | 1.16543152 02 | 9.66702594-02 | 4.53363896 02 | -4.93409959 03 |

| Approximants to $\mu_4(\sqrt{b_1})$ | | | | | | | |
|---|---|---|---|---|---|---|---|
| | | $\rho = 20$ | | | $\rho = 8/3$ | | |
| $n$ | $s$ | $C_{2s}^{(4)}$ | $C_{2s+1}^{(4)}$ | Levin's $\alpha_{10}$ | $C_{2s}^{(4)}$ | $C_{2s+1}^{(4)}$ | Levin's $\alpha_{10}$ |
| 5 | 1 | -0.93610469 | 8.90612782 | | -0.53538967 | 136.31549034 | |
| | 2 | 1.34724747 | -1.97692403 | | 5.53134645 | -112.64106417 | |
| | 3 | 1.06745260 | -0.54051031 | | 3.88053376 | -48.81089922 | |
| | 4 | 0.75012749 | -0.00028402 | 0.0737 | 0.96405927 | -34.28373213 | - |
| 10 | 1 | -0.53539963 | 2.21865732 | | -0.26983478 | 33.94202615 | |
| | 2 | 0.60458102 | 0.00003113 | | 2.67737082 | -24.08436984 | |
| | 3 | 0.50697730 | 0.24450694 | | 1.92159862 | -9.90761281 | |
| | 4 | 0.42171983 | 0.32719353 | 0.3717 | 0.67536150 | -6.70949550 | - |
| 20 | 1 | -0.37586503 | 0.55097063 | | -0.13710949 | 8.41869077 | |
| | 2 | 0.24893427 | 0.16280562 | | 1.256894444 | -4.28273836 | |
| | 3 | 0.22380979 | 0.19377369 | | 0.93926113 | -1.45555262 | |
| | 4 | 0.20946583 | 0.20230040 | 0.2068 | 0.48020973 | -0.83125252 | - |
| 50 | 1 | 0.70863662 | 0.08664264 | | -0.05765404 | 1.31683798 | |
| | 2 | 0.06307498 | 0.05934476 | | 0.42174203 | -0.11178200 | |
| | 3 | 0.06108589 | 0.06036345 | | 0.34462277 | 0.13436925 | |
| | 4 | 0.06060566 | 0.06052846 | 0.06057 | 0.26617854 | 0.18438226 | 0.2686 |
| 100 | 1 | 0.03367313 | 0.02118677 | | -0.03151877 | 0.31804481 | |
| | 2 | 0.01876786 | 0.01856157 | | 0.16328712 | 0.09748124 | |
| | 3 | 0.01862156 | 0.01860234 | | 0.14522029 | 0.12383916 | |
| | 4 | 0.01860669 | 0.01860576 | 0.0186062 | 0.13372557 | 0.12837767 | 0.1323 |
| 200 | 1 | 0.00599182 | 0.00514388 | | -0.01937831 | 0.07495189 | |
| | 2 | 0.00496522 | 0.00495781 | | 0.05411253 | 0.04862133 | |
| | 3 | 0.00495903 | 0.00495876 | | 0.05158350 | 0.05038411 | |
| | 4 | 0.00495880 | 0.00495880 | - | 0.05076397 | 0.05060132 | - |

| Preferred Assessments of $\mu_4(\surd b_1)$ | | |
|---|---|---|
| $n$ | $\rho = 20$ | $\rho = 8/3$ |
| 5 | -- | ??? |
| 10 | 0.4 | ?? |
| 20 | 0.207 | ? |
| 50 | 0.0606 | 0.27 |
| 100 | 0.018606 | 0.132 |
| 200 | 0.0049588 | 0.0507 |
| 250 | 0.003184314 (Levin) | 0.03504 (Levin) |
| 500 | -- | 0.00961153 (Levin) |

**Comments:**

(i) Notice the irregular sign patterns of the partial numerators in the c.f.s.

(ii) For small skewness the Stieltjes fraction works well for the mean, not quite so well for the variance; for the third and fourth central moments $n \geqslant 30$ or so is needed to produce some agreement with the Levin algorithms.

(iii) For moderately large skewness ($\surd \beta_1 = 1.5$) the mean and variance sequences seem acceptable for $n > 10$ or so. However the fourth moment sequences in particular are chaotic for small $n$ but stabilize for $n \geqslant 50$.

## 2.9 General Comments on Moment Series and c.f.s

There are the following components in the summation process:

- The population sampled. Limited ranges are easier to deal with than unlimited ranges. A linked property relates to the rate of increase of moments, which in turn relates to "tail" lightness. From a computational point of view, problems may arise if non-central moments only are known in closed form, the conversion to central moments involving variable signs and moment products leading to loss of accuracy (an example would be the moments of Johnson's $S_U$ system).

- The Statistic studied. Is it implicit or explicit? Explicit functions of sample moments (such as the skewness and kurtosis) present no particular problems although dimensionality exacerbate computer implementation. An example of an implicit statistic is the shape parameter of moment estimator $c^*$, where

  $$\Gamma(1 + 2/c^*)/\Gamma^2(1 + 1/c^*) = m_2/m_1'^2 .$$

  Faà di Bruno's derivative formula is needed here.

- The Algorithm Used. Generally, rational fractions work well provided the modulus increase in the coefficients is not too steep - inclusions toward the triple factorial tend to be problematical. For series of alternating signs, Levin's algorithm has proved quite useful in conjunction with a certain stopping rule in the sequences.

- Validation Studies. In general there are several possibilities - many arising from truncation or inversion procedures. In addition the Levin approach appears to be independent, but not always applicable. Simulation studies are always possible.

- Computer Facilities. Large numbers almost certainly appear and loss of accuracy in non-linear algorithms is difficult to avoid; alternatives in machine and even program may be advisable.

## Further Examples

### Further Example 2.1

### Low Order Terms and $E(\sqrt{m_2})$ in Exponential Sampling

If we use either of the truncated series

$$M_2(n) = 1 - \frac{1.5}{n} + \frac{6.125}{n^2}, \qquad M_3(n) = M_2(n) - \frac{151.3125}{n^3},$$

as approximations to $E(\sqrt{m_2})$ in sampling from an exponential distribution, then note that

$$\begin{aligned} M_2(2) &= 1.78\ (0.5), & M_3(2) &= -17.13,\\ M_2(3) &= 1.181\ (0.65), & M_3(3) &= -4.424,\\ M_2(4) &= 1.008\ (0.7271), & M_3(4) &= -1.356,\\ M_2(10) &= 0.9113\ (0.8767), & M_3(10) &= 0.7599, \end{aligned}$$

the parenthetic entries denoting preferred assessments (see Section 2.8.2). Thus the error in using three or four terms could be regarded as acceptable for $n > 10$, assuming we have a basis for error assessment for larger $n$. How often can the basic term or asymptotic be relied on more so than more complicated assessments? Note that in the case of the Stieltjes moment problem, when determined

$$\int_0^\infty \frac{d\sigma(t)}{n+t} = \frac{e_0}{n} - \frac{e_1}{n^2} + \cdots + \frac{(-1)^s e_s}{n^{s+1}} + \frac{(-1)^{s+1}}{n^{s+1}} \int_0^\infty \frac{t^{s+1} d\sigma(t)}{n+t}$$

so that the value, using the first two terms of the error in using $s+1$ terms of the series relates to the first term omitted $(e_{s+1}/n^{s+2})$. Hence for fixed $s$, this error will decrease as $n$ increases; this provides some insight into the reason for using rational fraction approximants for $n$ large enough. In the present example $1 - 1.5/n$ would be expected to have an error of less than $6.125/n^2$, or 0.01 when $n \geqslant 25$. This assumes that the c.f. is a Stieltjes form and converges to $1 - E(\sqrt{m_2})$. Our preferred assessment in this case is $E(\sqrt{m_2}) \sim 0.94597$.

**Further Example 2.2**

For the kurtosis $b_2$ under normality,

$$Var\,(b_2) \sim \frac{24}{n} - \frac{360}{n^2} + \frac{2976}{n^3} + \cdots - \frac{15816096}{n^8} + \cdots .$$

Note that the ratio of successive terms is 15, 8.2$\dot{6}$, 6.4, 5.7, 5.4, 5.2, 5.1. Moreover, there is the exact result

$$Var\,(b_2) = \frac{24n(n-2)(n-3)}{(n+1)^2(n+3)(n+5)}$$

with c.f. of four terms

$$\frac{24}{n+15+}\;\frac{101}{n-4.485149+}\;\cdots\;\frac{0.280691}{n+0.188049},$$

and partial fraction form

$$\frac{4}{n+1} - \frac{36}{(n+1)^2} - \frac{270}{n+3} + \frac{210}{n+5} .$$

Note the squared term, indicating the failure as a Stieltjes form. This also comes out when we consider the series with "moments"

$$\nu_1' = 15, \quad \nu_2' = 124, \quad \nu_3' = 798, \quad \text{with variance} \quad \nu_2 = -101 .$$

Some comparisons of series assessments with true values are:

| $n$ | True | $24/n$ | $24/n - 360/n^2$ | c.f. Convergent 1 | 2 | 3 |
|---|---|---|---|---|---|---|
| 2 | 0 | 12 | -78.0 | 1.41 | -1.02 | 0.06 |
| 5 | 0.25 | 4.8 | -9.6 | 1.20 | 0.11 | 0.252 |
| 10 | 0.5696 | 2.4 | -1.2 | 0.96 | 0.554 | 0.5697 |
| 20 | 0.5792 | 1.2 | 0.3 | 0.69 | 0.5782 | 0.5792 |
| 30 | 0.4904 | 0.8 | 0.4 | 0.53 | 0.4992 | 0.4904 |
| 50 | 0.3571 | 0.48 | 0.3360 | 0.369 | 0.3570 | 0.3571 |
| 100 | 0.2068 | 0.2490 | 0.2040 | 0.209 | 0.2068 | 0.2068 |
| 200 | 0.1114 | 0.1290 | 0.1110 | 0.1116 | 0.1114 | 0.1114 |

(Comment: How does using the "first term omitted" procedure work out? For small $n$ it is true but wide of the mark; for large $n$ it is acceptable but not true.)

**Further Example 2.3 (Continuation of Further Example 2.2)**

For the third central moment of the kurtosis,

$$\mu_3(b_2) \sim \frac{1728}{n^2} - \frac{63936}{n^3} + \frac{1297728}{n^4} - \cdots + \frac{30726248832}{n^8} - \cdots$$

with successive term ratios 37.6, 20.3, 15.1, 12.8, 11.5, 10.7 and exact value

$$\frac{1728n(n-2)(n-3)(n^2-5n+2)}{(n+1)^3(n+3)(n+5)(n+7)(n+9)}.$$

It does not have a Stieltjes form for the normed series since $\nu_1' = 37$, $\nu_2' = 751$, with $\nu_2 = -618$. How well does the basic term perform?

| $n$ | True | $1728/n^2$ | $-63936/n^3$ |
|---|---|---|---|
| 5 | 0.0357 | 69.1 | -511 |
| 10 | 0.6002 | 17.3 | -63.9 |
| 20 | 0.7660 | 4.32 | -7.99 |
| 50 | 0.3376 | 0.6912 | -0.511 |
| 100 | 0.1201 | 0.1728 | -0.064 |
| 250 | ? | 0.0276 | -0.0041 |

Is the missing value using the first two terms about 0.0235? Note that for small $n$, the first term omitted merely provides gross error assessments. As for the missing value it is 0.0233, but keep in mind that we are fortunate in knowing a correct value for making the decision.

**Further Example 2.4**

Padé-type sequences may be set up using c.f. remainders as follows, applications being for moderately diverging cases. Let

$$R(n;\sigma) = \int_0^\infty \frac{d\sigma(t)}{n+t}, \qquad \text{(Stieltjes moment problem determined)}$$

be a Stieltjes transform with c.f. expansions

$$= \frac{b_0}{n+a_1-}\ \frac{b_1}{n+a_2-}\ \cdots = \frac{c_1}{n+}\ \frac{c_2}{1+}\ \frac{c_3}{n+}\ \cdots,$$

and for the Stieltjes form convergents $\chi_s(n)/\omega_s(n)$. Then a summation algorithm for the series $e_0 + e_1/n + \cdots$ is

$$F_{2r}(n;\sigma) = n\sum_{s=0}^{r}\left\{\frac{\omega_{2r}(n)R(n;\sigma)-\chi_{2s}(n)}{c_1c_2\cdots c_{2s+1}}\right\}\omega_{2s}^{*}.$$

Here if $C$ is a (linear) operator such that

$$C(n^s) = e_s,$$

then

$$\omega_s^{*} = C\{\omega_s(n)\}.$$

Similarly, there is the "odd" part,

$$F_{2r+1}(n;\sigma) = e_0+\sum_{s=0}^{r}\left\{\frac{\chi_{2s+1}(n)-\omega_{2s+1}(n)R(n;\sigma)}{c_1c_2\cdots c_{2s+2}}\right\}\omega_{2s+1}^{*}.$$

If the J-fraction form is preferred, then

$$F_r^{(J)}(n;\sigma) = n\sum_{s=0}^{\infty}\left\{\frac{\omega_{2s}(n)R(n;\sigma)-\chi_{2s}(n)}{b_0b_1\cdots b_s}\right\}\omega_{2s}^{*}.$$

$$(\omega_0 = 1, \omega_2 = n+a_1)$$

For the Stieltjes form

$$R(n) = \frac{1}{k_1n+}\;\frac{1}{k_2+}\;\cdots, \qquad (k_s>0, \textstyle\sum k_s=\infty)$$

with convergents $\hat{\chi}_s(n)/\hat{\omega}_s(n)$, the algorithm is

$$F_{2r}^{(S)}(n) = n\sum_{s=0}^{r}k_{2s+1}\{\hat{\omega}_{2s}(n)R(n)-\hat{\chi}_{2s}(n)\}\hat{\omega}_{2s},$$

$$F_{2r+1}^{(S)} = e_0+\sum_{s=0}^{r}k_{2s+2}\{\hat{\chi}_{2s+1}(n)-\hat{\omega}_{2s+1}(n)R(n)\}\hat{\omega}_{2s+1}.$$

**Further Example 2.5 (Continuation of Further Example 2.4)**

Consider $E(\sqrt{m_2})$ and a uniform population (Tables 2.8). Let

$$R(n;\sigma) = \frac{1}{n+b-}\;\frac{b}{n+b+1-}\;\frac{2b}{n+b+2-}\;\frac{3b}{n+b+3-}\;\cdots,$$

where

$$R(n;\sigma) = e^{-b}\int_0^b x^{n-1}e^x\,dx\,/\,b^n\,, \qquad (b>0)$$

$$= \sum_{x=0}^{\infty}\frac{e^{-b}b^x}{x!(x+n)}. \qquad \text{(See Example 1.6)}$$

Take $b = 1, n = 5$, so that

$$R(5) = e^{-1}\int_0^1 x^4 e^x\,dx = 9 - 24e^{-1}\,,$$

$$F_0^{(J)}(n;\sigma) = 5R(5) = 0.8545\,,$$

$$\omega_1 = n+1\,, \qquad \omega_1^* = 1 - 0.6 = 0.4\,,$$

$$F_1^{(J)}(n;\sigma) = F_0^{(J)}(n;\sigma) + n(\omega_2 R - \chi_2)0.4 = 0.9052\,.$$

Other approximants to $E(\sqrt{m_2})$ in uniform sampling are:

| $r$ | $F_r^{(J)}(5;\sigma)$ | $r$ | $F_r^{(J)}(5;\sigma)$ |
|---|---|---|---|
| 2 | 0.8858 | 7 | 0.8618 |
| 3 | 0.8732 | 8 | 0.86135 |
| 4 | 0.8672 | 9 | 0.86108 |
| 5 | 0.8643 | 10 | 0.86096 |
| 6 | 0.8627 | 11 | 0.86094 |

For comparison we give the use of the basic $R(n)$ in the forms

(a) Gauss: $R(n) = \ln\left(1 + \frac{1}{n}\right)$

$$= \frac{1}{n+p_1-}\;\frac{p_1q_1}{n+p_2+q_1-}\;\frac{p_2q_2}{n+p_3+q_2-}\;\cdots.$$

$$(p_s = s/(4s-2),\ q_s = s/(4s+2))$$

(b) Gamma: $R(n) = \int_0^\infty \frac{e^{-t}\,dt}{n+t}$

$$= \frac{1}{n+1-}\;\frac{1^2}{n+3-}\;\frac{2^2}{n+5-}\;\cdots.$$

| Approximants to $E(\sqrt{m_2})$ in Uniform Sampling | | | | |
|---|---|---|---|---|
| | $n = 5$ | | $n = 10$ | |
| $r$ | Gauss | Gamma | Gauss | Gamma |
| 0 | 0.9116 | 0.8521 | 0.9531 | 0.9156 |
| 1 | 0.8950 | 0.8972 | 0.9440 | 0.9444 |
| 2 | 0.8706 | 0.8877 | 0.9371 | 0.9406 |
| 3 | 0.8633 | 0.8778 | 0.9360 | 0.9379 |
| 4 | 0.8610 | 0.8713 | 0.93578 | 0.9367 |
| 5 | 0.86030 | 0.8673 | 0.93575 | 0.9362 |
| 6 | 0.86030 | 0.8648 | 0.93575 | 0.9360 |
| 7 | - | 0.8633 | - | 0.9359 |
| 8 | - | 0.86228 | - | 0.93580 |
| 9 | - | 0.86165 | - | 0.93578 |
| 10 | - | 0.86125 | - | 0.93576 |
| 11 | - | 0.86102 | - | 0.935759 |
| 12 | - | 0.86090 | - | 0.935757 |
| 13 | - | 0.86086 | - | 0.935757 |
| Levin | 0.8616 | | 0.93576 | |

(Note: Loss of accuracy is a problem, so that the Gauss algorithm could only proceed as far as $r = 6$.)

**Further Example 2.6**

For the function

$$f(n) = 1 + \int_0^\infty \frac{e^{-t}\,dt}{1 + t^2/n}$$

and the corresponding series $1 + \sum_{s=0}^{\infty} (-1)^s (2s)!/n^s$ there is the c.f.

$$f(n) = \frac{2n}{n+}\;\frac{p_1}{1+}\;\frac{q_1}{n+}\;\cdots$$

with partial numerators

| $s$ | $p_s$ | $q_s$ | $s$ | $p_s$ | $q_s$ |
|---|---|---|---|---|---|
| 1 | 1.00000000 00 | 1.10000000 01 | 8 | 5.46925887 02 | 6.39522867 02 |
| 2 | 1.96363636 01 | 4.16969697 01 | 9 | 7.03961479 02 | 8.08183005 02 |
| 3 | 5.80910853 01 | 9.20618199 01 | 10 | 8.80747135 02 | 9.96571669 02 |
| 4 | 1.16330609 02 | 1.62122929 02 | 11 | 1.07728150 03 | 1.20469018 03 |
| 5 | 1.94340387 02 | 2.51893472 02 | 12 | 1.29356349 03 | 1.43253958 03 |
| 6 | 2.92112519 02 | 3.61380806 02 | 13 | 1.52959225 03 | 1.68012072 03 |
| 7 | 4.09642116 02 | 4.90589556 02 | 14 | 1.78536706 03 | |

For $n = 1$ and $n = 10$ convergent sequences are

| $n=1$ | | $n=10$ | |
|---|---|---|---|
| $s$ | $C_s$ | $s$ | $C_s$ |
| 1 | 2 | 1 | 2 |
| 2 | 1 | 2 | 1.8182 |
| 3 | 1.8462 | 3 | 1.9091 |
| 4 | 1.2104 | 4 | 1.8641 |
| 26 | 1.5051 | 5 | 1.8946 |
| 27 | 1.6823 | 6 | 1.8747 |
| 28 | 1.5110 | 11 | 1.8866 |
| | | 12 | 1.8817 |

($C_s$ refers to the $s$ th convergent.)

By quadrature $f(1) = 1.6214$, and $f(10) = 1.8843$.

**Further Example 2.7**

Find assessments for the sum of the series $f(n) \sim \sum_{s=0}^{\infty} \frac{(-1)^s (2s)!}{s!n^s}$ when $n = 1$, 5 and 10. Defining $x = 1/n$, check the coefficients in the approximants,

$$[1|2] = (1 + 10x)/(1 + 12x + 12x^2),$$

$$[2|2] = (1 + 18x + 32x^2)/(1 + 20x + 60x^2),$$

$$[2|3] = (1 + 28x + 132x^2)/(1 + 30x + 180x^2 + 120x^3),$$

$$[3|3] = (1 + 40x + 348x^2 + 384x^3)/(1 + 42x + 420x^2 + 840x^3),$$

$$[3|4] = \frac{(1 + 54x + 740x^2 + 2232x^3)}{(1 + 56x + 840x^2 + 3360x^3 + 1680x^4)},$$

$$[4|4] = \frac{(1 + 70x + 1380x^2 + 7800x^3 + 6144x^4)}{(1 + 72x + 1512x^2 + 10080x^3 + 15120x^4)},$$

$$[4|5] = \frac{(1 + 88x + 2352x^2 + 21120x^3 + 46320x^4)}{(1 + 90x + 2520x^2 + 25200x^3 + 75600x^4 + 30240x^5)},$$

$$[5|5] = \frac{(1 + 108x + 3752x^2 + 48720x^3 + 202320x^4 + 122880x^5)}{(1 + 110x + 3960x^2 + 55440x^3 + 277200x^4 + 332640x^5)}.$$

Compare the expansion (in ascending power of $x$) of each ratio with the series, noting the location and magnitude of the first discrepancy.

A formal solution is

$$f(n) = (n/2)^{1/2} e^{4/n} \int_{(n/2)^{1/2}} e^{-y^2/2} dy = (n/2)^{1/2} R\{(n/2)^{1/2}\}$$

where $R(\cdot)$ refers to Mills' ratio (1.61) for the normal integral. Hence

$$f(n) = \frac{N}{N+}\,\frac{1}{N+}\,\frac{2}{N+}\,\frac{3}{N+}\,\frac{4}{N+}\,\cdots. \qquad (N = \surd(n/2))$$

| | $n = 1$ | | $n = 5$ | | $n = 10$ | |
|---|---|---|---|---|---|---|
| $s$ | $[s\mid s]$ | $[s\mid s+1]$ | $[s\mid s]$ | $[s\mid s+1]$ | $[s\mid s]$ | $[s\mid s+1]$ |
| 1 | 0.71 | 0.440 | 0.818 | 0.773 | 0.8750 | 0.8621 |
| 2 | 0.630 | 0.486 | 0.795 | 0.783 | 0.86667 | 0.86486 |
| 3 | 0.593 | 0.510 | 0.790 | 0.786 | 0.86563 | 0.86528 |
| 4 | 0.574 | 0.523 | 0.7882 | 0.7870 | 0.86544 | 0.86537 |
| 5 | 0.564 | 0.531 | 0.7877 | 0.7873 | 0.865406 | 0.865385 |
| 6 | 0.558 | 0.535 | 0.78755 | 0.78736 | 0.865396 | 0.865390 |
| 7 | 0.5534 | 0.5384 | 0.78749 | 0.78741 | 0.8653938 | 0.8653919 |
| 8 | 0.5517 | 0.5405 | 0.787462 | 0.787423 | 0.8653930 | 0.8653923 |
| 9 | 0.5500 | 0.5419 | 0.787450 | 0.787431 | 0.8653927 | 0.8653925 |
| 10 | 0.5489 | 0.5429 | 0.787445 | 0.787435 | 0.86539264 | 0.86539255 |
| 11 | 0.5480 | 0.5436 | 0.787442 | 0.787437 | 0.86539261 | 0.86539257 |
| True value | 0.54564 | | 0.787439 | | 0.86539259 | |

(Note $[s\mid s]$, $[s\mid s+1]$ refer to the c.f.

$$\frac{1}{1+}\,\frac{1}{(n/2)+}\,\frac{2}{1+}\,\frac{3}{(n/2)+}\,\cdots;$$

thus $[1\mid 1] = (n+4)/(n+6)$,
and $[1\mid 2] = (n^2+10n)/(n^2+12n+12)$.)

**Further Example 2.8**

Develop a Stieltjes continued fraction for the series

$$f(x) \sim \sum_{s=0}^{\infty} (-1)^s (1+s!)x^s$$

in the form

$$\frac{2}{1+}\,\frac{xp_1}{1+}\,\frac{xq_1}{1+}\,\frac{xp_2}{1+}\,\frac{xq_2}{1+}\,\cdots.$$

Check the following:

| $s$ | $p_s$ | $q_s$ | $s$ | $p_s$ | $q_s$ |
|---|---|---|---|---|---|
| 1 | 1.00000000 | 0.50000000 | 14 | 13.4258808 | 14.4037842 |
| 2 | 2.50000000 | 1.80000000 | 15 | 14.7632325 | 15.0241089 |
| 3 | 2.75555556 | 3.47670251 | 16 | 16.1606078 | 15.6471439 |
| 4 | 3.34918523 | 4.56761836 | 17 | 17.5137252 | 16.3617194 |
| 5 | 4.57823153 | 5.10587392 | 18 | 18.7420772 | 17.2215529 |
| 6 | 6.14900736 | 5.57788318 | 19 | 19.8039984 | 18.2440012 |
| 7 | 7.61833244 | 6.30887386 | 20 | 20.6980534 | 19.4150704 |
| 8 | 8.72753039 | 7.38112090 | 21 | 21.4561154 | 20.6959106 |
| 9 | 9.47760910 | 8.72663673 | 22 | 22.1324271 | 22.0308705 |
| 10 | 10.0280335 | 10.1856180 | 23 | 22.7906015 | 23.3581018 |
| 11 | 10.5825738 | 11.5710579 | 24 | 23.4907336 | 24.6213314 |
| 12 | 11.2987698 | 12.7484438 | 25 | 24.2794687 | - |
| 13 | 12.2489497 | 13.6775305 | | | |

(Computed using double-double precision on IBM (33 significant digits) but truncated in this tabulation.)

The successive approximants when $x = ½$ are:

| $s$ | $C_{2s-4}$ | $C_{2s-3}$ | $s$ | $C_{2s-4}$ | $C_{2s-3}$ |
|---|---|---|---|---|---|
| 3 | 1.33333333 | 1.42857143 | 16 | 1.38932386 | 1.38932393 |
| 4 | 1.37931034 | 1.39423077 | 17 | 1.38932388 | 1.38932392 |
| 5 | 1.38692098 | 1.39035639 | 18 | 1.38932389 | 1.38932391 |
| 6 | 1.38874099 | 1.38962542 | 19 | 1.38932390 | 1.38932390 |
| 7 | 1.38917542 | 1.38942577 | 20 | 1.38932390 | 1.38932390 |
| 8 | 1.38928132 | 1.38935950 | 21 | 1.38932390 | 1.38932390 |
| 9 | 1.38930978 | 1.38933635 | 22 | 1.38932390 | 1.38932390 |
| 10 | 1.38931854 | 1.38932832 | 23 | 1.38932390 | 1.38932390 |
| 11 | 1.38932166 | 1.38932554 | 24 | 1.38932390 | 1.38932390 |
| 12 | 1.38932291 | 1.38932454 | 25 | 1.38932390 | 1.38932390 |
| 13 | 1.38932346 | 1.38932417 | 26 | 1.38932390 | 1.38932390 |
| 14 | 1.38932370 | 1.38932402 | | | |
| 15 | 1.38932381 | 1.38932396 | $\infty$ | 1.389323901 | |

Note that $p_s$ and $q_s$ are nearly equal to $s$ as might be expected from the factorial series and its c.f. Clearly

$$f(x) = \frac{1}{1+x} + \int_0^\infty \frac{e^{-t}\,dt}{1+xt}$$

and the distribution function $\sigma(t)$ consists of $\int_0^t e^{-x}\,dx$ plus the unit function $U(x-1)$, where $U(x) = 0$, $x < 0$, $U(x) = 1$, $x \geqslant 0$. Thus since Carleman's criterion is satisfied, the c.f. converges for all $x$ not on the negative half of the real axis.

**Further Example 2.9**

Show that for the series

$$f(x) \sim 1 - 2{\cdot}2!x + 3{\cdot}3!x^2 - 4{\cdot}4!x^3 + \cdots, \qquad \text{(see (2.12}a\text{))}$$

there is the Stieltjes continued fraction type development

$$\frac{1}{1+}\ \frac{p_1x}{1+}\ \frac{q_1x}{1+}\ \frac{p_2x}{1+}\ \frac{q_2x}{1+}\ \cdots$$

where

| $s$ | $p_s$ | $q_s$ | $s$ | $p_s$ | $q_s$ |
|---|---|---|---|---|---|
| 1 | 4.00000000 00 | 5.00000000-01 | 9 | 1.61823660 01 | 4.88915832 00 |
| 2 | 7.50000000 00 | -8.00000000-01 | 10 | 2.01212360 01 | 2.79674534 00 |
| 3 | -1.12000000 01 | 2.03571429 01 | 11 | 4.32191601 01 | -1.82707480 01 |
| 4 | 6.95488722-01 | 1.20284495 01 | 12 | -7.86227482 00 | 3.49967744 01 |
| 5 | 6.93753356-01 | 3.93134009 01 | 13 | 4.74383036 00 | 2.48270651 01 |
| 6 | -2.52752197 01 | -1.65292539 00 | 14 | 7.45641738 00 | 2.51763057 01 |
| 7 | 1.76317731 01 | 3.18543209 00 | 15 | 7.54200891 00 | |
| 8 | 1.50114164 01 | 4.66536504 00 | | | |

Derive the convergents for $x = 1$;

| $s$ | $C_{2s-4}$ | $C_{2s-3}$ | $s$ | $C_{2s-4}$ | $C_{2s-3}$ |
|---|---|---|---|---|---|
| 3 | 0.20000000 | 0.27272727 | 10 | 0.19268168 | 0.19322063 |
| 4 | 0.20930233 | 0.20207254 | 11 | 0.19271502 | 0.19283242 |
| 5 | 0.20993228 | 0.19191498 | 12 | 0.19272216 | 0.19273818 |
| 6 | 0.03186646 | 0.19057988 | 13 | 0.19272258 | 0.19270809 |
| 7 | 0.18966667 | 0.19055710 | 14 | 0.19272360 | 0.19269753 |
| 8 | 0.19207655 | 0.19011966 | 15 | 0.19273531 | 0.19269381 |
| 9 | 0.19255664 | 0.20127230 | 16 | 0.19257907 | 0.19269262 |

Show by quadrature, or otherwise, that the correct value is 0.19926947.

**Comments:** Note the anomalous sign pattern in the $p$'s and $q$'s, so that successive convergents do not necessarily provide bounds. However in this connection note that

$$f(x) = \frac{1}{x}\int_0^\infty \frac{(1-t)e^{-t}\,dt}{1+xt}$$

with a change of sign in the integrand. But the modified form

$$f(x) = \left(\frac{1}{x} + \frac{1}{x^2}\right)\int_0^\infty \frac{e^{-t}\,dt}{1+xt} - \frac{1}{x^2}$$

would produce a $(p,q)$ form Stieltjes sequence (monotonic increasing, decreasing) for $x > 0$). Note also that sequences from 2nd order c.f.s (see (3.64) for $F(z,z)$ and Section 4.5.1) can be constructed from $f(x) = \int_0^\infty te^{-t}/(1+xt)^2 dt$.

If (2.19) is used, as it may be as a check on any other algorithm for the c.f., then one uses the polynomials $p_s(k)$ based on the c.f.

$$\frac{1}{t+2-}\;\frac{1\cdot 2}{t+4-}\;\frac{2\cdot 3}{t+6-}\;\frac{3\cdot 4}{t+8-}\;\cdots . \qquad \text{(See (1.32) with } a = 2\text{)}$$

For example

$p_0(1) = 1,$ $\quad p_1(1) = -1,$ $\quad p_2(1) = 1,$
$p_3(1) = 1,$ $\quad p_4(1) = -19,$ $\quad p_5(1) = 151,$
$p_6(1) = -1091,$ $\quad p_7(1) = 7841,$ $\quad p_8(1) = -56519,$
$p_9(1) = 396271,$ $\quad p_{10}(1) = -2442439,$ $\quad p_{11}(1) = 7701409,$
$p_{12}(1) = 145269541.$

**Further Example 2.10**

A series for $\psi_3(z)$, the 4th derivative of $\ln\Gamma(z)$, is

$$f(z) \sim \frac{2}{z^3} + \frac{3}{z^4} + \frac{2}{z^5} - \frac{1}{z^7} + \frac{4}{3z^9} - \frac{3}{z^{11}} + \frac{10}{z^{13}} - \frac{691}{15z^{15}} + \frac{280}{z^{17}} - \frac{10851}{5z^{19}} + \frac{438670}{21z^{21}} - \frac{1222277}{5z^{23}} .$$

(i) Show that

$$f(z) = \frac{2}{z^3} + \frac{3}{z^4} + \frac{1}{z^3}\left(\frac{2}{z^2+}\;\frac{\frac{1}{2}}{1+}\;\frac{\frac{5}{6}}{z^2+}\;\frac{\frac{22}{15}}{1+}\;\frac{\frac{116}{55}}{z^2+}\;\frac{\frac{942}{319}}{1+}\;\cdots\right)$$

for which there are the approximants ($s$ th terms):

| $s$ | $\psi_3(1)$ | $\psi_3(2)$ | $\psi_3(3)$ |
|---|---|---|---|
| 1 | $7.0^+$ | $0.50000^+$ | $0.119341564^+$ |
| 2 | 6.333 | 0.49306 | 0.118908381 |
| 3 | $6.571^+$ | $0.49601^+$ | $0.118943316^+$ |
| 4 | 6.456 | 0.49388 | 0.118938764 |
| 5 | $6.516^+$ | $0.49404^+$ | $0.118939535^+$ |
| 6 | 6.481 | 0.49393 | 0.118939370 |
| Upper | 7.000 | 0.494547 | 0.118940482 |
| Lower | 6.000 | 0.493142 | 0.118937885 |
| True value | 6.494 | 0.493939 | 0.118939402 |

(Upper and Lower represent series sums to smallest term numerically and previous term.)

(ii) Omitting the second term $3/z^4$, show that

$$f(z) \sim \frac{3}{z^4} + \frac{1}{z}\left( \frac{2}{z^2+} \; \frac{p_1}{1+} \; \frac{q_1}{z^2+} \; \cdots \right)$$

where

| $s$ | $p_s$ | $q_s$ |
|---|---|---|
| 0 | - | 2.0 |
| 1 | -1.0. | 1.5. |
| 2 | 0.27 | 2.02 |
| 3 | 1.529670330 | 3.532398636 |
| 4 | 3.235549571 | 5.581033149 |
| 5 | 5.429380164 | 8.139302429 |

*Assuming a continuation of the pattern exhibited for the p*'s and $q$'s (i.e. increasing sequences), deduce

$$6.5007 > f(1) > 6.4911$$

$$0.49394025 > f(2) > 0.49393907$$

$$0.1189394036 > f(3) > 0.1189394019.$$

Can one prove that $1 - 2/(z^3 f(z) - 3)$ has a Stieltjes continued fraction expression $\dfrac{1}{z^2+} \; \dfrac{a_1}{1+} \; \dfrac{b_1}{z^2+} \; \cdots$? For some comments on c.f.s. for polygamma functions see Shenton and Bowman (1971).

We have proved that

$$\psi_3(z) = \frac{2}{z^3} + \frac{3}{z^4} + \left(\frac{2\pi}{z}\right)^3 \int_0^\infty \frac{(y + 4y^2 + y^3)x^{3/2}dx}{(x + z^2)(y-1)^4}.$$

$$(y = \exp(2\pi\sqrt{x}))$$

**Further Example 2.11**

If $I_m = \int_0^{\pi} \frac{\sin(mx)dx}{1+x^2}$, derive the result

$$I_m \sim A_m - B_m ,$$

where $A_m = \int_0^{\infty} \frac{\sin(mx)dx}{1+x^2} = \tfrac{1}{2}e^m E_1(m) + \tfrac{1}{2}e^{-m} Ei(m)$

(in terms of the exponential integrals);

$$B_m \sim \cos(m\pi) \sum_{s=0}^{\infty} \frac{(-1)^s (2s)! \sin\{(2s+1)\phi\} \sin^{2s+1}(\phi)}{m^{2s+1}} ,$$

where $\phi = \tan^{-1}(\pi^{-1})$.

Verify that for $m = 4$, summing the series to the smallest term, yields $I_4 = 0.260592$ in excellent agreement with a quadrature assessment.

**Further Example 2.12**

The series part of $\ln \Gamma(x)$, $1/12x - 1/360x^3 + \cdots$ with c.f. development $\frac{a_0}{x+} \, \frac{a_1}{x+} \, \frac{a_3}{x+} \cdots$, $(\mathrm{Re}(x) > 0)$ is mentioned in Bowman and Shenton (1988, p27). The following tabulation shows the discrepancies when compared to a note by Char (1980), who used MACSYMA and gave 41 coefficients.

| $s$ | $a_{2s}$ | $a_{2s+1}$ |
|---|---|---|
| 0 | 8.33333333333333333333333333333-02 | 3.33333333333333333333333333333-02 |
| 1 | 2.523809523809523809523809523809$\underline{6}$-01 | 5.256064690026954177897574123989$\underline{1}$-01 |
| 2 | 1.011523068126841711747372124730$\underline{9}$ 00 | 1.5174736491532873984284915194$\underline{943}$ 00 |
| 3 | 2.269488974204959960909150672$\underline{2130}$ 00 | 3.00991738325939817007314073$\underline{41994}$ 00 |
| 4 | 4.026887192343901226168875953$\underline{2133}$ 00 | 5.00276808075403005168850241$\underline{21603}$ 00 |
| 5 | 6.28391137081578218007266315$\underline{53301}$ 00 | 7.49591912238403392975235470$\underline{71934}$ 00 |
| 6 | 9.04066023436772669953113936$\underline{30158}$ 00 | 1.04893036545094822771883712$\underline{98496}$ 01 |
| 7 | 1.22971936103862058639894371$\underline{51873}$ 01 | 1.39828769539924301882597606$\underline{32250}$ 01 |
| 8 | 1.60535514167049354697156163$\underline{88392}$ 01 | 1.79766073998702775925694723$\underline{02198}$ 01 |
| 9 | 2.0309762027441653743805414$\underline{614520}$ 01 | 2.2470471639933132495517942$\underline{068346}$ 01 |
| 10 | 2.506584654894597202916339$\underline{8709867}$ 01 | 2.746445182502913360917556$\underline{3214289}$ 01 |
| 11 | 3.032182123167304712688258$\underline{9970321}$ 01 | 3.295853392997298721999408$\underline{2960162}$ 01 |
| 12 | 3.6077698931299242645153301$\underline{171880}$ 01 | 3.895270668231155573454438$\underline{5562496}$ 01 |

(Char (1980), using MACSYMA gives the first 41 coefficients. Underscored digits show the discrepancies in the two independent studies. Bowman calculations were carried out in 1978).

**Further Example 2.13**

Loss of accuracy, as shown in the previous example, is a constant source of concern when dealing with the large coefficients that occur in statistical moment series. We give two further examples.

(a) *Var* $\{\ln(m_1')\}$ in Exponential Sampling

This example is given as a clear warning that loss of accuracy in machine evaluation must always be carefully scrutinized.

Using the gamma population with density

$$f(x) = e^{-x} x^{\rho-1} / \Gamma(\rho), \qquad (\rho > 0)$$

with

$$E\{e^{\alpha \ln(m_1')}\} = \int_0^\infty e^{-x} x^{n\rho+\alpha-1} dx \,/\, \Gamma(n\rho), \qquad (\alpha + n\rho > 0)$$

it follows that

$$E\{\ln(m_1')\} = \psi(n\rho) - \ln(n),$$
$$E\{\ln^2(m_1')\} = \psi_1(n\rho) + \psi^2(n\rho) - 2\ln(n)\psi(n\rho) + \ln^2 n,$$

and

$$Var\{\ln(m_1')\} = \psi_1(n\rho)$$

in terms of the derivative of the psi function. A comparison for the case $\rho = 1$, using two precision forms is given in Table 2.17.

(b) In Sampling from a Poisson-Poisson Distribution with p.g.f

$$h(u) = e^{-\lambda_1} + \sum_{s=1}^{\infty} \frac{u^s}{s!} \frac{d^{s-1}}{du^{s-1}} \{\lambda_1 e^{\lambda_1(t-1)} e^{s\lambda_2(t-1)}\} \Big|_{t=0},$$
$$(\lambda_1 > 0, \; 0 \leqslant \lambda_2 < 1)$$

the series for two parameter points $(\lambda_1, \lambda_2)$ for $E(m_2'/m_1')^2$ are shown in Table 2.18.

Error analysis is based on an exact recursive formula for a general coefficient (Bowman and Shenton, 1985).

**Table 2.17** *Var* $\{\ln(m_1')\}$ **in Sampling from the Exponential Distribution: Loss of Accuracy in the Coefficient of** $n^{-s}$

| $s$ | Exact | RQ | RD |
|---|---|---|---|
| 7 | 0.0238095238095 | 0.023095238095 | 0.02380952378 |
| 8 | 0.0 | -0.41 Q-26 | 0.19 D-9 |
| 9 | -0.0333333333333 | -0.033333333333 | -0.033333305 |
| 10 | 0.0 | 0.62 Q-24 | 0.11 D-05 |
| 11 | 0.075757575758 | 0.075757575758 | 0.07578 |
| 12 | 0.0 | -0.14 Q-19 | 0.19 D-03 |
| 13 | -0.253113553136 | -0.253113553136 | -0.229 |
| 14 | 0.0 | 0.26 Q-17 | 0.17 D+01 |
| 15 | 1.16666666667 | 1.16666666667 | 0.145 D+03 |
| 16 | 0.0 | -0.58 Q-13 | 0.63 D+04 |
| 17 | -7.0921568627 | -7.0921568627 | 0.77 D+05 |
| 18 | 0.0 | -0.23 Q-09 | 0.54 D+07 |
| 19 | 54.9711779336 | 54.9711779336 | 0.50 D+09 |
| 20 | 0.0 | -0.20 Q-06 | 0.77 D+11 |
| 21 | -529.1242424 | -529.1242487 | 0.45 D+13 |
| 22 | 0.0 | 0.49 Q-03 | 0.21 D+15 |
| 23 | 6192.123188 | 6192.134205 | 0.42 D+16 |
| 24 | 0.0 | -0.24 Q+01 | 0.22 D+18 |

(RQ ~ recursive formula using double-double precision arithmetic,
RD ~ recursive formula using double precision arithmetic on IBM 3033.

Comment: The series is

$$\psi_1(n\rho) \sim \frac{1}{n\rho} + \frac{1}{2(n\rho)^2} + \sum_{k=1}^{\infty} \frac{B_{2k}}{(n\rho)^{2k+1}}. \qquad (\rho = 1)$$

where $(B_k)$ are Bernoulli's numbers. This provides further support for the validity of the COETS algorithm, but also brings out the important fact that loss of accuracy can be a serious factor.)

**Table 2.18 Series for** $E(m_2'/m_1')^2$ **in Sampling from the Poisson-Poisson Distribution**

| s | $\lambda_1 = 1.0, \quad \lambda_2 = 0.05$ | $\lambda_1 = 1.5, \quad \lambda_2 = 0.15$ |
|---|---|---|
| 0 | 4.6684724641462235556817397042687 00 | 9.9148717089115312316662875205038 00 |
| 1 | -1.1469654484735868800076248970660 00 | -2.2396621631615014751485536923789 00 |
| 2 | -5.2945761606632470119966506738393 00 | -1.0837876838384180279298217645696 01 |
| 3 | 4.4451120072831836112003984052557 00 | 2.0159763656170586355302335021848 01 |
| 4 | -4.8919996233668998337956002998060 00 | -2.7584539148574191690020694087118 01 |
| 5 | -5.8841100329160621340899389714361 00 | 3.4644358646124990022760205294044 01 |
| 6 | -3.5829661628675568402918805663149 01 | -5.1752488808926943864583942469691 01 |
| 7 | -1.8871192519682944870487838915774 02 | 5.6321936040529729944030246450658 01 |
| 8 | -1.2020008732559012856066671252525 03 | -1.6497621691907076565831238589677 02 |
| 9 | -8.7542537461815521992906710186232 03 | -1.1985901092295777436504075150587 02 |

**Table 2.18 continued**

| s | $\lambda_1 = 1.0, \quad \lambda_2 = 0.05$ | | $\lambda_1 = 1.5, \quad \lambda_2 = 0.15$ | |
|---|---|---|---|---|
| 10 | -7.2167417767966724052985350322902 | 04 | -1.9656605259245075336388416770901 | 03 |
| 11 | -6.6478953949222891594178521608638 | 05 | -9.8647838024794635551892065964812 | 03 |
| 12 | -6.7724926318795991241962645383471 | 06 | -7.3208717833573244368198864778015 | 04 |
| 13 | -7.5635495378075994316640695514994 | 07 | -5.4243090833030628463121804711910 | 05 |
| 14 | -9.1914377970476380925038561303664 | 08 | -4.4733172833467947793367356854044 | 06 |
| 15 | -1.2076822657380614161446972229519 | 10 | -3.9543066825446934682217963764883 | 07 |
| 16 | -1.7062373052276527270892901388954 | 11 | -3.7599829677040742600487956137645 | 08 |
| 17 | -2.5796250352677770642028957013281 | 12 | -3.8196160267015101537614452098524 | 09 |
| 18 | -4.1559493909451952548694234792492 | 13 | -4.1312827543419280699848361138038 | 10 |
| 19 | -7.1081644627835340961807074199896 | 14 | -4.7398529672953538962770471698605 | 11 |
| 20 | -1.2863896460360698370847992599010 | 16 | -5.7502355609998199970664427382872 | 12 |
| 21 | -2.4559526783376724121963107585907 | 17 | -7.3551624448719273660595528781414 | 13 |
| 22 | -4.9332453055431033134681739807129 | 18 | -9.8936013291680088540385290980339 | 14 |
| 23 | -1.0400493878738927595562988281250 | 20 | -1.3961893143804619521073341369629 | 16 |
| 24 | -2.2962684262184033351015117187500 | 21 | -2.0626473557044413023913574218750 | 17 |

(Set up by COETS algorithm. Underscored digits indicate loss of accuracy.)

## Further example 2.14

The series for $E(\sqrt{m_2})$ in sampling from a half Gaussian density ($f(x) = \sqrt{(2/\pi)}\exp(-x^2/2)$, $x > 0$, $f(x) = 0$, $x < 0$) is:

**Table 2.19** $E(\sqrt{m_2})$ **from the Half-Gaussian Distribution**

| $r$ | $e_r$ | $e_r / e_{r-1}$ | $r$ | $e_r$ | $e_r / e_{r-1}$ |
|---|---|---|---|---|---|
| 0 | 0.6028102749890869 00 | | 13 | -0.2043212738774243 12 | -32.9600 |
| 1 | -0.5176013324169388 00 | -0.8586 | 14 | 0.7400030996651372 13 | -36.2176 |
| 2 | 0.2428930392759523-01 | -0.0469 | 15 | -0.2922094039571403 15 | -39.4876 |
| 3 | -0.2247110244310805 00 | -9.2514 | 16 | 0.1249728132322679 17 | -42.7682 |
| 4 | 0.6951344282200770 00 | -3.0935 | 17 | -0.5756042652916438 18 | -46.0584 |
| 5 | -0.6827814981426405 01 | -9.8223 | 18 | 0.2841005053436587 20 | -49.3569 |
| 6 | 0.7145451055411003 02 | -10.4652 | 19 | -0.1496160584747404 22 | -52.6631 |
| 7 | -0.9906406217958537 03 | -13.8639 | 20 | 0.8374929383162668 23 | -55.9761 |
| 8 | 0.1683039595099694 05 | -16.9894 | 21 | -0.4965958497531367 25 | -59.2955 |
| 9 | -0.3381204030075665 06 | -20.0899 | 22 | 0.3109719800706051 27 | -62.6207 |
| 10 | 0.7875017229165029 07 | -23.2906 | 23 | -0.2050901919350491 29 | -65.9513 |
| 11 | -0.2086034414910586 09 | -26.4893 | 24 | 0.1421007775881540 31 | -69.2870 |
| 12 | 0.6199059626103409 10 | -29.7170 | | | |

(Coefficients evaluated using double-double precision on IBM System 360 Model 195.)

**Table 2.20 Levin on $E(\sqrt{m_2})$ from Half-Gaussian Distribution**

| r | $n=1$ | $n=2$ | $n=3$ | $n=4$ | $n=5$ | $n=10$ |
|---|---|---|---|---|---|---|
| 7 | .2800-1 | .3339 | .4274 | .47238 | .49890 | .551105 |
| 8 | -.5071-2 | .3310 | .4268 | .47222 | .49884 | .551103 |
| 9 | -.4081-1 | .3284 | .4264 | .47210 | .49880 | .551102 |
| 12 | -.1402 | .3237 | .4257 | .47197 | .49877 | .551102 |
| 13 | -.1392 | .3239 | .4258 | .47198 | .49877 | .551102 |
| 14 | -.7595-1 | .3257 | .4259 | .47201 | .49877 | .551102 |
| Exact | .0000 | .3305 | .4265** | -- | -- | -- |
| Padé* | -- | -- | .4259 | .4722 | .49881 | -- |

(*For the series $\sum e_S x^S$, the Padé sequence ($x = 1/n$) is derived from

$$\sum_1^6 e_S x^S + x^7 \hat{\pi}_5(x)/\ \pi_6(x),$$

where $\hat{\pi}_5(\cdot)$, $\pi_6(\cdot)$ are polynomials of degrees 5 and 6.
**derived from a Gauss formula for quadrature in three-dimensions.)

**Further Example 2.15**

Given

$$\frac{1}{2\pi}\int_0^4 \frac{\sqrt{(4/t-1)}dt}{z+t} = \frac{1}{z+1-}\ \frac{1}{z+2-}\ \frac{1}{z+2-}\ \cdots, \qquad (z>0)$$

with $s$th convergent $\chi_s(z)/\omega_s(z)$, $s = 0, 1, \cdots$, with ($\chi_0 = 0$, $\chi_1 = 1$; $\omega_0 = 1$, $\omega_1 = z+1$). Show that if

$$I(k) = \frac{1}{2\pi}\int_0^4 \frac{\sqrt{(4/t-1)}(t+k)dt}{z+t}, \qquad (k>0)$$

$$= \frac{Q_0}{z+}\ \frac{P_1}{1+}\ \frac{Q_1}{z+}\ \frac{P_2}{1+}\ \frac{Q_2}{z+}\ \cdots, \qquad (Q_0 = k+1)$$

then

$$\begin{cases} P_s = 1 + \dfrac{1}{\chi_s(k)\omega_s(k)}, \\ Q_s = 1 - \dfrac{1}{\chi_{s+1}(k)\omega_s(k)}. \end{cases} \qquad (s = 1, 2, \cdots)$$

(Note that $\chi_{s+1}(k)$, $\omega_s(k)$ are polynomials of degree $s$ with positive coefficients.)

In particular

$$I(1) = \frac{2}{z+}\;\frac{\frac{3}{2}}{1+}\;\frac{\frac{5}{6}}{z+}\;\frac{\frac{16}{15}}{1+}\;\frac{\frac{39}{40}}{z+}\;\frac{\frac{105}{104}}{1+}\;\frac{\frac{272}{273}}{z+}\;\frac{\frac{715}{714}}{1+}\;\frac{\frac{1869}{1870}}{z+}\;\cdots,$$

$$I(2) = \frac{3}{z+}\;\frac{\frac{4}{3}}{1+}\;\frac{\frac{11}{12}}{z+}\;\frac{\frac{45}{44}}{1+}\;\frac{\frac{164}{165}}{z+}\;\frac{\frac{616}{615}}{1+}\;\frac{\frac{2295}{2296}}{z+}\;\cdots,$$

$$I(3) = \frac{4}{z+}\;\frac{\frac{5}{4}}{1+}\;\frac{\frac{19}{20}}{z+}\;\frac{\frac{96}{95}}{1+}\;\frac{\frac{455}{456}}{z+}\;\frac{\frac{2185}{2184}}{1+}\;\frac{\frac{10464}{10465}}{z+}\;\cdots.$$

Indication of a proof is to show first that

$$\begin{cases} \chi_s(k) = (\theta_1^{2s} - \theta_2^{2s})/(\theta_1^2 - \theta_2^2), \\ \omega_s(k) = (\theta_1^{2s+1} + \theta_2^{2s+1})/(\theta_1 + \theta_2), \end{cases}$$

where

$$\theta_1 = \{\sqrt{(k+4)} + \sqrt{k}\}/2,$$
$$\theta_2 = \{\sqrt{(k+4)} - \sqrt{k}\}/2. \qquad (\theta_1\theta_2 = 1)$$

Then from Section 2.7 show that

$$P_s Q_s = \omega_{s-1}(k)\omega_{s+1}(k)/\omega_2^2(k).$$

Now consider the odd-part of the basic c.f. defined from

$$\frac{1}{z+1-}\;\frac{1}{z+2-}\;\frac{1}{z+2-}\;\cdots = \frac{1}{z} - \frac{1}{z}\left(\frac{1}{z+2-}\;\frac{1}{z+2-}\;\cdots\right).$$

Notice that the denominators of the convergents are now $\chi_2(z)$, $\chi_3(z)$, $\cdots$. But if

$$\frac{1}{z+2-}\;\frac{1}{z+2-}\;\cdots = \frac{1}{z+P_1+Q_1-}\;\frac{P_2Q_1}{z+P_2+Q_2-}\;\frac{P_3Q_2}{z+P_3+Q_3-}\;\cdots$$

then

$$P_{s+1}Q_s = \chi_s(k)\chi_{s+2}(k)/\chi_{s+1}^2(k).$$

Hence

$$\frac{P_{s+1}}{P_s} = \frac{\chi_s(k)\chi_{s+2}(k)}{\chi_{s+1}^2(k)} \cdot \frac{\omega_s^2(k)}{\omega_{s-1}(k)\omega_{s+1}(k)}$$

leading to

$$P_s = \frac{\chi_{s+1}(k)\omega_{s-1}(k)}{\chi_s(k)\omega_s(k)}, \qquad Q_s = \frac{\chi_s(k)\omega_{s+1}(k)}{\chi_{s+1}(k)\omega_s(k)}.$$

But

$$\chi_{s+1}(k)\omega_s(k) - \chi_s(k)\omega_{s+1}(k) = 1,$$
$$\chi_{s+1}(k)\omega_{s-1}(k) - \chi_s(k)\omega_s(k) = 1,$$

and the result follows.

**Further Example 2.16**

**A Stieltjes type fraction for $\ln^2(1+x)$**

(a) Genesis

Using a symbolic programming language and integer arithmetic, a c.f. has been found for $\ln^2(1+x)$, taking the form

$$\ln^2(1+x) = \frac{x^2}{1+} \frac{p_1 x}{1+} \frac{q_1 x}{1+} \frac{p_2 x}{1+} \frac{q_2 x}{1+} \cdots, \tag{2.91}$$

as far as $p_{12}$. An interesting pattern emerged,

$$\begin{aligned}
p_1 &= 1, & q_1 &= -1/12, \\
p_2 &= 1/12, & q_2 &= -3/5, \\
p_3 &= 8/5, & q_3 &= -23/504, \\
p_4 &= 23/504, & q_4 &= -632/529, \\
p_5 &= 1161/529, & q_5 &= -13237/448404, \\
p_6 &= -q_5, & q_6 &= -44376903/24905465, \\
p_7 &= 1-q_6, & q_7 &= -204831433/8380316304, \\
p_8 &= -q_7, & q_8 &= -4399255846592/1861490525033, \\
p_9 &= 1-q_8, & q_9 &= -\frac{281973151659310 73}{141625537374411846 0}, \\
p_{10} &= -q_9, & q_{10} &= -\frac{2135167424130848106537 5}{726111990465101800932 1}, \\
p_{11} &= 1-q_{10}, & q_{11} &= -\frac{35504168154466332100977960 97}{21113007118747373509999745186 4}, \\
p_{12} &= -q_{10}.
\end{aligned}$$

This suggests that in general

$$p_{2s} + q_{2s-1} = 0, \qquad p_{2s+1} + q_{2s} = 1. \qquad (s = 1, 2, \cdots)$$

Consideration of the even part of the c.f., prompted by the sign pattern in (2.91), suggests that the c.f.

$$\ln^2(1+x) = \frac{1}{y+}\ \frac{-p_1q_1}{1+}\ \frac{-p_2q_2}{y+}\ \cdots \tag{2.92}$$

where $y = (1+x)/x^2$ and the partial numerators (as far as they have been calculated) are positive, is a Stieltjes fraction. This is in turn suggests a Stieltjes transform

$$\ln^2(1+x) = \int_0^\infty \frac{d\psi(u)}{u+y}, \qquad (y = (1+x)/x^2) \tag{2.93}$$

and a series of the form

$$\ln^2(1+x) = A_1y^{-1} + A_2y^{-2} + \cdots. \tag{2.94}$$

The first few coefficients in (2.94) are

$$A_1 = 1, \quad A_2 = -1/12, \quad A_3 = 1/90, \quad A_4 = -1/560,$$
$$A_5 = 1/3150,$$

and it is not difficult to conjecture that the series is

$$\ln^2(1+x) = \sum_{n=0}^\infty \frac{(-1)^n n!n!y^{-n-1}}{(n+1)(2n+1)!}. \tag{2.95}$$

If the conjecture is correct, then (after some manipulation)

$$\ln^2(1+x) = 2\int_0^{1/4} \ln\left\{\frac{1+(1-4u)^{1/2}}{1-(1-4u)^{1/2}}\right\}\frac{du}{u+(1+x)/x^2}, \tag{2.96}$$

and so in (2.93), $\psi(u)$ is constant for $u < 0$ and $u > 1/4$.

(b) Shenton and Kemp (1989) have proved the conjectures in (2.92), (2.95), and (2.96). They also show that the results are true for $x > -1$.

There are the inequalities

$$\ln^2(1+x) < x^2/(1+x), \tag{2.97}$$

$$\ln^2(1+x) > x^2/(1+x+x^2/12), \tag{2.98}$$

$$\ln^2(1+x) < \frac{x^2(1+x+x^2/20)}{(1+x)(1+x+2x^2/15)}, \tag{2.99}$$

$$\ln^2(1+x) > \frac{x^2(1+x+31x^2/252)}{(1+x)^2+13x^2(1+x)/63+23x^4/3780}, \quad \text{etc.}$$

An inequality similar to (2.97) appears in Mitrinović (1970, 3.6.15 and 3.6.18), who attributes it to Karamata (1960); the others would seem to be new. They have the advantage of holding for $x > -1$, compared with the inequalities

$$\frac{2x}{2+x} < \frac{6x+3x^2}{6+6x+x^2} < \ln(1+x) < \frac{x(30+21x+x^2)}{30+36x+9x^2} < \frac{x(6+x)}{6+4x}$$

where $x > 0$, obtained from the Padé approximants to $z\ln(1+1/z)$, $z > 0$, by taking $x = 1/z$, see e.g. Luke (1975) page 41. It is elementary though tedious to show that, whilst

$$\ln(1+x) < x(6+x)/(6+4x), \qquad (x > 0)$$

gives a poorer upper bound than (2.97) only when $x > 3$, on the other hand

$$(6x+3x^2)/(6+6x+x^2) < \ln(1+x), \qquad (x > 0)$$

always gives a poorer lower bound than (2.98), and

$$\ln(1+x) < x(30+21x+x^2)/(30+36x+9x^2), \qquad (x > 0)$$

always gives a poorer upper bound than (2.99).

**Further Example 2.17**

For the c.f.

$$\phi(n) = \frac{n}{n+}\;\frac{1}{1+}\;\frac{1}{n+}\;\frac{2}{1+}\;\frac{2}{n+}\;\frac{3}{1+}\;\frac{3}{n+}\;\frac{4}{1+}\;\frac{4}{n+}\;\cdots$$

derive the convergents

$$c_1 = n/n, \qquad c_2 = n/(n+1),$$
$$c_3 = (n^2+n)/(n^2+2n), \qquad c_4 = (n^2+3n)?(n^2+4n+2),$$
$$c_5 = (n^3+5n^2+2n)/(n^3+6n^2+6n),$$
$$c_6 = (n^3+8n^2+11n)/(n^3+9n^2+18n+6),$$
$$c_7 = (n^4+11n^3+26n^2+6n)/(n^4+12n^3+36n^2+36n),$$
$$c_8 = (n^4+15n^3+58n^2+50n)/(n^4+16n^3+72n^2+108n+24),$$
$$c_9 = \frac{(n^5+19n^4+102n^3+154n^2+24n)}{(n^5+20n^4+120n^3+252n^2+168n)},$$

$$c_{10} = \frac{(n^5 + 14n^4 + 177n^3 + 444n^2 + 274n)}{(n^5 + 25n^4 + 200n^3 + 600n^2 + 600n + 120)}.$$

(a) Can you find a formula for (i) $C_{2s} - C_{2s-1}$, (ii) $C_{2s+1} - C_{2s}$ ?

(b) If $\phi(n) \sim a_0 + a_1/n + a_2/n^2 + \cdots$, set up formulas to obtain $a_0, a_1, \cdots$.

**Further Example 2.18**

Being given

$$J(x) = \int_0^\infty \frac{te^{-xt}\,dt}{\sinh(t)} = \frac{1}{x+}\;\frac{\frac{1^4}{1\cdot 3}}{x+}\;\frac{\frac{2^4}{3\cdot 5}}{x+}\;\cdots, \qquad \text{(Rogers, 1907)}$$

use the Bernoulli numbers

$$\frac{te^{t/2}}{e^t - 1} = \sum_{n=0}^{\infty} \frac{t^n B_n(0.5)}{n!},$$

to show that

$$J(x) \sim \frac{1}{x} - \frac{1}{3x^3} + \frac{7}{15x^5} - \frac{31}{21x^7} + \frac{127}{15x^9} - \frac{2555}{33x^{11}} + \frac{1414477}{1365x^{13}} - \frac{57344}{3x^{15}} + \cdots$$

$$\sim \sum_{n=0}^{\infty} \frac{(2^n - 2)B_n}{x^{n+1}}.$$

Hence $J(x)$ has alternating signs when expanded in descending powers of $x$, diverges as fast as $2(2n)!/\pi^{2n}$, but the partial numerators of its c.f. expansion only increase as fast as $n^2/4$, $n \to \infty$. Note that

(i) $J(x) = \frac{1}{2}\psi\left(\frac{x+1}{2}\right)$, $\quad (x > 0)$

(ii) the series coefficients decrease at first, but soon increase without limit;

(iii) the c.f. converges for $x > 0$ but very slowly for small $x$. Thus for $x = 1/4$, the first few approximants are 4.0, 0.63, 3.1, 0.85 and 2.76. The best approximant from the series in this case is the first term 4.0.

**Further Example 2.19**

Geary (1936) studied the statistic

$$\omega_n = n^{-1/2}\sum |x_i| / \sqrt{(\sum x_i^2)}$$

where the $x_i \in N(0,1)$ as a test of normality. Reacting to remarks of E. S. Pearson, he (1936) studied the modified statistic

$$\omega_n' = (n+1)^{-1/2}\sum_{j=1}^{n+1} |x_j - \bar{x}| / \sqrt{\{\sum_{j=1}^{n+1}(x_j - \bar{x})^2\}}.$$

Apart from a few misprints (Shenton, Bowman, and Lam, 1979), he deduced through some intricate mathematics the first four moments. In fact, he found exact expressions for $\mu_1'$, $\mu_2$, $\mu_3$ of $\omega_n'$, using the independence of $\omega_n'$ and $s_x^2$. However, a problem arises with $\mu_4$, for $Z_n = E|z_1 z_2 z_3 z_4|$ is required ($z_i = x_i - \bar{x}$); Geary expresses this as the sum of 16 apparently intractable integrals, resulting in the series involving $a_s$ (Table 2.21a). We are indebted to Dr. Bob Byers for a check on the coefficients and the extension to $a_{14}$ using a FORMAC package on expressions such as $E(\pm z_1 z_2 \pm z_1 z_3 \pm z_1 z_4 \pm z_2 z_3 \pm z_2 z_4 \pm z_3 z_4)^s$, $s = 3$ to 14. Notice the ratios $a_s / a_{s-1}$, and the Shanks' (1955) five point extrapolate $3.19 \cdots$; does the ratio tend to $\pi$ or to 3? Now, a remakable serendipitous event occurs, at least looking at the work with a time lag of some 40 years. For Geary had in mind his work on $\omega_n$, so was disposed to compare the moments with those of $\omega_n'$; the transformation to the series in $b_s$ now takes place, for which the coefficient ratios seem to tend to unity. Since, in Geary's notation,

$$\mu_4'(\omega_n') = \left(\frac{n+1}{n}\right)^2 \frac{E(d^4)}{E(s^4)}, \qquad (d = \sum_{i=1}^{n+1} |z_i| / (n+1))$$

he constructs the series for $E(d^4)$, a factor in it being $Z_n = E|z_1 z_2 z_3 z_4|$. The series for $(1 + 1/n)^2 E(d^4)$ (Table 2.21b) is remarkable for the smallness of the coefficients, although the new terms ($c_7 - c_{14}$) hint at divergent tendencies. Notice that

$$\mu_4 \sim \frac{n}{n+2}\sum_{s=0}^{\infty} \frac{c_s}{n^s}$$

and this series, using terms up to $s = 14$, seems reasonable to use for $n \geqslant 3$. As checks on the fourth moments we use

$$(n+1)^3 E(d^4) = A_1 + A_2 + A_3 + A_4 + A_5 \tag{2.100}$$

where

$$A_1 = 3(1 + 1/n)^{-2},$$

$$A_2 = [8n^3 / \{\pi(n+1)^2\}]\{3n^{-1}\arcsin(n^{-1}) + (2 + n^{-2})\sqrt{(1 - n^{-2})}\},$$

$$A_3 = 3n^3(n+1)^{-2}(1 + 2n^{-2}),$$

$$A_4 = \left\{\frac{12(n-1)}{\pi(n+1)^2}\right\}\{n(n-2)\arcsin(n^{-1}) + (n^2+2)\surd(n^2-1)\},$$

$$A_5 = n(n-1)(n-2)Z_n ;$$

and

$$Z_n = \frac{n(n+1)a^4}{(n-1)(n-2)} B_n , \qquad \text{with} \qquad B_n = \sum(-1)^s \frac{b_s}{n^s} .$$

The series for $B_n$ (Table 2.21a) may be summed using (1) direct summation with Shanks' acceleration, (2) Levin's U-algorithm, and (3) series mimicry (Table 2.22).

For (1) we set up the partial sums and successively use Shanks' 3-point formula (Table 2.23).

For (3) we use the approximation

$$B_n^* = \frac{c_{14}}{d_{14}}\left\{c\left(1+\frac{p}{n}\right)^{-\alpha} - \sum_{s=0}^{13}\frac{a_s}{n^s}\right\} + 1 - \frac{6}{n} + \cdots - \frac{b_{13}}{n^{13}}$$

where $c(1+p/n)^{-\alpha}$ is determined from $c_{11}$, $c_{12}$, and $c_{13}$. We find

$$\begin{cases} p = 0.927856214 \\ c = 3.977732609 \\ \alpha = 4.836679845 . \end{cases}$$

For $n = 3$, fitting 4-moment Pearson curves, the effect on the percentage points is shown in Table 2.24.

For the corresponding case with $n = 4$, there is at most 2 units difference in the fourth decimal place, and for $n \geqslant 5$, the three approaches agree to within at most a unit in the fourth decimal place.

For those interested in mathematical curiosities, consider $Z_3$, which is not defined by the $a_s$-series but may be summed from the $b_s$-series (Table 2.21a). Levin's U-algorithm yields 0.280985 for 12 and 0.280963 for 13 terms; a mimicry algorithm (fitting $c(1+p/n)^{-\alpha}$ to the 10th, 11th and 12th terms) gave 0.280951. But $Z_3 = E|z_1z_2z_3(z_1+z_2+z_3)|$ which we have failed to evaluate exactly, now yields by quadrature (5 runs with Gauss-Legendre with intervals such as 1/5, 1/6, 9/80, 2/15, and range $-6$ to 6 in each dimension) values between 0.280891 to 0.280982.

This example clearly points out the advantage of being able to spot dominant singularities in a slowly convergent or divergent series, although when it was done by Geary, one surmises it was mainly a matter of remarkable scientific instinct.

**Table 2.21a Geary's $\omega_n'$**

| $s$ | $a_s$ | $a_s / a_{s-1}$ | $b_s$ | $b_s / b_s$ |
|---|---|---|---|---|
| 0 | 1 | -- | 1 | -- |
| 1 | 0 | -- | 6 | 6 |
| 2 | 12 | -- | 20 | 3.3333 |
| 3 | 32 | 2.6667 | 56 | 2.8000 |
| 4 | 172 | 5.3750 | 133 | 2.3750 |
| 5 | 640 | 3.7209 | 258 | 1.9399 |
| 6 | 12736/5 | 3.9800 | 2136/5 | 1.6558 |
| 7 | 47104/5 | 3.6985 | *3208/5 | 1.5019 |
| 8* | 1204992/35 | 3.6545 | *31498/35 | 1.4027 |
| 9* | 4289536/35 | 3.5598 | *42124/35 | 1.3374 |
| 10* | 45109504/105 | 3.5054 | *4652/3 | 1.2884 |
| 11* | 22255616/15 | 3.4536 | *68016/35 | 1.2532 |
| 12* | $114247139 \cdot 2^8/5775$ | 3.4134 | *13742274/5775 | 1.2245 |
| 13* | $1300699889049 \cdot 2^{13}/13!$ | 3.3787 | 16514004/5775 | 1.2017 |
| 14* | $304979663213568 \cdot 2^{14}/14!$ | 3.3496 | 254507804/75075 | 1.1855 |

(Here
$$E\,|z_1 z_2 z_3 z_4| \sim \frac{a^4(n+1)^{1/2}(n-3)^{7/2}}{(n-2)^4} \sum_{s=0}^{\infty} \frac{(-1)^s a_s}{(n-2)^s}$$
$$\sim \frac{a^4 n(n+1)}{(n-1)(n-2)} \sum_{s=0}^{\infty} \frac{(-1)^s b_s}{n^s}. \qquad (a^2 = 2/\pi)$$

Shank's extrapolate (last five entries) for $a_s / a_{s-1}$ is 3.191789; for $b_s / b_{s-1}$ it is 1.090596.)

**Table 2.21b Series for $(1 + 1/n)^2 E(d^4)$ Geary's Expression (22)**

| $s$ | $c_s$ (exact) | $c_s$ |
|---|---|---|
| 0 | $a^4$ | 0.405285 |
| 1 | $6a^2 - 6a^4$ | 1.388010 |
| 2 | $3 - 16a^2 + 20a^4$ | 0.919778 |
| 3 | $-6 + 45a^2 - 56a^4$ | -0.048055 |
| 4 | $15 - 108a^2 + 133a^4$ | 0.147934 |
| 5 | $-30 + (845/4)a^2 - 258a^4$ | -0.077535 |
| 6 | $51 - 352a^2 + (2136/5)a^4$ | 0.047479 |
| 7 | $-78 + (21233/40)a^2 - (3208/5)a^4$ | -0.096995 |
| 8 | $111 - (7473/10)a^2 + (31498/35)a^4$ | -0.012854 |
| 9 | $-150 + (2243603/2240)a^2 - (42124/35)a^4$ | -0.133783 |
| 10 | $195 - (362207/280)a^2 + (4652/3)a^4$ | -0.067535 |
| 11 | $-246 + (21817811/13440)a^2 - (68016/35)a^4$ | -0.139523 |
| 12 | $303 - (6689057/3360)a^2 + (13742274/5775)a^4$ | 0.044680 |
| 13 | $-366 + (1416880403/591360)a^2 - (16514004/5775)a^4$ | 0.382254 |
| 14 | $435 - (14989081/5280)a^2 + (254507804/75075)a^4$ | 1.672062 |

$(a^2 = 2/\pi)$

Table 2.22 Assessments of $B_n \sim 1 - \frac{6}{n} + \frac{20}{n^2} - \frac{56}{n^3} + \frac{133}{n^4} - \cdots$

| $n$ | 3 | 4 | 5 | 6 | 7 |
|---|---|---|---|---|---|
| Mimicry (M) | 0.115537221 | 0.21798024 | 0.303182804 | 0.3730427424 | 0.4307809471 |
| Levin's U(L) | 0.115536590 | 0.21798023 | 0.303182803 | 0.3730427424 | 0.4307809473 |
| Shanks (S) | 0.115536180 | 0.21798023 | 0.303182803 | 0.3730427424 | 0.4307809472 |
| Direct Sum Corrected | 0.115537142 | 0.21798024 | 0.303182804 | 0.373042730 | 0.4307809476 |

(Direct sum entry derived from (2.100) using series in ($c_s$) in Table 2.21a for the $E(d^4)$ term and evaluating $B_n$ from $Z_n$.)

Table 2.23 $B_3$ Using Shanks ($n = 3$)

| $s$ | Partial Sum | $s^{(1)}$ | $s^{(2)}$ | $s^{(3)}$ |
|---|---|---|---|---|
| 8 | 0.158199 | | | |
| 9 | 0.097053 | | | |
| 10 | 0.123313 | 0.115423 | | |
| 11 | 0.112343 | 0.115576 | | |
| 12 | 0.116821 | 0.115523 | 0.11553645 | |
| 13 | 0.115027 | 0.115540 | 0.11553695 | |
| 14 | 0.115736 | 0.115535 | 0.11553637 | 0.11553618 |

(Successive columns are iterated Shanks' 3-point approximants; see Van Dyke (1974). Entries are reduced from the original output.)

Table 2.24 Percentage Points of $\omega_n'$ Comparison ($n = 3$)

| | Percentage Points | | | | | |
|---|---|---|---|---|---|---|
| | 1% | 5% | 10% | 90% | 95% | 99% |
| M | 0.7577 | 0.7702 | 0.7823 | 0.9591 | 0.9811 | 1.0104 |
| L | 0.7553 | 0.7694 | 0.7823 | 0.9585 | 0.9811 | 1.0126 |
| S | 0.7583 | 0.7706 | 0.7823 | 0.9593 | 0.9811 | 1.0097 |

(M $\equiv$ Mimicry series, L $\equiv$ Levin's algorithm, and S $\equiv$ Direct sum on the series).

**Further Example 2.20**

Noting that

$$\int_0^1 \frac{(1-t)^2 dt}{n+t} = -n - \frac{3}{2} + (n-1)^2 \ln(1 + \frac{1}{n}), \qquad (n > 0)$$

show that

$$\ln(1+\frac{1}{n}) = \frac{1}{n+1} + \frac{1}{2(n+1)^2} + \frac{1}{(n+1)^2}$$
$$\times\left[\frac{1}{3n+}\;\frac{1\cdot 3}{4+}\;\frac{1\cdot 3}{5n+}\;\frac{2\cdot 4}{6+}\;\frac{2\cdot 4}{7n+}\;\frac{3\cdot 5}{8+}\;\frac{3\cdot 5}{8n+}\;\cdots\right].$$

Similarly if

$$I_n(k) = \int_0^1 \frac{(1-t)^k\,dt}{n+t}, \qquad (k = 1, 2, \cdots)$$

deduce that

$$\ln(1+\frac{1}{n}) = \frac{1}{n+1} + \frac{1}{2(n+1)^2} + \cdots + \frac{1}{k(n+1)^k}$$
$$+\frac{1}{(n+1)^k}\left[\frac{1}{(k+1)n+}\;\frac{1\cdot(k+1)}{k+2+}\;\frac{1\cdot(k+!)}{(k+3)n+}\;\frac{2\cdot(k+2)}{k+4+}\right.$$
$$\left.\frac{2\cdot(k+2)}{(k+5)n+}\;\cdots\right]. \qquad (n>0;\; k = 1, 2, \cdots)$$

For $n = 1$, $k = 5$, approximants to $\ln(2)$ are

$$\begin{array}{ll} < & 0.693750 \\ > & 0.693099 \\ < & 0.693155 \\ > & 0.6931463 \\ < & 0.69314732 \\ > & 0.693147163\,. \end{array}$$

**Further Example 2.21**

Let $J(n) = \ln\left[\frac{n+1}{n-1}\right]$, so that

$$\frac{J(n)}{n} = \int_0^1 \frac{t^{-1/2}dt}{n^2-t}. \qquad (n>1)$$

By considering

$$\int_0^1 \frac{t^{-1/2}(n^2-t+1-n^2)^k\,dt}{n^2-t} - (1-n^2)^k \int_0^1 \frac{t^{-1/2}dt}{n^2-t},$$

show that

$$\frac{1}{n}\ln\left|\frac{n+1}{n-1}\right| = \frac{2\cdot 0!}{n^2-1} - \frac{2^2\cdot 1!}{1\cdot 3(n^2-1)^2} + \frac{2^3\cdot 2!}{1\cdot 3\cdot 5(n^2-1)^3} + \cdots$$

$$+ \frac{(-1)^{k-1}2^k(k-1)!}{1\cdot 3\cdots(2k-1)(n^2-1)^k} + \frac{(-1)^k\Gamma(½)\Gamma(k+1)}{\Gamma(k+½)(n^2-1)^k}$$

$$\times \left\{ \frac{1}{n^2(k+\frac{1}{2})-}\;\frac{\frac{1}{2}(k+\frac{1}{2})}{k+\frac{3}{2}-}\;\frac{\frac{2}{2}(k+\frac{2}{2})}{n^2(k+\frac{5}{2})-}\;\frac{\frac{3}{2}(k+\frac{3}{2})}{k+\frac{7}{2}-}\;\cdots \right\}.$$

The c.f. may be written

$$\frac{2}{n^2(2k+1)-}\;\frac{1\cdot(2k+1)}{2k+3-}\;\frac{2\cdot(2k+2)}{n^2(2k+5)-}\;\frac{3(2k+3)}{2k+7-}\;\frac{4(2k+4)}{n^2(2k+9)-}\;\cdots .$$

For $n = 3$ and $k = 1$

$$\begin{aligned}\frac{1}{3}\ln(2) &< 0.2\\ &< 0.231481\\ &< 0.231060\\ &< 0.231049383\\ &< 0.231049069\\ &< 0.231049060 \quad .\end{aligned}$$

**Further Example 2.22**

Using an algebraic computer language we have

$$\ln\left|\frac{e^x-1}{x}\right| = \frac{x}{2} + \frac{x^2}{24+}\;\frac{2x^2}{10+}\;\frac{19x^2}{252+}\;\frac{551x^2}{361+}\;\cdots$$

so that the c.f. arises from

$$\ln\left\{\frac{\sinh(x/2)}{x/2}\right\} = \sum_{n=1}^{\infty}\frac{(-1)^n 2^{2n-1}B_{2n}x^{2n}}{n(2n)!}. \qquad (|x|<\pi)$$

Note that

$$\ln\left\{\frac{\sinh(x)}{x}\right\} = 4\int_0^\infty \frac{\sin^2(xt)dt}{t(e^{2\pi t}-1)}.$$

**Further example 2.23**

If $f_n(x) = \ln\{\prod_{j=1}^{n}(1+jx)\}$, show, for $x > 0$, that

$$f_3(x) = \frac{6x}{1+} \; \frac{\frac{7}{6}x}{1+} \; \frac{\frac{23}{42}x}{1+} \; \frac{\frac{165}{161}x}{1+} \; \frac{\frac{357}{575}x}{1+} \; \cdots,$$

$$f_8(x) = \frac{36x}{1+} \; \frac{\frac{17}{6}x}{1+} \; \frac{\frac{143}{102}x}{1+} \; \frac{\frac{6177}{2431}x}{1+} \; \frac{\frac{34527}{20735}x}{1+} \; \cdots.$$

Note that

$$f_n(x) = \int_0^x \left\{ \frac{1}{1+t} + \frac{2}{1+2t} + \cdots + \frac{n}{1+nt} \right\} dt = \sum_{j=1}^{n} \int_0^{jx} \frac{dt}{1+t}.$$

**Further Example 2.24**

If $L(x) = \dfrac{\ln(1+x)}{1+x} = \dfrac{x}{1+} \; \dfrac{a_1 x}{1+} \; \dfrac{a_2 x}{1+} \; \cdots$ then

| $s$ | $a_s$ | | | $s$ | $a_s$ | | |
|---|---|---|---|---|---|---|---|
| 1 | 3/2 | = | 1.5 | 7 | 4940/14147 | = | 0.133 |
| 2 | -5/18 | = | -0.28 | 8 | 5633/34827 | = | 0.162 |
| 3 | 17/45 | = | 0.378 | 9 | 31960/97071 | = | 0.329 |
| 4 | 6/85 | = | 0.07 | 10 | 17537/97988 | = | 0.179 |
| 5 | 129/340 | = | 0.379 | 11 | 4192/13277 | = | 0.316 |
| 6 | 799/6020 | = | 0.133 | | | | |

Thus, is $\{x - (1 + a_1 x) L(x)\} / \{x - L(x)\}$ a Stieltjes transform?

By contrast,

$$\frac{x \ln(1+x)}{x+2} = \frac{1}{w+} \; \frac{1\cdot 1}{3+} \; \frac{1\cdot 1}{5w+} \; \frac{2\cdot 3}{7+} \; \frac{2\cdot 3}{9w+} \; \frac{3\cdot 5}{11+} \; \frac{3\cdot 5}{13w+} \; \cdots$$

$$= \int_0^{1/4} \frac{1}{\sqrt{(1-4t)}} \, \frac{dt}{t+w}. \qquad (w = (1+x)/x^2)$$

**Further Example 2.25**

Set up the first few terms of

$$L^*(x) = \frac{2\ln(1+x) + \ln^2(1+x)}{6 + 4\ln(1+x)}.$$

Thus

$$L^*(x) = \frac{x/3}{1+}\ \frac{2x/3}{1+}\ \frac{x/4}{1+}\ \frac{x/2}{1+}\ \frac{11x/40}{1+}\ \frac{487x/1320}{1+}$$
$$\frac{0.232148x}{1+}\ \frac{0.304172x}{1+}\ \frac{0.223951x}{1+}\ \frac{0.284238x}{1+}\ \frac{0.226385x}{1+}$$
$$\frac{0.276010x}{1+}\ \frac{0.229666x}{1+}\ \frac{0.271316x}{1+}\ \frac{0.23237x}{1+}\ \frac{0.268144x}{1+}$$
$$\frac{0.234492x}{1+}\ \frac{0.265824x}{1+}\ \frac{0.236159x}{1+}\ \frac{0.263926x}{1+}\ \frac{0.239894x}{1+}$$
$$\frac{0.233943x}{1+}\ \cdots .$$

**Further Example 2.26**

(a) $$\int_0^\infty \frac{e^{-t/x}\,dt}{(1+t)^n} = \frac{x}{1+}\ \frac{nx}{1+}\ \frac{x}{1+}\ \frac{(n+1)x}{1+}\ \frac{2x}{1+}\ \cdots$$
$$= \frac{x}{1+nx-}\ \frac{nx^2}{1+(n+2)x-}\ \frac{2(n+1)x^2}{1+(n+4)x-}\ \frac{3(n+2)x^2}{1+(n+6)x-}\ \cdots .$$

(b) $$\int_0^\infty \frac{\cosh(at)e^{-t/x}\,dt}{\cosh(t)}$$
$$= \frac{x}{1+}\ \frac{(1^2-a^2)x^2}{1+}\ \frac{2^2x^2}{1+}\ \frac{(3^2-a^2)x^2}{1+}\ \frac{4^2x^2}{1+}\ \cdots .$$

$$\int_0^\infty \frac{\sinh(at)e^{-t/x}\,dt}{\cosh(t)}$$
$$= \frac{ay}{1+}\ \frac{(2^2-a^2)y}{1+}\ \frac{2^2y}{1+}\ \frac{(4^2-a^2)y}{1+}\ \frac{4^2y}{1+}\ \cdots .$$
$$(y = x^2/(1-x^2))$$

(c) $$\exp\left(\int_0^\infty \frac{\tanh(t)e^{-t/x}\,dt}{t}\right) = \frac{1}{1-x+}\ \frac{x^2}{2(1-x)+}\ \frac{3x^2}{2(1-x)+}\ \cdots .$$

(d) $$\exp\left\{\int_0^\infty \left(1-\frac{\cosh(2at)}{\cosh(2t)}\right)\frac{e^{-t/x}\,dt}{t}\right\}$$
$$= 1+\frac{2(1-a^2)x^2}{1+}\ \frac{(3^2-a^2)x^2}{1+}\ \frac{(5^2-a^2)x^2}{1+}\ \cdots .$$

(e) $$\tanh\left\{\frac{1}{2}\int_0^\infty \frac{\sinh(2at)e^{-t/x}\,dt}{t\cosh(t)}\right\}$$

$$= \frac{ax}{1+}\;\frac{(1^2-a^2)x^2}{1+}\;\frac{(2^2-a^2)x^2}{1+}\;\frac{(3^2-a^2)x^2}{1+}\;\cdots.$$

(f) $$\tanh\left\{\int_0^\infty \frac{\sinh(at)e^{-t/x}\,dt}{t\cosh(t)}\right\}$$

$$= \frac{ax}{1+}\;\frac{1^2x^2}{1+}\;\frac{(2^2-a^2)x^2}{1+}\;\frac{3^2x^2}{1+}\;\frac{(4^2-a^2)x^2}{1+}\;\cdots.$$

(Rogers 1907a, 1907b)

**Further Example 2.27**

Find the first few terms of the c.f. in $y = (1+x)/x^2$, for $\exp\{\ln^2(1+x)\}$. If

$$\exp\{\ln^2(1+x)\} = 1 + \frac{1}{y-a_1+}\;\frac{a_1a_2}{y-a_3+a_2+}\;\frac{a_3a_4}{y-a_5+a_4+}\;\cdots$$

and $a_{2s} = p_{2s}p_{2s+1}$, $a_{2s+1} = (1-p_{2s+1})p_{2s}$, $(a_1 = p_1p_2)$ then

| $s$ | $p_s$ |
|---|---|
| 1 | 1 |
| 2 | 5/12 |
| 3 | 57/125 |
| 4 | 32545/162792 |
| 5 | 26218012/42367081 |
| 6 | 197650491007/1602119481644 |
| 7 | 13213240867501299/17980504099598215 5 |

Investigate $[\exp\{\ln^2(1+x)\}-1]/x^2$ in the form

$$\frac{1}{y-}\;\frac{r_1}{1+}\;\frac{r_2}{y-}\;\frac{r_3}{1+}\;\frac{r_4}{y-}\;\cdots.$$

What can be said about $r_1, r_2, \cdots$ ?

**Further Example 2.28**

In (2.21) the J-fraction is given for

$$H_n(k;\sigma) = \int_0^\infty \frac{(t+k)d\,\sigma(t)}{n+t}.$$

From Section 2.7.2, the Stieltjes form may be constructed after minor modifications. Thus

$$H_n(k;\sigma) = \frac{Q_0(k)}{n+}\;\frac{P_1(k)}{1+}\;\frac{Q_1(k)}{n+}\;\cdots, \qquad (n,k>0)$$

where

$$\text{(a)}\quad P_s(k) = p_s\left\{\frac{\omega_{s+1}(k) - p_{s+1}\omega_s(k)}{\omega_s(k) - p_s\omega_{s-1}(k)}\right\}\frac{\omega_{s-1}(k)}{\omega_s(k)},$$

$$Q_s(k) = q_s\left\{\frac{\omega_s(k) - p_s\omega_{s-1}(k)}{\omega_{s+1}(k) - p_{s+1}\omega_s(k)}\right\}\frac{\omega_{s+1}(k)}{\omega_s(k)},$$

$$Q_0(k) = b_0\omega_1(k); \qquad (s = 1, 2, \cdots)$$

$$\text{(b)}\quad \int_0^\infty \frac{d\,\sigma(t)}{n+t} = \frac{b_0}{n+}\;\frac{p_1}{1+}\;\frac{q_1}{n+}\;\cdots.$$

Note that if $k = n\ (>0)$ there is the identity

$$1 = \frac{\omega_1(n)}{n+}\;\frac{P_1(n)}{1+}\;\frac{Q_1(n)}{n+}\;\frac{P_2(n)}{1+}\;\frac{Q_2(n)}{n+}\;\cdots$$

where for example

$$\omega_1(k) = k + p_1,$$

$$\omega_2(k) = (k + p_2 + q_1)\omega_1(k) - p_1q_1\omega_0(k). \qquad (\omega_0(k) = 1)$$

**Further Example 2.29**

$$1 - \frac{x}{(1+x)\ln(1+x)} = \frac{(1/2)x}{1+}\;\frac{(5/6)x}{1+}\;\frac{(1/15)x}{1+}\;\frac{(2/5)x}{1+}\;\frac{(3/20)x}{1+}$$

$$\frac{(47/140)x}{1+}\;\frac{(60/329)x}{1+}\;\frac{(131/423)x}{1+}$$

$$\frac{(235/1179)x}{1+}\;\cdots,$$

$$\sim \frac{x}{2} - \frac{5x^2}{112} + \frac{3x^3}{8} - \frac{251x^4}{720} + \frac{95x^5}{288}$$

$$- \frac{19087x^6}{60480} + \frac{5257x^7}{17280} - \cdots.$$

**Further Example 2.30**

$$\ln\left(\frac{24+18x+x^2}{24+6x}\right) = \frac{\frac{x}{2}}{1+}\;\frac{\frac{5x}{12}}{1+}\;\frac{\frac{x}{12}}{1+}\;\frac{\frac{x}{5}}{1+}\;\frac{\frac{9x}{40}}{1+}\;\frac{\frac{389x}{3240}}{1+}\;\frac{\frac{182269x}{1102815}}{1+}\;\frac{\frac{530036613x}{1985273948}}{1+}\;\cdots .$$

It turns out that from using an algebraic program, the first fifteen partial numerators are positive. What is the explanation and can it be proved that this holds in general.

**Further Example 2.31**

The Johnson $S_B$ distribution (Bowman and Shenton, 1988, p95) is defined by the real transformation

$$X = \zeta + \lambda y\,, \qquad (y = (\theta e^{-z/\delta}+1)^{-1};\; z \in N(0,1);\; \theta = e^{\gamma/\delta},\; \delta > 0)$$

so that for the mean value of the variate $y$,

$$E(y) = \frac{1}{\sqrt{(2\pi)}}\int_{-\infty}^{\infty}\frac{e^{-z^2/2}dz}{(\theta e^{-z/\delta}+1)}\,.$$

The substitution $t = \exp(-z/\delta)$ leads to a Stieltjes transform. A series (with lacunas), and continued fraction may be set up; thus

$$\begin{aligned} E(y) &\sim 1-\theta q+\theta^2q^4-\theta^3q^9+\theta^4q^{16}-\cdots \\ &\sim \frac{1}{1+}\;\frac{q\theta}{1+}\;\frac{q(q^2-1)\theta}{1+}\;\frac{q^5\theta}{1+}\;\frac{q^3(q^4-1)\theta}{1+}\;\frac{q^9\theta}{1+} \\ &\quad\frac{q^5(q^6-1)\theta}{1+}\;\frac{q^{13}\theta}{1+}\;\cdots, \qquad (q=\exp\{1/(2\delta^2)\}>1,\ \theta>0) \end{aligned}$$

the c.f. being of the form studied by Heine (see Perron, 1957, p125). Bounds are provided by the c.f.; for example, when $\delta = 4.241$, $\gamma = 0$, $0.493 < 0.4998 < 0.4999 < E(y) < 0.50001 < 0.50023 < 0.507 < 1$. However they deteriorate as $\delta$ decreases.

From theorem 12.11f (Henrici, 1977) there are the bounds for the difference between successive convergents

$$|C_{2s}-C_{2s-1}| \leqslant \{\tfrac{1}{2}(1+\sqrt{5})\}^{1/2}\prod_{k=1}^{s}\left(1+\frac{2\sigma}{q^{2k-3/2}}\right)^{-1/2}\prod_{k=1}^{s-1}\left(1+\frac{2\sigma}{q^{k-1/2}\sqrt{(q^{2k}-1)}}\right)^{-1/2}$$

where $\sigma = 1/\sqrt{\theta}$, and a similar expression for $|C_{2s+1}-C_{2s}|$ (see Appendix III). The product terms ultimately converge to a positive quantity. Thus the series for $E(y)$ diverges by infinite oscillation, whereas the c.f. provides valid bounds but without closure.

# 3.

# SECOND-ORDER CONTINUED FRACTIONS

## 3.1 Introduction

Stieltjes showed (Shohat and Tamarkin, 1943, p. 75) that when there is a solution to the Stieltjes' moment problem mentioned in 1.7, then there is a unique $\psi(t)$ such that

$$F(z,\psi) - \frac{\chi_{2s}}{\omega_{2s}} = \min_{P_s} \int_0^\infty \frac{P_s^2(t)}{z+t}\, d\psi(t), \qquad (z > 0) \tag{3.1}$$

$$z\left\{\frac{\chi_{2s+1}}{\omega_{2s+1}} - F(z,\psi)\right\} = \min_{P_s} \int_0^\infty \frac{P_s^2(t)}{z+t}\, t\, d\psi(t),$$

where

$$F(z,\psi) = \int_0^\infty \frac{d\psi(t)}{z+t},$$

and the minimum is over all polynomials $P_s(t)$ of degree $\leqslant s$ which assume the value unity at $t = -z$; $\chi_s$, $\omega_s$ being the numerator and denominator of the $s$ th convergent of the Stieltjes fraction (1.21).

## 3.2 Generalization

A change of emphasis arises from

$$\min_{\lambda} \int_0^\infty C_n(t)\{A(t)/C_n(t) - \sum_{s=0}^{r} \lambda_s P_s(t)\}^2\, d\psi(t) \tag{3.2}$$

where

(i) $\quad C_n(t) = \prod_{s=1}^{n}(t + z_s) > 0$ for $t > 0$,

(ii) $\{P_s(t)\}$ is a polynomial base ($t^s$ for example, or the orthogonal system associated with $d\psi(t)$, or the orthogonal system associated with $C_n(t)d\psi(t)$),

(iii) the minimum is over $\{\lambda_s\}$,

(iv) $A(t)$ is a real function; generally a polynomial of finite degree.

The formalization leads to approximants (rational fractions in the symmetric functions $\sum z_s$, $\sum_{s\neq t} z_s z_t$, etc.) to the function

$$F_A(z) = \int_0^\infty \frac{A^2(t)d\psi(t)}{C_n(t)}, \tag{3.3}$$

which for $A(t) = 1$, $n = 1$ reduces to the c.f. for $F(z,\psi)$. A more general form would point to sequences of approximants to the inner-product type integral

$$F_{A,B}(z) = \int_0^\infty \frac{A(t)B(t)d\psi(t)}{C_n(t)}. \tag{3.4}$$

Let it be pointed out immediately that generalizations such as (3.3) are natural extensions of the matrix approach in Section 1.2. For the terminating c.f.

$$\phi(z) = \frac{b_0}{a_1+z-}\,\frac{b_1}{a_2+z-}\cdots\frac{b_{n-1}}{a_n+z} \tag{3.5}$$

is the value of $y_0$ for the system of $n$ equations

$$\begin{aligned}
(a_1+z)y_0 + b_1y_1 \qquad\qquad &= b_0\\
y_0 + (a_2+z)y_1 + b_2y_2 &= 0\\
\cdot\quad\cdot\quad\cdot\quad\cdot\quad\cdot\qquad &\\
y_{n-2} + (a_n+z)y_{n-1} &= 0.
\end{aligned} \tag{3.6}$$

Thus $y_0/b_0$ is the leading element in the inverse of the matrix $(z\underset{\sim}{I} + \underset{\sim}{M})$, where $\underset{\sim}{I}$ is a unit matrix of order $n$ and

$$\underset{\sim}{M} = \begin{bmatrix} a_1 & b_1 & \cdot & \cdot & \\ 1 & a_2 & b_2 & \cdot & \\ & & & & \\ & & 1 & a_{n-1} & b_{n-1} \\ & & & 1 & a_n \end{bmatrix}. \tag{3.7}$$

So $\phi(z)$ arises from

$$(z\underset{\sim}{I} + \underset{\sim}{M})\underset{\sim}{y}' = \underset{\sim}{U}' \tag{3.8}$$

where $\underset{\sim}{y} = [y_0, y_1, \cdots, y_{n-1}]$, $\underset{\sim}{U} = [b_0, 0, 0, \cdots]$.

At this point it is advantageous in view of developments to consider $\phi(z)$ as the value of $y_0$ in the system

$$(z\underset{\sim}{I} + \underset{\sim}{M}^*)\underset{\sim}{y}' = \underset{\sim}{U}', \tag{3.9}$$

where $\underset{\sim}{M}^*$ is now the symmetric matrix

$$\begin{bmatrix} a_1 & \sqrt{b_1} & \cdot & \cdot & \\ \sqrt{b_1} & a_2 & \sqrt{b_2} & \cdot & \\ & & & & \\ \cdot & & & \cdot & \sqrt{b_{n-1}} \\ \cdot & & & \sqrt{b_{n-1}} & a_n \end{bmatrix},$$

the question of the signs of $b_1, b_2, \cdots$ being of little importance.

An extension to several variables is now obvious. The *second order* case considers

$$(z_1\underset{\sim}{I} + \underset{\sim}{M}^*)(z_2\underset{\sim}{I} + \underset{\sim}{M}^*)\underset{\sim}{y}' = \underset{\sim}{U}' \tag{3.10}$$

so that

$$\underset{\sim}{y}' = \{(z_1\underset{\sim}{I} + \underset{\sim}{M}^*)(z_2\underset{\sim}{I} + \underset{\sim}{M}^*)\}^{-1}\underset{\sim}{U}', \tag{3.11}$$

and since the matrices are commutative,

$$(z_2 - z_1)\underset{\sim}{y}' = \left\{\frac{\underset{\sim}{I}}{z_1\underset{\sim}{I} + \underset{\sim}{M}^*} - \frac{\underset{\sim}{I}}{z_2\underset{\sim}{I} + \underset{\sim}{M}^*}\right\}\underset{\sim}{U}' \tag{3.12}$$

Hence the leading element in $\underset{\sim}{y}'$ is $\{\phi(z_1) - \phi(z_2)\}/(z_2 - z_1)$; the confluent case being that the leading element in $(z\underset{\sim}{I} + \underset{\sim}{M}^*)^{-2}\underset{\sim}{U}'$ is $-d\,\phi(z)/dz$.

This is a version of an almost forgotten paper of E. T. Whittaker (1916), who was attempting to remove certain obvious deficiencies (at least those conceptualized at the time) of continued fraction mathematics; for example, the deficiencies of non-additivity, of non-differentiability and so on, displaying an unfavorable comparison to the additivity, differentiability, etc. of Taylor series. It was to be several decades later that the positive attributes, such as the gain in the domain of convergence (often the split plane compared to the unit circle), the power to sum slowly convergent or seemingly divergent series, were to become part of conventional thinking.

In a personal communication (Shenton, 1959), A. C. Aitken had these comments: "Whittaker had an excellent idea in continued fractions - I mean about the leading element in $(xI+M)^{-1}$ - but went about it in a rather old-fashioned way. I have somewhere in my private MS notes all that, much more simply, and a good deal more; I have known it ever since 1923, when first I came to this country and secured an off-print of his paper". It is of interest to recall that linear algebra was a very fashionable branch of pure mathematics in the 1920's and 1930's, and Scotland boasted the three eminent scientists A. C. Aitken, H. W. Turnbull, and E. T. Whittaker. The unmistakable underlying preoccupation with so-called "old-fashioned" mathematics is discernible in a contribution to Whittaker's paper by Turnbull (1932), who sets out the basic tie-up between the leading element in a matrix inverse and continued fractions. For Aitken's published remarks on the Whittaker approach see Appendix II.

For example, Turnbull (1932) (according to some, the innovator of modern matrix notation and its popularization in Britain) makes a point of deriving a continued fraction for the derivative of

$$f(x) = \frac{1}{b_0+x-}\ \frac{a_1}{b_1+x-}\ \frac{a_2}{b_2+x-}\ \cdots\ \frac{a_n}{b_n+x}.$$

However, his generalization, consisting of a study of the leading element in the inverse $C^{-1}$ of $C = A+Bx$ (where $A$ and $B$ are each of the type $M$), does not seem to have been fruitful, so far. He gives an expression for $dC/dx$, namely, under certain conditions

$$\frac{dC}{dx} = -(A+Bx)^{-1}A_1(A+Bx)^{-1}BA_1^{-1}, \qquad (|A_1| \neq 0)$$

when $C = A_1 + B(x-x_1)$, and $C$ is non-singular at $x = x_1$.

It is not known to us how far Aitken was informed of Turnbull's work; he did not mention it in his communication.

To appreciate the structure of the generalized form in (3.11) let the determinant of $(z_1\underset{\sim}{I} + \underset{\sim}{M}^*)(z_2\underset{\sim}{I} + \underset{\sim}{M}^*)$ be $|P_{ij}|_{i,j=0}^{n-1}$, so that, for example

$$\begin{aligned} P_{00} &= (a_1+z_1)(a_1+z_2)b_1, \\ P_{01} &= P_{10} = (z_1+z_2+a_1+a_2)\sqrt{b_1}, \\ P_{02} &= P_{20} = \sqrt{(b_1b_2)}, \qquad\qquad (3.13) \\ P_{11} &= (a_2+z_1)(a_2+z_2)+b_1+b_2, \end{aligned}$$

$$P_{12} = P_{21} = (z_1 + z_2 + a_2 + a_3)\sqrt{b_2}\,,$$
$$P_{13} = P_{31} = \sqrt{(b_2 b_3)}\,.$$

The determinant is symmetric and $P_{ij} = P_{ji}$, where in general

$$\begin{aligned}
P_{i,i} &= (a_{i+1} + z_1)(a_{i+1} + z_2) + b_i^* + b_{i+1}\,, \qquad (i = 0, 1, \cdots, n-1)\\
P_{i,i+1} &= (z_1 + z_2 + a_{i+1} + a_{i+2})\sqrt{b_{i+1}}\,,\\
P_{i,i+2} &= \sqrt{(b_{i+1} b_{i+2})}\,;\\
P_{i,j} &= 0\,, \qquad (j > i+2);\\
b_i^* &= b_i\,, \ (i > 0); \qquad b_0^* = 0;\\
b_i &= 0, \ (i > n-1); \qquad a_i = 0, \ (i > n).
\end{aligned} \tag{3.14}$$

Approximants to $\{\phi(z_1) - \phi(z_2)\}/(z_2 - z_1)$ are now

$$\begin{aligned}
P[0\,|\,1] &= b_0 / P_{00}\,,\\
P[1\,|\,2] &= b_0 P_{11} / (P_{00} P_{11} - P_{10}^2)\,,\\
P[2\,|\,3] &= \frac{b_0 (P_{11} P_{22} - P_{12}^2)}{(P_{00} P_{11} P_{22} - P_{00} P_{21}^2 - P_{11} P_{20}^2 - P_{22} P_{10}^2 + 2 P_{10} P_{20} P_{21})}\,,
\end{aligned}$$

the final one having $|P_{ij}|$ itself as denominator, and $b_0$ times its principal minor as numerator.

The extension to three and more matrix products is clear. For example, for the *third order* case

$$\underset{\sim}{y}' = \{(z_1 \underset{\sim}{I} + \underset{\sim}{M}^*)(z_2 \underset{\sim}{I} + \underset{\sim}{M}^*)(z_3 \underset{\sim}{I} + \underset{\sim}{M}^*)\}^{-1} \underset{\sim}{U}' \tag{3.15}$$

will have approximants arising from the leading term in the matrix product reciprocal, the related determinant $|P_{ij}|$ having elements ($P_{ij} = P_{ji}$)

$$\begin{aligned}
P_{i,i} &= (a_{i+1} + z_1)(a_{i+1} + z_2)(a_{i+1} + z_3) + p\,(b_i^* + b_{i+1}) + b_{i+1} a_{i+2}\\
&\qquad + b_i^* a_i + 2 a_{i+1} (b_i^* + b_{i+1})\,,\\
P_{i,i+1} &= \{q + p\,(a_{i+1} + a_{i+2}) + a_{i+2}^2 + a_{i+1} a_{i+2} + a_{i+1}^2\\
&\qquad + b_i^* + b_{i+1} + b_{i+2}\}\sqrt{b_{i+1}}\,,\\
P_{i,i+2} &= \{p + a_{i+1} + a_{i+2} + a_{i+3}\}\sqrt{(b_{i+1} b_{i+2})}\,,\\
P_{i,i+3} &= \sqrt{(b_{i+1} b_{i+2} b_{i+3})}\,;\\
b_i^* &= b_i\,, \ (i > 0); \ \ b_0^* = 0; \ \ b_i = 0, \ (i > n-1); \ \ a_i = 0, \ i > n\,;\\
(x + z_1)(x + z_2)(x + z_3) &\equiv x^3 + p x^2 + q x + r\,.
\end{aligned} \tag{3.16}$$

The approximants finally take the value

$$\begin{vmatrix} 1 & 1 & 1 \\ z_1 & z_2 & z_3 \\ \phi(z_1) & \phi(z_2) & \phi(z_3) \end{vmatrix} \div \begin{vmatrix} 1 & 1 & 1 \\ z_1 & z_2 & z_3 \\ z_1^2 & z_2^2 & z_3^2 \end{vmatrix} .$$

Note that approximants in the second-order case (3.10) are ratios of five-diagonal determinants, those for the third-order (3.15) are seven-diagonal determinants, whereas those for the traditional first-order case (3.8) are three-diagonal determinants.

We next consider that happens when infinite orders are involved. Whittaker gave details of the determinant ratio in the second-order confluent case (see his (19), p. 254) and no other illustration. Quoting from his introduction, "The ultimate purpose of this work lies in its application to function theory and the solution of differential equations; the continued fractions are then non-terminating, and the determinants associated with them are of infinite order. But as the present paper is occupied only with the formal algorithms, the order may for convenience be supposed to be finite, and questions relating to the convergence of infinite processes need not here be considered."

## 3.3 Infinite Continued Fractions

These can be envisaged for the formulations discussed in Section 2.1, and we can assume that $\phi(z)$ is the even part of a Stieltjes fraction. Thus, in the first-order case

$$F(z) = \frac{b_0}{z+a_1-}\;\frac{b_1}{z+a_2-}\;\cdots \tag{3.17}$$

is associated with

$$F(z) = \int_0^\infty \frac{d\psi(t)}{z+t};$$

in the second-order case

$$F(z_1, z_2) = \int_0^\infty \frac{d\psi(t)}{(z_1+t)(z_2+t)} \tag{3.18}$$

has a sequence of approximants $\{t_s\}$, where

$$t_s = \frac{b_0 K_{s-1}(\gamma_1, \beta_1, \alpha_1)}{K_s(\gamma_0, \beta_0, \gamma_0)}, \tag{3.19}$$

where $K_s(\gamma_0, \beta_0, \gamma_0)$ is a five-diagonal symmetric determinant of order $s$ with elements

$$\gamma_i = P_{i,i}, \quad \beta_i = P_{i,i+1}, \quad \alpha_i = P_{i,i+2}$$

in the notation of (3.13) and (3.14), and $K_{s-1}(\gamma_1, \beta_1, \alpha_1)$ is the principal minor determinant of $K_s(\gamma_0, \beta_0, \alpha_0)$.

Similarly convergents of the third-order c.f. relating to

$$F(z_1, z_2, z_3) = \int_0^\infty \frac{d\psi(t)}{(t+z_1)(t+z_2)(t+z_3)}, \tag{3.20}$$

based on the parameters in the J.f. in (3.17), are defined by

$$t_s = \frac{b_0 K_{s-1}(\delta_1, \gamma_1, \beta_1, \alpha_1)}{K_s(\delta_0, \gamma_0, \beta_0, \alpha_0)} \tag{3.21}$$

where

$$\delta_i = P_{i,i}, \quad \gamma_i = P_{i,i+1}, \quad \beta_i = P_{i,i+2}, \quad \alpha_i = P_{i,i+3}$$

in the notation of (3.16). The numerator and denominator terms are 7-diagonal determinants, the numerator being the principal minor of the denominator, and related to the leading term in segments of $\{(z_i \underset{\sim}{I} + \underset{\sim}{M}^*)(z_2 \underset{\sim}{I} + \underset{\sim}{M}^*)(z_3 \underset{\sim}{I} + M^*)\}$, $\underset{\sim}{M}^*$ being the modified form of $\underset{\sim}{M}^*$ in (3.9).

**Example 3.1**

$$F(z) = \int_0^\infty \frac{e^{-t}\,dt}{z+t}$$

$$= \frac{1}{z+1-}\;\frac{1^2}{z+3-}\;\frac{2^2}{z+5-}\;\frac{3^2}{z+7-}\;\cdots .$$

<u>Matrix</u>

$$\underset{\sim}{M}^* + i\underset{\sim}{I} = \begin{bmatrix} 1+i & 1 & 0 & \\ 1 & 3+i & 2 & \cdot \\ 0 & 2 & 5+i & \cdot \\ & \cdot & \cdot & \cdot \\ & & \cdot & \cdot \end{bmatrix}. \tag{3.22}$$

Matrix Product

$$(\underset{\sim}{M}^* + i\underset{\sim}{I})(\underset{\sim}{M}^* - i\underset{\sim}{I}) = \begin{bmatrix} 3 & 4 & 2 & 0 & 0 & \\ 4 & 15 & 16 & 6 & 0 & \cdot \\ 2 & 16 & 39 & 36 & 12 & \cdot \\ 0 & 6 & 36 & 75 & 49 & \cdot \\ 0 & 0 & 12 & 49 & 123 & \cdot \\ & \cdot & \cdot & \cdot & \cdot & \cdot \end{bmatrix}. \tag{3.23}$$

Approximants $\{t_s\}$ to

$$\frac{F(i) - F(-i)}{(-2i)} = \int_0^\infty \frac{e^{-t}\,dt}{1+t^2} \sim 0.6214\,.$$

| Sequence of Approximants | | | | |
|---|---|---|---|---|
| 1 | 2 | 3 | 4 | 5 |
| 1/3 | 15/29 | 329/559 | 10743/17553 | 498973/806651 |
| 0.3333 | 0.5172 | 0.5886 | 0.6120 | 0.6186 |

**Example 3.2**

$$F(z,z) = \int_0^\infty \frac{e^{-t}\,dt}{(z+t)^2}$$

and is $-dF/dz$ from Example 3.1.

Matrix

$$\underset{\sim}{M}^* + \underset{\sim}{I} = \begin{bmatrix} 2 & 1 & 0 & 0 & \\ 1 & 4 & 2 & 0 & \cdot \\ 0 & 2 & 6 & 3 & \cdot \\ 0 & 0 & 3 & 8 & \cdot \\ & \cdot & \cdot & \cdot & \cdot \end{bmatrix}. \tag{3.24}$$

Matrix Product

$$(\underset{\sim}{M}^* + I)^2 = \begin{bmatrix} 5 & 6 & 2 & 0 & 0 & & \\ 6 & 21 & 20 & 6 & 0 & \cdot & \\ 2 & 20 & 49 & 42 & 12 & \cdot & \cdot \\ 0 & 6 & 42 & 89 & 72 & \cdot & \cdot \\ 0 & 0 & 12 & 72 & 141 & \cdot & \cdot \\ & \cdot & \cdot & \cdot & \cdot & \cdot & \cdot \end{bmatrix}. \tag{3.25}$$

Approximants $\{t_s\}$ to $F(1,1) \sim 0.4037$ .

| Sequence of Approximants | | | | |
|---|---|---|---|---|
| 1 | 2 | 3 | 4 | 5 |
| 1/5 | 21/69 | 629/1777 | 27253/72113 | 1634721/4192941 |
| 0.2 | 0.3043 | 0.3540 | 0.3779 | 0.3899 |

C.f. approximants from $\frac{1}{1+}\;\frac{2}{1+}\;\frac{1}{1+}\;\frac{3}{1+}\;\frac{2}{1+}\;\frac{4}{1+}\;\frac{3}{1+}\;\cdots$ .

| | Increasing | | Decreasing | |
|---|---|---|---|---|
| 1 | 1/3 | 0.3333 | 1/1 | 1.0 |
| 2 | 5/13 | 0.3846 | 2/4 | 0.5 |
| 3 | 29/73 | 0.3973 | 9/21 | 0.4286 |
| 4 | 201/501 | 0.4012 | 56/136 | 0.4118 |
| 5 | 1631/4051 | 0.4026 | 425/1045 | 0.4067 |

**Example 3.3**

$$F(z) = \frac{1}{2\pi}\int_0^4 \frac{\sqrt{(4/t-1)}\,dt}{z+t}$$

$$= \frac{1}{z+1-}\;\frac{1}{z+2-}\;\frac{1}{z+2-}\cdots ,$$

$$F(z) = -\frac{1}{2}+\frac{1}{2}\sqrt{(1+4/z)}. \qquad (z>0)$$

Matrix Product

$$(\underset{\sim}{M}^* + i\underset{\sim}{I})(\underset{\sim}{M}^* - i\underset{\sim}{I}) = \begin{vmatrix} 3 & 3 & 1 & 0 & 0 & & \\ 3 & 7 & 4 & 1 & 0 & \cdot & \\ 1 & 4 & 7 & 4 & 1 & \cdot & \cdot \\ 0 & 1 & 4 & 7 & 4 & \cdot & \cdot \\ 0 & 0 & 1 & 4 & 7 & \cdot & \cdot \\ & \cdot & \cdot & \cdot & \cdot & \cdot & \cdot \end{vmatrix} . \tag{3.26}$$

Approximants $\{t_s\}$ to $F(i,-i) = \frac{1}{2}\sqrt{\{\frac{1}{2}(\sqrt{17}-1)\}} \sim 0.6248$ .

| Sequence of Approximants | | | | | |
|---|---|---|---|---|---|
| 1 | 2 | 3 | 4 | 5 | 6 |
| 1/3 | 7/12 | 33/53 | 144/231 | 624/1000 | 2704/4329 |
| 0.3333 | 0.5833 | 0.6226 | 0.6234 | 0.6240 | 0.6246 |

After some algebra, it can be shown that

$$F(ix,-ix) = \frac{1}{2x}\sqrt{\left\{\frac{1}{2}\sqrt{\left(1+\frac{16}{x^2}\right)}-\frac{1}{2}\right\}}, \qquad (x>0)$$

with approximants $\{t_s\}$, where $t_s = k_s^*/k_s$, given by

$$k_s^* = \left(\frac{x^{2s-2}}{2^{4s+3}r}\right)\{r_1^s(R_{11}^{2s+2}+R_{12}^{2s+2})+r_2^s(R_{21}^{2s+2}+R_{22}^{2s+2})\},$$

$$k_s = \left(\frac{x^{2s}}{2^{4s+7/2}r}\right)\{r_1^{s+1}(R_{11}^{2s+2}-R_{12}^{2s+2})-ir_2^{s+1/2}(R_{21}^{2s+2}-R_{22}^{2s+2})\},$$

and

$$r_1 = \sqrt{(r+1)}, \quad r_2 = \sqrt{(r-1)}, \quad r = \sqrt{(1+16/x^2)};$$
$$R_{11} = r_1+\sqrt{2}, \quad R_{12} = r_1-\sqrt{2},$$
$$R_{21} = r_2+i\sqrt{2}, \quad R_{22} = r_2-i\sqrt{2}.$$

*Hint.* Set up the recurrence for the $s$ th order five diagonal determinant in (3.26).

## 3.4 Convergence Questions

What are the conditions under which these higher order c.f.s converge? The general question of convergence of ordinary c.f.s has proved a difficult field, partly because the structure of the partial numerators and denominators can be so diversified (for example, Perron (1957) discusses c.f.s $\frac{a_1}{b_1+}\ \frac{a_2}{b_2+}\ \cdots$ where $a_s = a+bs+cs^2$, $b_s = d+es$). It is perhaps true to say that a unified theory of convergence (if there is one) is still lacking. However, our interests lie mainly with c.f.s associated with definite integrals (relating to the moment problems discussed in Section 1.7), and especially those defined on the positive real axis. Again, we point out that the use of c.f.s (which embrace the Padé table complex) in summing series is often applied when only a finite number of terms is available. Thus, convergence questions although not entirely irrelevant, are not as important as error analysis; it becomes more a question of seeking out subsets of sequences pointing to an assessment lying in a certain interval. Needless to say, one is well aware of the pitfalls of relying on a finite number of terms, and of the danger of drawing conclusions from the closeness of contiguous approximants.

We consider the Stieltjes' case, and in particular the mean square formulation in (3.2), for the $n$ th order c.f. for

$$F_Z(z) = \int_0^\infty \frac{A^2(t)d\psi(t)}{(z_1+t)(z_2+t)\cdots(z_n+t)} \tag{3.27}$$

where the denominator $C_n(t) > 0$ for $0 < t < \infty$. The key evidently lies in Parseval's formula.

First of all for a function $f(x)$ belonging to $L^2(a,b)$, and $\psi(x)$ a non-decreasing function with infinitely many points of increase in the interval $(a,b)$ with moments

$$\mu_s' = \int_a^b t_s d\psi(t), \qquad (s = 0, 1, \cdots)$$

of all orders, the minimum values of the mean square deviation

$$S = \int_a^b \{f(t) - \pi_s(t)\}^2 d\psi(t) \tag{3.28}$$

for the polynomials $\pi_s(t)$, is given by the choice

$$\pi_s(t) = \sum_{r=0}^{s} f_r P_r(t). \tag{3.29}$$

Here $\{P_r(t)\}$ is the orthogonal set associated with $d\psi(t)$ and $f_r$ is the Fourier coefficient

$$f_r = \int_a^b f(t)P_r(t)d\psi(t). \tag{3.30}$$

The minimum is

$$S^* = \sum_0^s f_r^2\phi_r \tag{3.31}$$

with

$$\phi_r = \int_a^b P_r^2(t)d\psi(t).$$

Parseval's formula (Szegö, 1939, p. 39) asserts that

$$\int_a^b \{f(t)\}^2 d\psi(t) = \sum_0^\infty f_r{}^2 \phi_r \tag{3.32}$$

for $f(t)$ of the class $L_\alpha^2(a,b)$. (For a definition of this class see Szegö, section 1.5 on closure.)

Now turn to solutions $\psi(\cdot)$ of the Stieltjes' moment problem (see Section 1.7 and (1.63)). If the $\alpha$'s are positive and $\Sigma\alpha_j = \infty$, then $\psi(\cdot)$ is uniquely determined. If, however, the $\alpha$'s are positive but $\Sigma\alpha_j$ converges there will be many solutions $\psi(\cdot)$, including two called extremal solutions, which have the property that $\{\omega_{2s}(-z)\}$ and $\{\omega_{2s+1}(-z)\}$, $\{\omega_s(z)\}$ being a denominator convergent of the c.f. in (1.63), are orthogonal systems with respect to them. But Riesz (Shohat and Tamarkin (1950), pp. 61-76) has shown that if $\psi(\cdot)$ is an extremal solution or unique solution of the Stieltjes' moment problem, then Parseval's formula applies to all $f(\cdot) \in L_\psi^2$. This is readily extended to cover our present requirements; if $\psi(\cdot)$ is a unique solution of the Smp, then Parseval's formula applies to $f(\cdot) \in L_{\hat\psi}^2$ where

$$\hat\psi(x) = \int_0^x C(t)\,d\psi(t), \qquad x \geqslant 0,$$

for a non-negative polynomial $C(t)$ of fixed degree which is not identically zero. Clearly $\hat\psi(x)$ is bounded and non-decreasing (in view of the nature of $\psi(t)$), and the moments are given by

$$\mu_s' = \int_0^\infty t^s C(t)\,d\psi(t), \qquad (s = 1, 2, \cdots)$$

and so exist.

## 3.5 Convergence of 2nd- and Higher Order c.f.s

To avoid complication, consider

$$F(z_1, z_2) = \int_0^\infty \frac{d\psi(t)}{(z_1+t)(z_2+t)}$$

where $\psi(\cdot)$ is a unique solution of the Stieltjes' moment problem, and

$$\begin{aligned} C_2(t) &= (z_1+t)(z_2+t) \\ &= t^2 + p_1 t + p_2 \end{aligned}$$

is non-negative on $(0, \infty)$. Then Parseval's formula holds, and we can set up approximants to $F(z_1, z_2)$ from the minimization of

$$S_r = \int_0^\infty C_2(t)d\psi(t)\{1/C_2(t) - \sum_{s=0}^{r} \lambda_s P_s(t)\}^2, \tag{3.33}$$

which is (3.2) with $A(t) = 1$.

The equations for the minimum set $\{\hat\lambda_s\}$ are

$$\int_0^\infty P_l(t)\{\sum_{s=0}^{r} \hat\lambda_s P_s(t)\}C_2(t)d\psi(t) = \int_0^\infty P_l(t)d\psi(t). \tag{3.34}$$

$$(l = 0, 1, \cdots, r; \quad r = 0, 1, \cdots)$$

But from the recurrence for the orthogonal polynomials in (1.11)

$$\begin{aligned} C_2(t)P_{r-1}(t) = {} & P_{r+1}(t) + (p_1 + a_r + a_{r+1})P_r(t) \\ & + (C_2(a_r) + b^*_{r-1} + b_r)P_{r-1}(t) + b_{r-1}(p_1 + a_{r-1} + a_r)P_{r-2}(t) \\ & + b_{r-1}b_{r-2}P_{r-3}(t), \end{aligned} \tag{3.35}$$

for $r = 1, 2, \cdots,$ with $P_1(t) = t - a_1$, $P_0(t) = 1$, $P_r(t) = 0$ if $r < 0$, and $b^*_r = b_r$, $r > 0$, $b^*_0 = 0$. Hence using (1.10), we have the system of equations for $r = 2$;

$$\begin{aligned} &\{C_2(a_1) + b_1\}\phi_0\hat\lambda_0 + (p + a_1 + a_2)\phi_1\hat\lambda_1 + \phi_2\hat\lambda_2 = \phi_0 \\ &b_1(p + a_1 + a_2)\phi_0\hat\lambda_0 + \{C_2(a_2) + b_1 + b_2\}\phi_1\hat\lambda_1 + (p + a_2 + a_3)\phi_2\hat\lambda_2 = 0 \\ &b_1b_2\phi_0\hat\lambda_0 + b_2(p + a_2 + a_3)\phi_1\hat\lambda_1 + \{C_2(a_3) + b_2 + b_3\}\phi_2\hat\lambda_2 = 0, \end{aligned} \tag{3.36}$$

and the minimum of $S_2$ is

$$S^*_2 = F(z_1, z_2) - \phi_0\hat\lambda_0;$$

i.e., the third approximant to $F$ is $\hat\lambda_0\phi_0$ where $\hat\lambda_0$ is the ratio of the first minor of

$$\begin{vmatrix} C_2(a_1)+b_1 & p+a_1+a_2 & 1 \\ b_1(p+a_1+a_2) & C_2(a_2)+b_1+b_2 & p+a_2+a_3 \\ b_1b_2 & b_2(p+a_2+a_3) & C_2(a_3)+b_2+b_3 \end{vmatrix} \tag{3.37}$$

to the determinant itself. Multiplying the second and third columns by $\sqrt{b_1}$, $\sqrt{(b_1b_2)}$, and dividing the second and third rows by $\sqrt{b_1}$, $\sqrt{(b_1b_2)}$ brings the result into line with (3.13).

For general $r$ in (3.33), the equations $\hat{\lambda}_0, \hat{\lambda}_1, \cdots, \hat{\lambda}_r$ are

$$\begin{aligned}\hat{\lambda}_{t-2}b_t b_{t-1}\phi_{t-2} &+ b_t(p + a_t + a_{t+1})\hat{\lambda}_{t-1}\phi_{t-1} \\ &+ \{C_2(a_{t+1}) + b_t^* + b_{t+1}\}\hat{\lambda}_t\phi_t \\ &+ (p + a_{t+1} + a_{t+2})\hat{\lambda}_{t+1}\phi_{t+1} + \hat{\lambda}_{t+2}\phi_{t+2} = \delta_{t,0}\phi_0\end{aligned} \tag{3.38}$$

where

$$t = 0, 1, \cdots, r; \quad \hat{\lambda}_t = 0 \text{ if } t > r \text{ or } t < 0;$$
$$b_t^* = b_t, \ t > 0; \quad b_0^* = 0.$$

The $(r+1)$th approximate to $F(z_1, z_2)$ is $\hat{\lambda}_0\phi_0$ and from (3.38) $\hat{\lambda}_0$ is the ratio of the principal minor of the determinant of the system to the determinant itself (strictly speaking $\hat{\lambda}_s$ in (3.33) should be described as $\hat{\lambda}_s^{(r)}$, but this would make the notation unduly cumbersome). To bring the approximants into line with (3.14), so that they involve symmetric determinants, we multiply the $m$ th column of the determinant (of the system) by $\sqrt{(b_1 b_2 \cdots b_{m-1})}$ and divide the $m$ th row by the same factor, $m = 2, 3, \cdots, r$; similarly for the first minor in the numerator. *Thus for a unique solution $\psi(\cdot)$ to the Stieltjes' moment problem, for $C_2(t) > 0$ for $0 \leqslant t < \infty$, we have shown that approximants $\{t_r\}$ exists such*

$$F(z_1, z_2) = \underset{r \to \infty}{l.i.s} \frac{b_0 K_{r-1}(\gamma_1, \beta_1, \alpha_1)}{K_r(\gamma_0, \beta_0, \alpha_0)}, \tag{3.39}$$

*where in view of the positivity of the terms in Parseval's formula, we write l.i.s. for limit of increasing sequence.*

The extension to the case where $C_n(t)$ is a positive polynomial of degree $n$ presents no new problem, the approximants being ratios (principal minor of det./det.) of symmetric $(2n+1)$ diagonal determinants. The structure for $n = 3$ is given in (3.16).

Note that in the process of making the determinants symmetrical, we have used divisors involving $\sqrt{b_1}, \sqrt{b_2}, \cdots$; in the Stieltjes situation these are nonzero (the sign being of little consequence), but more general cases might involve zeros in which case the situation would require more scrutiny.

We see then that Examples 3.1-3.3 present 2nd order convergent c.f.s, the sequences being monotonic increasing. Moreover in the case of Example 3.3, $\psi(t)$ is constant for $t > 4$.

The extension to the inner product integral in (3.4) proceeds by considering Parseval expansions for $\{A(t) \pm B(t)\}/C(t)$, leading to

$$\int_0^\infty \frac{A(t)B(t)d\,\psi(t)}{(z_1+t)(z_2+t)} = \lim_{r\to\infty} t_r \tag{3.40}$$

where $t_r = \hat{K}_r / K_r$. $\hat{K}_r$ is now a bordered determinant of order $r+1$, and

$$\hat{K}_r = \begin{vmatrix} 0 & A_0 & A_1 & A_2 & \cdot & \cdot & \cdot & A_{r-1} \\ B_0 & \gamma_0 & \beta_0 & \alpha_0 & \cdot & \cdot & \cdot & \\ B_1 & \beta_0 & \gamma_1 & \beta_1 & \alpha_1 & \cdot & & \\ B_2 & \alpha_0 & \beta_1 & \gamma_2 & \beta_2 & \cdot & & \\ \cdot & \cdot & \cdot & \cdot & \cdot & \cdot & & \\ \cdot & \cdot & & & & & & \alpha_{r-3} \\ \cdot & & & & & & & \beta_{r-2} \\ B_{r-1} & & & & & \alpha_{r-3} & \beta_{r-2} & \gamma_{r-1} \end{vmatrix},$$

where

$$A_s = \int_0^\infty P_s(t)A(t)d\,\psi(t),$$

$$B_s = \int_0^\infty P_s(t)B(t)d\,\psi(t).$$

If $A(t)$, $B(t)$ are polynomials of degree $\leqslant r$, then $A_s$, $B_s$, $s > r$, are zero; in particular, if $A(t) = B(t) = 1$, then (3.41) reduces to (3.39). In general $A(t)$, $B(t)$ will be real polynomials, in which case $\{t_r\}$ will not necessarily form an increasing sequence. Also, in more general cases $A(t)$ and $B(t)$ must belong to the class $L_\psi^2$.

*Note that if* $(t+z_1)(t+z_2) = t^2 + pt + q$ *where* $q - p^2/4 > 0$, *then a sequence of decreasing bounds may be set up from the relation*

$$(q - p^2/4)F(z_1, z_2) = b_0 - \int_0^\infty \frac{(t+p/2)^2 d\,\psi(t)}{(t+z_1)(t+z_2)}.$$

If the $r$th approximant to $F(z_1, z_2)$ using this form is $t_r^*$, and using (3.39) is $t_r$, then it can be shown that

$$t_r^* = t_r + \frac{\prod_{\lambda=0}^{r} b_\lambda}{(q - p^2/4)K_r(\gamma_0, \beta_0, \alpha_0)}. \tag{3.41}$$

(Details are given in Shenton, 1957, p. 162.)

## 3.6 Recurrence Relations for Second-Order c.f.s

### 3.6.1 A fifth order recurrence

It will be abundantly clear from Examples 3.1-3.3 that the numerical evaluation of higher order c.f.s become arduous and demanding even with the facilities of digital computers. On the other hand since three-diagonal determinants follow a recurrence of order two, one hopes for a similar relation for higher cases. Note that it is the balanced form of the 5-diagonal determinants which causes some of the complication; determinants with one and only one super-diagonal follow easily derivable recurrences, the order depending solely on the number of sub-diagonals (they are usually called recurrents (see Muir, 1920, p208)).

Let $K_s(h_1, g_1, f_1)$ be a symmetric 5-diagonal determinant with elements $h_1, h_2, \cdots$, in the main diagonal, $g_1, g_2, \cdots$, in the first super-diagonal, and $f_1, f_2, \cdots$, in the second super-diagonal. Expand $k_s$ by its last row and column. Then

$$K_s = h_s K_{s-1} - g_{s-1}^2 K_{s-2} + 2g_{s-1} f_{s-2} K_{s-2}^* - f_{s-2}^2 h_{s-1} K_{s-3} + f_{s-2}^2 f_{s-3}^2 K_{s-4}^* \qquad (s = 4, 5) \tag{3.42}$$

where

$$K_s^* = \begin{vmatrix} & K_{s-1} & & \vdots & \\ & & & \vdots & f_{s-2} \\ \cdots & \cdots & \cdots & \vdots & g_{s-1} \\ & & & f_{s-1} & g_s \end{vmatrix}$$

and $K_{s-1}$ is the matrix consisting of the elements of $K_{s-1}(h_1, g_1, f_1)$. For example,

$$K_2^* = \begin{bmatrix} h_1 & g_1 \\ f_1 & g_2 \end{bmatrix}, \qquad K_3^* = \begin{bmatrix} h_1 & g_1 & f_1 \\ g_1 & h_2 & g_2 \\ & f_2 & g_3 \end{bmatrix} .$$

But expanding $K_s^*$ by its last row, we have

$$K_s^* = g_s K_{s-1} - f_{s-1} K_{s-1}^* . \qquad (s = 3, 4, \cdots) \tag{3.43}$$

Eliminating $K^*$ from (3.42) and (3.43), we find

$$\begin{aligned} g_{s-2}K_s &= (h_s g_{s-2} - f_{s-2}g_{s-1})K_{s-1} \\ &\quad - (g_{s-1}g_{s-2} - h_{s-1}f_{s-2})(g_{s-1}K_{s-2} - g_{s-2}f_{s-2}K_{s-3}) \\ &\quad - f_{s-3}^2 f_{s-2}(h_{s-2}g_{s-1} - f_{s-2}g_{s-2})K_{s-4} \\ &\quad + f_{s-2}f_{s-3}^2 f_{s-4}^2 g_{s-1}K_{s-5}. \end{aligned} \tag{3.44}$$

$$(s = 3, 4, \cdots; \quad K_{-2} = K_{-1} = 0; \quad K_0 = 1)$$

This recurrence relation satisfied by $K_s(h_1, g_1, f_1)$ is of order five. Note that if one or more of the terms $g_1, g_2, \cdots$, happened to be zero, the recurrence would break down because of $g_{s-2}$ on the left side. Special consideration would be needed in this case. In any event we develop alternatives in the sequel.

By a slight modification of this approach, it may be shown that the recurrence relation for the asymmetric determinant

$$K_s \left( h_1, \begin{matrix} g_1, & f_1 \\ g_1', & f_1' \end{matrix} \right),$$

(where now elements $g$, $f$ in the sub-diagonal are replaced by elements $g'$, $f'$) is of order six, but we do not require it in the present context.

### 3.6.2 Special Cases

There are three interesting special cases (see Muir, 1884):

(a) $f_j = 0$, when (3.44) reduces to

$$K_s = h_s K_{s-1} - g_{s-1}^2 K_{s-2}, \tag{3.45}$$

as we should expect since $K_s(h_1, g_1, 0)$ is now a "continuant" type of determinant with three main diagonals.

(b) $g_j = 0$, when (3.44) becomes

$$K_s = h_s K_{s-1} - h_{s-1} f_{s-2}^2 K_{s-3} + f_{s-2}^2 f_{s-3}^2 K_{s-4}, \tag{3.46}$$

which is the recurrence relation for the product of two "continuants," and indeed (Muir, 1884)

$$\begin{aligned} K_{2s}(h_1, 0, f_1) &= K_s(h_1^*, f_1^*)K_s(\hat{h}_2, \hat{f}_2), \\ K_{2s+1}(h_1, 0, f_1) &= K_{s+1}(h_1^*, f_1^*)K_s(\hat{h}_2, \hat{f}_2), \end{aligned} \tag{3.47}$$

where

$$h_s^* = h_{2s-1}, \qquad f_s^* = f_{2s-1},$$
$$\hat{h}_s = h_{2s}, \qquad \hat{f}_s = f_{2s}.$$

A third case of reducibility occurs for the recurrence associated with $S$-fractions (1.20b) and we defer this.

### 3.6.3 General Cases

Now applying (3.44) to (3.19) we may write the 2nd-order c.f. as

$$F(z_1, z_2) = \underset{r\to\infty}{l.i.s.} \frac{K_{s-1}(\gamma_1, \beta_1, \alpha_1)}{K_s(\gamma_0, \beta_0, \alpha_0)} \cdot \phi_0 \tag{3.48}$$

where $u_s = K_{s-1}(\gamma_1, \beta_1, \alpha_1)$, and $v_s = K_s(\gamma_0, \beta_0, \alpha_0)$ satisfy the recurrence (i.e., $y_s = u_s$ or $y_s = v_s$)

$$\begin{aligned} \beta'_{s-3} y_s &= (\gamma_{s-1}\beta'_{s-3} - b_{s-1}\beta'_{s-2}) y_{s-1} \\ &\quad - b_{s-1}(\beta'_{s-2}\beta'_{s-3} - \gamma_{s-2})(\beta'_{s-2} y_{s-2} - b_{s-2}\beta'_{s-3} y_{s-3}) \\ &\quad - b_{s-1} b_{s-2} b_{s-3}(\gamma_{s-3}\beta'_{s-2} - b_{s-2}\beta'_{s-3}) y_{s-4} \\ &\quad + b_{s-1} b_{s-2} b_{s-3}^2 b_{s-4} y_{s-5}, \qquad (s = 3, 4, \cdots) \end{aligned}$$

and $\beta'_s \sqrt{b_{s+1}} = \beta_s$, with $\alpha_s = P_{s,s+2}$, $\beta_s = P_{s,s+1}$, and $\gamma_s = P_{s,s}$ as given in (3.14). In addition,

$$\begin{cases} u_0 = 0, \quad u_1 = 1, \quad u_2 = \gamma_1; \\ v_0 = 1, \quad v_2 = \gamma_0, \quad v_2 = \gamma_0\gamma_1 - \beta_0^2; \\ \mu_s, \ v_s = 0 \ \text{if} \ s < 0. \end{cases}$$

**Example 3.4**

$$F(z) = \int_0^1 \frac{dt}{z + t}$$

$$= \frac{1}{z + 1/2 -} \ \frac{a_1}{z + 1/2 -} \ \frac{a_2}{z + 1/2 -} \ \frac{a_3}{z + 1/2 -} \ \cdots, \qquad (z > 0)$$

where $a_s = s^2/(16s^2 - 4)$.

$$F(iz, iz) = \int_0^1 \frac{dt}{z^2 + t^2} = z^{-1}\tan^{-1}(z^{-1}), \qquad (z \text{ real})$$

$$= \underset{r\to\infty}{l.i.s.} \frac{u_s}{v_s}\phi_0. \qquad (\phi_0 = 1)$$

$u_s, v_s$ follow;

$$a_s y_s = (z^2 + a_s + 1/4)y_{s-1} + (z^2 + a_{s-1} + a_{s-2} - 3/4)(y_{s-2} - y_{s-3}) - (z^2 + a_{s-3} + 1/4)y_{s-4} + a_{s-3}y_{s-5}. \qquad (s = 3, 4, \cdots)$$

Initial values;

$$u_0 = 0, \quad u_1 = 12, \quad u_2 = 3(60z^2 + 24), \quad u_s = 0,\ s < 0;$$
$$v_0 = 0, \quad v_1 = 12z^2 + 4, \quad v_2 = 3(60z^4 + 44z^2 + 3), \quad v_s = 0,\ s < 0;$$

For Example, using (3.48)

$$\begin{cases} 3u_3 = 16(525z^4 + 410z^2 + 45), \\ 3v_3 = 16(525z^6 + 585z^4 + 135z^2 + 3), \end{cases}$$

$$\begin{cases} 3u_4 = 132{,}300z^6 + 153{,}300z^4 + 41{,}300z^2 + 1{,}800, \\ 3v_4 = 132{,}300z^8 + 197{,}400z^6 + 80{,}640z^4 + 8{,}100z^2 + 75. \end{cases}$$

Numerical:

$$z = 1.5, \qquad z^{-1}\tan^{-1}(z^{-1}) = 0.392001731,$$
$$u_1/v_1 = 0.3871, \qquad u_2/v_2 = 0.39187,$$
$$u_3/v_3 = 0.3919938, \qquad u_4/v_4 = 0.392001731.$$

(The second-order c.f. is not the same, apart from the first convergent, as the even part of the hypergeometric c.f.

$$z^{-1}\tan^{-1}(z^{-1}) = \frac{1}{z^2+}\ \frac{b_1}{1+}\ \frac{b_2}{z^2+}\ \frac{b_3}{1+}\ \cdots, \qquad (b_s = \frac{s^2}{4s^2-1})$$

for which $\chi_2/\omega_2 = 1/(z^2 + 1/3)$, $\chi_4/\omega_4 = 5(21z^2 + 11)/(105z^4 + 90z^2 + 9)$, the latter giving 0.39196 for $z = 1.5$, nearer the true value than $u_2/v_2$; however, $u_2/v_2 > \chi_4/\omega_4$, and therefore nearer the true value for $0 < x^2 < 51/35$).

## 3.7 A Reducible Case of the Fifth Order Recurrence

The relation (3.48) reduces to a fourth order one when the basic c.f. is a S.f. of the form

$$F(z) = \frac{b_0}{z-}\ \frac{b_1}{z-}\ \frac{b_2}{z-}\ \cdots. \tag{3.49}$$

Formally, writing $b_0K_{s-1}(\gamma_1, \beta_1, \alpha_1) = T_s^*$, $K_s(\gamma_0, \beta_0, \alpha_0) = T_s$, we find the recurrence for $T_s^*$ and $T_s$ may be written

$$\Phi_s(y) = b_{s-1}\Phi_{s-1}(y), \tag{3.50}$$

where

$$\begin{aligned}\Phi_s(y) = {} & y_s - (q + b_s - b_{s-1})y_{s-1} - b_{s-1}(2b_{s-1} - p^2 + 2q)y_{s-2} \\ & - b_{s-1}b_{s-2}(q + b_{s-2} - b_{s-1})y_{s-3} + b_{s-1}b_{s-2}^2b_{s-3}y_{s-4}\end{aligned}$$

and $p = z_1 + z_2, \quad q = z_1z_2.$

But $\Phi_2(T) = 0$, assuming $T_s = 0$, $s < 0$. Hence from (3.50),

$$\begin{aligned}T_s = {} & (q + b_s - b_{s-1})T_{s-1} + b_{s-1}(2b_{s-1} - p^2 + 2q)T_{s-2} \\ & + b_{s-1}b_{s-2}(q + b_{s-2} - b_{s-1})T_{s-3} - b_{s-1}b_{s-2}^2b_{s-3}T_{s-4},\end{aligned}$$

$$(s = 2, 3, \cdots)$$

with $T_0 = 1, \quad T_1 = q + b_1, \quad T_s = 0$ for $s < 0$.

Similarly it will be found that if

$$\Psi_s(T^*) = \Phi_s(T^*) - 2\prod_{\lambda=0}^{s-1} b_\lambda$$

then

$$\Psi_s(T^*) = b_{s-1}\Psi_{s-1}(T^*),$$

and $\Psi_2(T^*) = 0$ assuming $T_s^* = 0, \quad s \leqslant 0$. Hence

$$\begin{aligned}T_s^* = {} & (q + b_s - b_{s-1})T_{s-1}^* + b_{s-1}(2b_{s-1} - p^2 + 2q)T_{s-2}^* \\ & + b_{s-1}b_{s-2}(q + b_{s-2} - b_{s-1})T_{s-3}^* - b_{s-1}b_{s-2}^2b_{s-3}T_{s-4}^* \\ & + 2\prod_{\lambda=0}^{s-1} b_\lambda, \qquad (s = 2, 3, \cdots)\end{aligned} \tag{3.52}$$

with $T_1^* = b_0, \quad T_s^* = 0, \quad s \leqslant 0.$

There is now the question as to the validity of

$$\frac{F(z_1) - F(z_2)}{z_2 - z_1} = \lim_{s \to \infty} \frac{T_s^*}{T_s}. \tag{3.53}$$

This can be related to the theory of the Hamburger moment problem. Let

$$F(z) \sim \mu_0/z + \mu_2/z^3 + \mu_4/z + \cdots$$

and assume

$$\begin{cases} b_s > 0, & (s = 0, 1, \cdots) \\ \sum_{s=0}^{\infty} \mu_{2s}^{-1/(2s)} = \infty. & \end{cases} \tag{3.54}$$

Then there exists a unique bounded non-decreasing function $\psi(t)$ in the interval $(-\infty, \infty)$ such that

$$\begin{cases} \int_0^{\infty} t^{2s} \, d\,\psi(t) = \mu_{2s}, \\ \int_0^{\infty} t^{2s+1} d\,\psi(t) = 0. \qquad (s = 0, 1, \cdots) \end{cases}$$

(See Shohat and Tamarkin, 1950, p. 5 and p. 19; also p. 62 for a theorem of M. Riesz.)

All we have to do now is justify Parseval's expansion for

$$\int_{-\infty}^{\infty} \{(t + z_1)(t + z_2)\}^{-2} d\,\hat{\psi}(t)$$

where

$$\hat{\psi}(t) = \int_{-\infty}^{t} (z_1 + x)(z_2 + x) d\,\psi(x),$$

the argument following that of Section 3.3. It turns out that using the Riesz theorem regarding a *determined* Hamburger moment problem that under the conditions in (3.54), that

$$F(z_1, z_2) = \int_{-\infty}^{\infty} \frac{d\,\psi(t)}{(t + z_1)(t + z_2)} = \underset{r \to \infty}{l.i.s.} \frac{T_r^*}{T_r} \tag{3.55}$$

where $z_1 = x + iy$, $z_2 = x - iy$, $(t + z_1)(t + z_2) > 0$ for all real $t$. A sequence of decreasing bounds can be set up using

$$y^2 F(z_1, z_2) = b_0 - \int_{-\infty}^{\infty} \frac{(t + x)^2 d\,\psi(t)}{(t + x)^2 + y^2}$$

in conjunction with (3.41).

**Example 3.5**

$$F(z) = \tan(z^{-1}/2), \qquad b_0 = 1, \qquad b_s = 1/(16s^2 - 4),$$

(see Perron (1957), p. 157; $m = 1/2$, $n = z$).

$\mu_{2s} = 2B_{s+1}(2^{2s+2} - 1)/(2s+2)!$, $B_s$ a Bernoulli number with $B_1 = 1/6$, $B_2 = 1/30$, etc.

$\mu_{2s}^{1/(2s)} \sim 1/\pi, \quad s \to \infty.$

$$F(x+iy, x-iy) = \frac{\sinh(Y)}{Y\{\cosh(Y) + \cos(X)\}} = \underset{r \to \infty}{l.i.s.} \frac{T_r^*}{T_r}$$

where $T_r^*$, $T_r$ are defined in (3.51) and (3.52), and $X = x/(x^2+y^2)$, $Y = y/(x^2+y^2)$, $y \neq 0$. As a numerical example

$$\frac{T_4^*}{T_4} = \frac{398499385800}{19896118681110}$$

which gives correct to 14 decimal places

$$\frac{\sinh(0.16)}{0.16\{\cosh(0.16) + \cos(0.12)\}} = 0.02002900124325996\,.$$

Using the upper bounds indicated in (3.41), the error in this approximant cannot exceed 1.6E–15.

**Example 3.6**

$$F(z) = 2z - \cot(z^{-1}/2)$$
$$= \frac{1/6}{z-}\;\frac{1/(6\cdot 10)}{z-}\;\frac{1/(10\cdot 14)}{z-}\;\cdots, \qquad (z > 0)$$

$$(b_s = 1/\{(4s+2)(4s+6)\},\quad s > 0)$$

$$F(x+iy, x-iy) = \frac{\sinh(Y)}{Y\{\cosh(Y) - \cos(X)\}} - 2$$
$$= \underset{r \to \infty}{l.i.s.} \frac{T_r^*}{T_r}. \qquad (y \neq 0)$$

(Refer to Example 3.5 for $X$, $Y$, and the recurrences.)

**Example 3.7**

$$F_1(z) = \frac{1}{2}\int_{-\infty}^{\infty} \frac{\operatorname{sech}(\tfrac{1}{2}\pi t)dt}{t+z}$$

$$= \frac{1}{z-}\,\frac{1^2}{z-}\,\frac{2^2}{z-}\,\cdots . \qquad (\operatorname{Im}(z)\neq 0)$$

$$F_2(z) = \frac{1}{2}\int_{-\infty}^{\infty} \frac{t\operatorname{cosech}(\tfrac{1}{2}\pi t)dt}{t+z}$$

$$= \frac{1}{z-}\,\frac{1\cdot 2}{z-}\,\frac{2\cdot 3}{z-}\,\cdots . \qquad (\operatorname{Im}(z)\neq 0)$$

(Stieltjes Correspondance, 1905, letter 175).

The Hamburger moment problem is determined in each case; see Wall (1948), p. 366, Example 2. Also recall that the Hamburger moment problem

$$\mu_s = \int_{-\infty}^{\infty} t^s e^{-by}\,dt , \qquad (y = |t|^{\alpha}, \alpha > 1, b > 0)$$

is determined.

The second order of c.f.s (see (3.49), (3.51) and (3.52)) for

$$\frac{1}{2}\int_{-\infty}^{\infty} \frac{\operatorname{sech}(\tfrac{1}{2}\pi t)dt}{(t+x)^2+y^2}, \qquad \frac{1}{2}\int_{-\infty}^{\infty} \frac{\operatorname{cosech}(\tfrac{1}{2}\pi t)dt}{(t+x)^2+y^2},$$

will provide convergent increasing sequences for $y \neq 0$.

**Example 3.8**

<u>The c.f. for Mills' Ratio (see Section 1.6.1)</u>

$$F(z) = \int_{-\infty}^{\infty} \frac{g(t)dt}{z+t}, \qquad (g(t) = e^{-t^2/2}/\sqrt{(2\pi)},\ \operatorname{Im}(z)\neq 0)$$

$$= \frac{1}{z-}\,\frac{1}{z-}\,\frac{2}{z-}\,\frac{3}{z-}\,\cdots ,$$

$$\int_{-\infty}^{\infty} \frac{g(t)dt}{(t+x)^2+y^2} = \underset{r\to\infty}{l.i.s.}\ \frac{T_r^*}{T_r} .$$

Recurrence

$$\begin{aligned} T_r^* &= (x^2+y^2)T_{r-1}^* + 2(r-1)(r-1+y^2-x^2)T_{r-2}^* \\ &\quad + (r-1)(r-2)(x^2+y^2-1)T_{r-3}^* - (r-1)(r-2)^2(r-3)T_{r-4}^* \\ &\quad + 2(r-1)!\,. \qquad\qquad (r = 2, 3, \cdots) \end{aligned}$$

$T_r$ the same recurrence omitting $2(r-1)!$

$$T_r^* = 0, \quad r < 1, \quad T_1^* = 1;$$
$$T_r = 0, \quad r < 0, \quad T_0 = 1, \quad T_1 = x^2+y^2+1.$$

Numerical $x = 1/2 \;\; y = 3/2$ Integral $= 0.762826$

| $r$ | $T_r^*$ | $T_r$ | $T_r^*/T_r$ | $\hat{T}_r/T_r$ |
|---|---|---|---|---|
| 0 | 0 | 1 | 0.0000 | |
| 1 | 1 | 2 | 0.5000 | 1.1667 |
| 2 | 4 | 7 | 0.5714 | 0.9524 |
| 3 | 22 | 34 | 0.6471 | 0.8824 |
| 4 | 140 | 203 | 0.6897 | 0.8473 |
| 5 | 1048 | 1486 | 0.7052 | 0.8129 |
| 6 | 9076 | 12457 | 0.7286 | 0.8057 |
| 7 | 88136 | 120422 | 0.7319 | 0.7877 |
| 8 | 962932 | 1293049 | 0.7447 | 0.7863 |
| 9 | 11528104 | 15468918 | 0.7453 | 0.7765 |
| 10 | 151848908 | 201821591 | 0.7524 | 0.7764 |

($T_r^*/T_r$ is an increasing sequence; $\hat{T}_r/T_r$ is a decreasing sequence from (3.41).)

## 3.8 Formula for Numerators and Denominators

In the first order c.f. (1.16) with convergent $P_s(n)/Q_s(n)$, it is readily seen that (Szegö, 1939, p. 54)

$$P_s(n) = (-1)^{s-1}\int_a^b \frac{(P_s(t) - P_s(-n))}{t+n}\, d\psi(t)\,, \tag{3.56}$$

where $Q_s(n) = (-1)^s P_s(-n)$; in the notation of Chapter 1, $\omega_s(n) = Q_s(n)$. For the most part the integral limits are $a = 0$, $b = \infty$.

Does this formula have an extension to the case of higher order c.f.s? First of all, a word about the denominators of 2nd order c.f.s (the general case is deferred to Chapter 4). Consider

$$Q_s(z_1, z_2) = Q_s(\underset{\sim}{z}) = \begin{vmatrix} Q_s(z_1) & Q_s(z_2) \\ Q_{s+1}(z_1) & Q_{s+1}(z_2) \end{vmatrix} \div \begin{vmatrix} 1 & 1 \\ z_1 & z_2 \end{vmatrix}. \tag{3.57}$$

Demonstrably

$$Q_0(\underset{\sim}{z}) = 1, \qquad Q_1(\underset{\sim}{z}) = K_1(\gamma_0, \beta_0, \alpha_0), \qquad Q_2(\underset{\sim}{z}) = K_2(\gamma_0, \beta_0, \alpha_0).$$

With some algebra, it can now be shown that $Q_s(\underset{\sim}{z})$ follows the recurrence (3.48).

Now consider

$$P_s(z_1, z_2) = P_s(\underset{\sim}{z}) = \int_a^b \frac{P_s(t, z_1, z_2)\,d\,\psi(t)}{(z_2 - z_1)(t + z_1)(t + z_2)} \tag{3.58}$$

where

$$P_s(t, z_1, z_2) = -\begin{vmatrix} 1 & 1 & 1 \\ P_s(t) & P_s(-z_1) & P_s(-z_2) \\ P_{s+1}(t) & P_{s+1}(-z_1) & P_{s+1}(-z_2) \end{vmatrix}.$$

Again,

$$P_0(\underset{\sim}{z}) = 0, \qquad P_1(\underset{\sim}{z}) = 1, \qquad P_2(\underset{\sim}{z}) = K_1(\gamma_1, \beta_1, \alpha_1).$$

Note that the denominator in the integral (3.58) may be written

$$(z_2 - z_1)(t + z_1)(t + z_2) = -\begin{vmatrix} 1 & 1 & 1 \\ t & -z_1 & -z_2 \\ t^2 & z_1^2 & z_2^2 \end{vmatrix}.$$

Since the formulae (3.57) and (3.58) are mainly of theoretical interest (and are dealt with in the general case in Chapter 4), we give here only an indication of proofs. For $P_s(\underset{\sim}{z})$, there are three components similar to $Q_s(\underset{\sim}{z})$, so that we need only consider this. Use the recurrence on $Q_{s+1}(\cdot)$ and express $Q_s(\underset{\sim}{z})$ in terms of $Q_{s-1}(\underset{\sim}{z})$ and a new term $Q_s(z_1)Q_s(z_2)$. Using this as the new pivot, the recurrence for $Q_s(\cdot)$ on each term produces a $Q_{s-1}(z_1)Q_{s-1}(z_2)$ term and another new term $Q_s(z_1)Q_{s-1}(z_2)$, or $Q_{s-1}(z_1)Q_s(z_2)$ using $Q_{s-1}(\underset{\sim}{z})$. But $Q_s(z_1)Q_{s-1}(z_2)$ recurses to $Q_{s-1}(z_1)Q_{s-1}(z_2)$ and a like term of lower order. We now have a closed system yielding finally the fifth order expression in (3.48).

We deduce a formula for the remainder, the analogue of (1.46). Thus

$$K_s(\gamma_0, \beta_0, \alpha_0)F(z_1, z_2) - b_0 K_{s-1}(\gamma_1, \beta_1, \alpha_1) = \int_a^b \frac{P_s^*(t, z_1, z_2)\,d\psi(t)}{(t+z_1)(t+z_2)(z_2-z_1)}, \tag{3.59}$$

where

$$P_s^*(t, z_1, z_2) = \begin{vmatrix} 0 & 1 & 1 \\ P_s(t) & P_s(-z_1) & P_s(-z_2) \\ P_{s+1}(t) & P_{s+1}(-z_1) & P_{s+1}(-z_2) \end{vmatrix},$$

the confluent case modifying the third column to derivatives. Note that whereas for the first order c.f.s $F_s(n) - \chi_s(n)/\omega_s(n) - n^{-2s-1}$ (c.f. Wall, 1948, p. 196; also (1.46)), 2nd order c.f.s have a lower order of asymptotic closeness. In fact from (3.59), if

$$r_s = F(z_1, z_2) - b_0 F_{s-1}(\gamma_1, \beta_1, \alpha_1)/\, F_s(\gamma_0, \beta_0, \alpha_0)$$

and $z_1 = -z_2 = i\sqrt{n}$, then $r_{2s} \sim n^{-2s-1}$, $r_{2s+1} \sim n^{-2s-3}$, much lower asymptotic orders than for $F_s(n)$.

Example 3.9

$$F(n) = \int_0^\infty \frac{e^{-t}\,dt}{n+t^2}.$$

<u>2nd Order c.f.</u>

$$P[0\,|\,1] = \frac{1}{n+2},$$

$$P[1\,|\,2] = \frac{n+14}{n^2+16n+12},$$

$$P[2\,|\,3] = \frac{n^2+52n+276}{n^3+54n^2+360n+144},$$

$$P[3\,|\,4] = \frac{n^3+126n^2+2792n+7824}{n^4+128n^3+3024n^2+11520n+2880},$$

$$P[4\,|\,5] = \frac{n^4+248n^3+13924n^2+17808n+306720}{n^5+250n^4+14400n^3+201600n^2+504000n+86400}.$$

Padé Approximants

$$P_a[0|1] = \frac{1}{n+2},$$

$$P_a[1|2] = \frac{5n+158}{5n^2+168n+216},$$

$$P_a[2|3] = \frac{149n^2+19712n+346236}{149n^3+20010n^2+38680n+392400},$$

$$P_a[3|4] = \frac{26825n^3+pn^2+qn+r}{26825n^4+an^3+bn^2+cn+d}.$$

$p = 9162382$, $q = 659127416$, $r = 8435409648$,
$a = 9216032$, $b = 676915680$, $c = 9587370240$, $d = 8482723200$.

| $s$ | $P[s\|s+1]$ | $P_a[s\|s+1]$ | $n$ | True Value |
|---|---|---|---|---|
| 1 | 0.5712 | 0.4190 | | |
| 2 | 0.5886 | 0.4604 | 1 | 0.6214 |
| 3 | 0.6120 | 0.4854 | | |
| 1 | 0.3333 | 0.2937 | | |
| 2 | 0.3529 | 0.3118 | 2 | 0.3574 |
| 3 | 0.3564 | 0.3216 | | |
| 1 | 0.1624 | 0.1550 | | |
| 2 | 0.16408 | 0.1588 | 3 | 0.1644 |
| 3 | 0.16410 | 0.1606 | | |

Remarks. $P[s|s+1]$ is nearer the true value than $P_a[s|s+1]$, $s = 1-3$. However, this trend may not be maintained for larger $s$; in any case, it breaks down for larger $n$, and $P_a[2|3] > P[2|3]$ for $n > 40$, $P_a[3|4] > P[3|4]$ for $n > 252$ approximately.

Note that

$$P[0|1] \sim 1/n - 2!/n^2 + 4/n^3 - \cdots,$$

$$P[1|2] \sim 1/n - 2!/n^2 + 20/n^3 - \cdots,$$

$$P[2|3] \sim 1/n - 2!/n^2 + 4!/n^3 - 6!/n^4 + 30528/n^5 - \cdots,$$

$$P[3|4] \sim 1/n - 2!/n^2 + 4!/n^3 - 6!/n^4 + 39744/n^5 - \cdots,$$

$$P[4|5] \sim 1/n - 2!/n^2 + 4!/n^3 - 6!/n^4 + 8!/n^5.$$

$$- 10!/n^6 + 459820800/n^7 - \cdots$$

### 3.9 A Recurrence with Even and Odd Parts

For 2nd order c.f.s there is an analogue to the Stieltjes c.f. in (1.21). For this c.f.

$$\omega_{2s} = \omega_{2s-1} + c_{2s}\omega_{2s-2}, \qquad (3.60)$$

$$\omega_{2s+1} = n\,\omega_{2s} + c_{2s+1}\omega_{2s-1}.$$

Consider then

$$K_s(\gamma_0, \beta_0, \alpha_0) = Q_s(z_1, z_2) = k_{2x}$$

say, where $Q_s(z) = \omega_{2s}(z)$ in terms of (1.21). Using (3.60),

$$k_{2s} = y_{2s} + c_{2s+1}c_{2s}k_{2s-2} \qquad (3.61)$$

where $y_s = \omega_s(z_1)\omega_s(z_2)$.

Similarly,

$$k_{2s-1} = \frac{\omega_{2s-1}(z_1)\omega_{2s+1}(z_2) - \omega_{2s-1}(z_2)\omega_{2s+1}(z_1)}{z_2 - z_1}$$

$$= y_{2s-1} + c_{2s}c_{2s-1}k_{2s-3}. \qquad (3.62)$$

But

$$y_{2s} = y_{2s-1} + c_{2s}^2 y_{2s-2} + c_{2s}\Theta_{2s-1}, \qquad (3.63)$$

where

$$\Theta_{2s-1} = \omega_{2s-1}(z_1)\omega_{2s-2}(z_2) + \omega_{2s-1}(z_2)\omega_{2s-2}(z_1)$$
$$= (z_1+z_2)y_{2s-2} + c_{2s-1}\{\omega_{2s-2}(z_1)\omega_{2s-3}(z_2) + \omega_{2s-2}(z_2)\omega_{2s-3}(z_1)\}.$$

Hence

$$\Theta_{2s-1} = (z_1+z_2)y_{2s-2} + 2c_{2s-1}y_{2s-3} + c_{2s-1}c_{2s-2}\Theta_{2s-3}$$

and so from (3.63)

$$y_{2s} - y_{2s-1} - c_{2s}^2 y_{2s-2} = c_{2s}(z_1+z_2)y_{2s-2} + 2c_{2s}c_{2s-1}y_{2s-3} + c_{2s}c_{2s-1}(y_{2s-2} - y_{2s-3} - c_{2s-2}^2 y_{2s-4}),$$

from which using (3.61) and (3.62) we deduce a sixth-order recurrence for $k_{2s}$. Similarly starting with

$$y_{2s+1} = z_1 z_2 y_{2s} + c_{2s+1}^2 y_{2s-1} + c_{2s+1}\Phi_{2s}$$

where

$$\Phi_{2s} = z_1\omega_{2s}(z_1)\omega_{2s-1}(z_2) + z_2\omega_{2s}(z_2)\omega_{2s-1}(z_1)$$

a recurrence is found for $k_{2s+1}$. *We thus have; if the Stieltjes' moment problem is determined and*

$$F(z) = \int_0^\infty \frac{d\,\psi(t)}{t+z} = \frac{b_1}{z+}\;\frac{b_2}{1+}\;\frac{b_3}{z+}\;\frac{b_4}{1+}\;\frac{b_5}{z+}\;\frac{b_6}{1+}\;\cdots, \tag{3.64}$$

*then*

$$\underset{s\to\infty}{l.i.s.}\frac{k^*_{2s}}{k_{2s}} = \lim_{s\to\infty}\frac{k^*_{2s}}{k_{2s}} = \frac{F(z_2)-F(z_1)}{z_1-z_2} = F(z_1,z_2),$$

*where*

(i) $k^*_s$ *and* $k_s$ *follow, for* $s = 2, 3, \cdots,$

$$\omega_{2s-1} = z_1z_2\omega_{2s-2} + \alpha_{2s-1}\omega_{2s-3} - \beta_{2s-1}\omega_{2s-5} - z_1z_2\gamma_{2s-1}\omega_{2s-6} + \delta_{2s-1}\omega_{2s-7},$$

$$\omega_{2s} = \omega_{2s-1} + \alpha_{2s}\,\omega_{2s-2} - \beta_{2s}\,\omega_{2s-4} - \gamma_{2s}\,\omega_{2s-5} + \delta_{2s}\,\omega_{2s-6},$$

(ii) $k^*_0 = 0,\quad k^*_1 = k^*_2 = b_1,\quad k^*_s = 0,\ s<0,$

$k_0 = 1,\quad k_1 = z_1z_2,\quad k_2 = z_1z_2 + b_2(z_1+z_2+b_3+b_2),$

$k_s = 0,\ s<0,$

(iii) $\alpha_s = b_s(z_1+z_2+b_{s+1}+b_s+b_{s-1}),\quad \beta_s = b_sb_{s-2}\alpha_{s-1},$

$\gamma_s = b_sb_{s-1}b_{s-2}b_{s-3},\quad \delta_s = b_sb_{s-1}b_{s-2}^2b_{s-3}b_{s-4},$

(iv) $(x+z_1)(x+z_2) > 0$ for $x \geqslant 0$.

Note: It has been assumed throughout that $z_1 \neq z_2$, but it is easily shown the result holds if $z_1 = z_2$ and $(t+z_1)^2 > 0$ for $t \geqslant 0$.

**Example 3.10**

$$F(z) = \frac{1}{2\pi}\int_0^4 \frac{\sqrt{(4/t-1)}dt}{t+z} = \frac{1}{z+}\;\frac{1}{1+}\;\frac{1}{z+}\;\frac{1}{1+}\;\cdots,$$

$$F(z,z) = \{z\sqrt{(z^2+4z)}\}^{-1} = \lim_{s\to\infty} k^*_s/k_s,$$

$$= \frac{1}{2\pi}\int_0^4 \frac{\sqrt{(4/t-1)}dt}{(t+z)^2} = -\frac{dF(z)}{dz}. \qquad (z>0)$$

Recurrence for $k_s^*$, $k_s$

$$w_{2s-1} = z^2 w_{2s-2} + (2z+3)w_{2s-3} - (2z+3)w_{2s-5} - z^2 w_{2s-6} + w_{2s-7},$$
$$w_{2s} = w_{2s-1} + (2z+3)w_{2s-2} - (2z+3)w_{2s-4} - w_{2s-5} + w_{2s-6}.$$
$$(s = 2, 3, \cdots)$$

$$k_1^* = 1, \quad k_2^* = 1, \qquad k_s^* = 0, \quad s < 0,$$
$$k_1 = z^2, \quad k_2 = z^2 + 2z + 2, \quad k_s = 0, \quad s < 0.$$

$F(1,1) = 1/\sqrt{5} = 0.4472136$

| $s$ | $k_s^*$ | $k_s$ | $k_s^*/k_s$ |
|---|---|---|---|
| 1 | 1 | 1 | 1.0 |
| 2 | 1 | 5 | 0.2* |
| 3 | 6 | 10 | 0.6 |
| 4 | 11 | 30 | 0.37* |
| 5 | 36 | 74 | 0.49 |
| 9 | 1590 | 3540 | 0.449 |
| 10 | 4140 | 9276 | 0.4463 |
| 18 | 9153025 | 20466835 | 0.4472125* |
| 19 | 23963005 | 53582855 | 0.4472140 |
| 20 | 62735880 | 140281751 | 0.4472134* |

(* increasing sequence)

Note that from elementary considerations,

$$F(z,z) = \frac{1}{z}\left(\frac{1}{z+}\;\frac{2}{1+}\;\frac{1}{z+}\;\frac{1}{1+}\;\frac{1}{z+}\;\frac{1}{1+}\;\cdots\right)$$

and this provides slightly sharper bounds than the 2nd order c.f. For example, when $z = 1$ the 19th and 20th convergents are $4181/9349 = 0.44721361$ and $6765/15127 = 0.44721359$.

**Example 3.11**

$$F(z) = \frac{1}{2\pi\lambda}\int_0^{4\lambda} \frac{\sqrt{\{t(4\lambda - t)\}}dt}{t(t+z)}, \qquad (\lambda > 0, \quad z > 0)$$

$$= \frac{1}{z+}\;\frac{\lambda}{1+}\;\frac{\lambda}{z+}\;\frac{\lambda}{1+}\;\cdots.$$

With $\lambda = 1/4$, we consider $F(\sqrt{2}, -\sqrt{2})$ with $t_s = X_s/Y_s$ where $k_s^* = X_s/4^s$, $k_s = Y_s/4^s$.

Recurrence for $X_x$ and $Y_s$

$$w_{2s-1} = -8w_{2s-2} + 3w_{2s-3} - 3w_{2s-5} + 8w_{2s-6} + w_{2s-7},$$
$$w_{2s} = 4w_{2s-1} + 3w_{2s-2} - 3w_{2s-4} - 4w_{2s-5} + w_{2s-6}, \qquad (s = 2, 3, \cdots)$$
$$X_1 = 4, \quad X_2 = 16, \quad X_s = 0, \quad s \leqslant 0;$$
$$Y_0 = 1, \quad Y_1 = -8, \quad Y_2 = 30, \quad Y_s = 0, \quad s < 0.$$

$-F(\surd 2, -\surd 2) = \tfrac{1}{2}\{\surd(2+\surd 2) - \surd(2-\surd 2)\} = 0.54119610$

| $s$ | 1 | 2 | 3 | 4 | 5 | 12 | 13 |
|---|---|---|---|---|---|---|---|
| $-k_s^*/k^s$ | 0.5 | 0.53 | 0.537 | 0.5396 | 0.5408 | 0.54119602 | 0.54119608 |

**Example 3.12** (see Example 3.1)

$$F(z,a) = \frac{1}{\Gamma(a)} \int_0^\infty \frac{e^{-t} t^{a-1} dt}{t+z}, \qquad (a, z > 0)$$
$$= \frac{1}{z+}\;\frac{a}{1+}\;\frac{1}{z+}\;\frac{a+1}{1+}\;\frac{2}{z+}\;\frac{a+2}{1+}\;\frac{3}{z+}\;\frac{a+3}{1+}\;\cdots.$$

Using (3.64) derive approximants to

$$F(i,-i) = \int_0^\infty \frac{e^{-t}\,dt}{t^2+1} = 0.62145$$

| $s$ | 2 | 4 | 6 | 8 | 10 | 12 |
|---|---|---|---|---|---|---|
| $k_{2s}^*/k_{2s}$ | 0.74 | 0.653 | 0.6284 | 0.6227 | 0.6217 | 0.6216 |

($k_{2s}^*/k_{2s}$ is 1 reduced by an even convergent for the integral using $t^2 e^{-t}$ as weight function.)

## 3.10 Further Comments on Recurrences

It is clear that the fifth order recurrence for 2nd order c.f.s given in (3.48), the reducible case in (3.51)-(3.52), and that in (3.64) relating directly to a Stieltjes fraction, are considerable advances on the straightforward evaluation of $K_s(\gamma_0, \beta_0, \alpha_0)$, and $K_s(\gamma_1, \beta_1, \alpha_1)$. Indeed some of the examples given in this chapter would involve the evaluation of determinants of order ten or larger, a task not likely to be undertaken lightly without a computer (incidently, most of the computations have been carried out or checked on a H.P. calculator). The evaluation of 3rd or 4th order c.f.s involving 7 and 9 diagonal determinants, for example, is obviously becoming too arduous for the technique to have any appeal (large computers are not always available to the researcher when that person has the need). We spent a considerable time trying to discover the recurrence for the 7-diagonal case, and ultimately gave it up, realizing that there appeared to be no logical strategy for its unfolding. In the meantime H. D. Ursell became aware of the problem and not only solved it, but solved a more general problem relating to systems of linear simultaneous equations in several unknowns (Ursell, 1958). For a determinant with $p$ superdiagonals and $p$ subdiagonals, Ursell proved that the recurrence relations were of order

| $p$ | 0 | 1 | 2 | 3 | 4 | 5 | 6 |
|---|---|---|---|---|---|---|---|
| Symmetrical Case | 1 | 2 | 5 | 15 | 49 | 169 | 604 |
| Asymmetrical Case | 1 | 2 | 6 | 20 | 70 | 252 | 924 |

One must hasten to add that the table, in general, refers to the existence of the recurrence, the recurrence itself still having to be constructed. So the usefulness of this type of evaluation for 3rd and higher order c.f.s is practically nil.

Fortunately, simultaneous recurrences of lower order can be constructed and these, described in Chapter 4, make numerical assessment for higher order c.f.s a reasonable undertaking.

## Further Examples

### Further Example 3.1

For Example 3.9 and approximations to $\int_0^\infty e^{-t}\,dt\,/\,(n+t^2)$ show that

$$P_a[1\,|\,2] = \frac{0.978626}{n+1.339082} + \frac{0.0213738}{n+32.260918}.$$

i.e. point masses of 0.978626 at 1.339082, and 0.0213738 at 32.260918 with moments $\mu_0'=1$, $\mu_1'=2!$, and $\mu_2'=4!$. For the second order c.f.

$$P[1\,|\,2] = \frac{n+14}{n^2+16n+12} = \frac{0.916025}{n+0.788897} + \frac{0.0839749}{n+15.211103}$$

so that the point masses are located nearer the origin.

Now look at the [2|3] c.f. approximants, with

$$P_a[2\,|\,3] = \frac{0.9566457}{n+1.086643} + \frac{0.0431556}{n+21.742711} + \frac{0.198701\times10^{-4}}{n+111.465948}$$

compared to the 2nd order representation

$$P[2\,|\,3] = \frac{0.807756}{n + 0.427152} + \frac{0.185709}{n + 7.709826} + \frac{0.653501\times10^{-3}}{n + 46.290174}$$

showing the same phenomenon.

Lastly, for [3|4] c.f. approximant, we have

$$P_a[3\,|\,4] = \frac{0.938728}{n + 0.947329} + \frac{0.0604553}{n + 17.316320} + \frac{0.815282\times10^{-3}}{n + 77.927758} + \frac{0.128777\times10^{-5}}{n + 247.369897}$$

against

$$P[3\,|\,4] = \frac{0.713822}{n + 0.268743} + \frac{0.260180}{n + 4.373881} + \frac{0.0256814}{n + 24.880074} + \frac{0.316949\times10^{-3}}{n + 98.477301}.$$

Note. Convergents of Stieltjes c.f. have partial fraction form

$$\sum_{r=1}^{p} \frac{L_r}{z + x_r},$$

where $L_r > 0$, $0 \leqslant x_1 < x_2 < \cdots < x_p$ (see Wall, 1948, p120; Shohat and Tamarkin, 1943, p37).

# 4.

# GENERALIZED CONTINUED FRACTIONS AND SIMULTANEOUS RECURRENCE RELATIONS

## 4.1 A General Formula

Our aim is to derive simultaneous recurrences for higher order c.f.s defined in (3.3) and (3.4). Our first objective is to establish a general result for the convergents in terms of the convergents of the J.f. in (1.20a).

Consider the inner-product type integral

$$(f_l, f_m; C_n) = \int_0^\infty \frac{f_l(t) f_m(t) d\psi(t)}{C_n(t)}, \tag{4.1}$$

where

(i) $f_l(t) = \prod_{\lambda=1}^{l}(t + x_\lambda), \qquad (l < n)$

$f_m(t) = \prod_{\lambda=1}^{m}(t + y_\lambda), \qquad (m \leqslant n)$

$C_n(t) = \prod_{\lambda=1}^{n}(t + z_\lambda), \qquad (n = 2, 3, \cdots)$

are real polynomials.

(ii) $$F(z) = \frac{b_0}{z + a_1 -}\ \frac{b_1}{z + a_2 -}\ \cdots \quad \left[= \frac{P_s(z)}{Q_s(z)},\ \text{as } s \to \infty\right]$$

$$= \int_0^\infty \frac{d\psi(t)}{t + z},$$

where $\psi(t)$ is a unique solution to the Stieltjes moment problem.

Let $\{q_r(t)\}$ be the orthogonal system associated with $C_n(t)d\psi(t)$, $q_r$ having highest coefficient unity, and integral square $\phi_r$ ; thus

$$\int_0^\infty q_r(t)q_s(t)C_n(t)d\psi(t) = \delta_{rs}\phi_r . \tag{4.2}$$

Similarly let $\{p_r(t)\}$ be the orthogonal system associated with $\psi(t)$, $p_r$ having highest coefficient unity, and

$$p_r(t) = (t - a_r)p_{r-1}(t) - b_{r-1}p_{r-2}(t), \qquad (r = 1, 2, \cdots)$$
$$p_0(t) = 1, \qquad p_r(t) = 0, \qquad (r < 0)$$

and

$$\int_0^\infty p_r(t)p_s(t)d\psi(t) = b_r b_{r-1} \cdots b_0 \delta_{r,s} . \tag{4.3}$$

We further assume that $C_n(t) > 0$ for $t \geqslant 0$, and has distinct zeros; the confluent case however presents no limiting process problem in the final result.

Since $\psi(t)$ is a unique solution to the Stieltjes moment problem, if in addition the integrals in (4.1) converge, then (3.3) shows that Parseval's theorem applies to the functions $f_l / C_n$ and $f_m / C_n$, and

$$(f_l, f_m : C_n) = \sum_{j=0}^\infty A_{lj} A_{mj} \phi_j \tag{4.4}$$

where

$$A_{kj}\phi_j = \int_0^\infty f_k(t)q_j(t)d\psi(t). \qquad (k = l, m; j = 0, 1, \cdots)$$

The partial sum of the first $r$ terms gives the $r$th convergent of the $n$th order c.f. associated with the integral; i.e.

$$\sum_{j=0}^{r-1} A_{lj} A_{mj} \phi_j = \frac{\chi_r(\underset{\sim}{z}, f_l, f_m)}{\omega_r(\underset{\sim}{z}, f_l, f_m)} . \tag{4.5}$$

We need now to introduce an abbreviation for some determinants which arise. We define

$$|A_r(x_1), A_s(x_2), A_t(x_3)| = \begin{vmatrix} A_r(x_1) & A_r(x_2) & A_r(x_3) \\ A_s(x_1) & A_s(x_2) & A_s(x_3) \\ A_t(x_1) & A_t(x_2) & A_t(x_3) \end{vmatrix}$$

an alternant type of determinant, with obvious extensions to higher orders, and without ambiguity. There could however be ambiguity with $|A_r(x_1), B_s(x_2), C_t(x_3)|$. So we define

$$^{+}|A_r(x_1), B_s(x_2), C_t(x_3)| = \begin{vmatrix} A_r(x_1) & A_r(x_2) & A_r(x_3) \\ B_s(x_1) & B_s(x_2) & B_s(x_3) \\ C_t(x_1) & C_t(x_2) & C_t(x_3) \end{vmatrix}$$

where the function symbol (such as A) can not be separated from its subscript ($r$ in this case). The extension to higher orders is obvious. Similarly,

$$|A_r(x_1), B_s(x_2), C_t(x_3)|^{+} = \begin{vmatrix} A_r(x_1) & B_r(x_2) & C_r(x_3) \\ A_s(x_1) & B_s(x_2) & C_s(x_3) \\ A_t(x_1) & B_t(x_2) & C_t(x_3) \end{vmatrix}$$

in which the function symbol and argument are tied.

Now introduce the polynomials

$$\Delta_j(t) = {}^{+}|f_j(t), p_r(-z_1), p_{r+1}(-z_2), \cdots, p_{r+n-1}(-z_n)| . \tag{4.6}$$

$$(j = l, m; r = 1, 2, \cdots)$$

Set

$$\Delta_j(t) = C_n(t) \sum_{s=0}^{r-1} \overline{A}_{sj}(r)\, q_s(t), \tag{4.7}$$

so that, in the notation of (4.1),

$$(\Delta_l, \Delta_m : C_n) = \sum_{j=0}^{r-1} \overline{A}_{lj}(r)\overline{A}_{mj}(r)\phi_j , \qquad (r = 1, 2, \cdots) \tag{4.8}$$

where

$$\overline{A}_{kj}(r) = |p_r(-z_1), p_{r+1}(-z_2), \cdots, p_{r+n-1}(-z_n)|\, A_{kj} .$$

$$(k = l, m; j = 0, 1, \cdots, r-1)$$

But since $l < n$, and the zeros of $C_n(t)$ are distinct, we have

$$\frac{f_l(t)}{f_n(t)} = (-1)^{n-1} \frac{{}^{+}| z_1^0, z_2^1, \cdots, z_{n-1}^{n-2}, h_l(z_n) |}{| z_1^0, z_2^1, \cdots, z_n^{n-1} |}, \tag{4.9}$$

where $h_l(z_n) = f_l(-z_n)/(t + z_n)$. For example

$$\frac{(t+x_1)(t+x_2)| z_1^0, z_2^1, z_3^2 |}{(t+z_1)(t+z_2)(t+z_3)} =$$

$$\begin{vmatrix} 1 & 1 & 1 \\ z_1 & z_2 & z_3 \\ \dfrac{(x_1-z_1)(x_2-z_1)}{t+z_1} & \dfrac{(x_1-z_2)(x_2-z_2)}{t+z_2} & \dfrac{(x_1-z_3)(x_2-z_3)}{t+z_3} \end{vmatrix}.$$

Now recall that in terms of the $s$th convergent of (4.1), (ii), the orthogonal polynomials are related as follows:

$$p_s(-z) = (-1)^s Q_s(z),$$

$$\int_0^\infty \frac{p_s(t) - p_s(-z)}{t+z} \, d\psi(t) = (-1)^{s-1} P_s(z). \tag{4.10}$$

From (4.5), (4.8), (4.9), and (4.10) we now have an $n$th order c.f. given by

$$(f_l, f_m : C_n) = \lim_{r\to\infty} \frac{\chi_r(\underset{\sim}{z}, f_l, f_m)}{\omega_r(\underset{\sim}{z})}, \qquad (\underset{\sim}{z} \equiv z_1, z_2, \ldots, z_n) \tag{4.11}$$

where

(i) $\omega_r(\underset{\sim}{z}) = | Q_r(z_1), Q_{r+1}(z_2), \cdots, Q_{r+n-1}(z_n) | / \Delta,$ (4.11a)

$(\Delta = | z_1^0, z_2^1, \cdots, z_n^{n-1} |);$

(ii) $\chi_r(\underset{\sim}{z}, f_l, f_m) =$ (4.11b)

$$\frac{(-1)^n}{\Delta} \begin{vmatrix} -(f_l\, f_m) & W_r(f_l) & W_{r+1}(f_l) & \cdots & W_{r+n-1}(f_l) \\ f_m(-z_1) & Q_r(z_1) & Q_{r+1}(z_1) & \cdots & Q_{r+n-1}(z_1) \\ \cdot & \cdot & \cdot & \cdots & \cdot \\ \cdot & \cdot & \cdot & \cdots & \cdot \\ f_m(-z_n) & Q_r(z_n) & Q_{r+1}(z_n) & \cdots & Q_{r+n-1}(z_n) \end{vmatrix}$$

in which

$$(f_l, f_m) = \begin{vmatrix} 1 & 1 & \cdots & 1 \\ z_1 & z_2 & \cdots & z_n \\ \cdot & \cdot & \cdots & \cdot \\ z_1^{n-2} & z_2^{n-2} & \cdots & z_n^{n-2} \\ \beta_1 & \beta_2 & \cdots & \beta_n \end{vmatrix} \div \Delta \tag{4.11c}$$

with

$$\beta_s = f_l(-z_s)\int_0^\infty \frac{(f_m(t) - f_m(-z_s))d\psi(t)}{t + z_s}, \tag{4.11d}$$

$$(s = 1, 2, \cdots, n)$$

(iii)

$$W_s(f_l) = \begin{vmatrix} 1 & 1 & \cdots & 1 \\ z_1 & z_2 & \cdots & z_n \\ \cdot & \cdot & \cdots & \cdot \\ z_1^{n-2} & z_2^{n-2} & \cdots & z_n^{n-2} \\ \gamma_1^{(s)} & \gamma_2^{(s)} & \cdots & \gamma_n^{(s)} \end{vmatrix} \div \Delta, \tag{4.11e}$$

with

$$\gamma_r^{(s)} = P_s(z_r)f_l(-z_r).$$

The elements in (4.11) may be real or complex, depending on the zeros of $C_n(t)$, $f_l(t)$, and $f_m(t)$. However the numerator and denominator (equivalent to $K_{r-1}(\gamma_1, \beta_1, \alpha_1)$, and $K_r(\gamma_0, \beta_0, \alpha_0)$ for a special case with $n = 2$, $f_l = f_m = 1$; see (3.19)) are real. To transform them into transparently real forms is our next objective.

## 4.2 Rational Forms

Define $\Delta U_s^{(\lambda)}$ to be the product of $\Delta$ and the determinant formed from the array

$$\begin{vmatrix} z_1^0 & z_2^0 & z_3^0 & \cdots & z_n^0 \\ z_1^1 & z_2^1 & z_3^1 & \cdots & z_n^1 \\ \cdot & \cdot & \cdot & \cdots & \cdot \\ z_1^{n-1} & z_2^{n-1} & z_3^{n-1} & \cdots & z_n^{n-1} \\ Q_s(z_1) & Q_s(z_2) & Q_s(z_3) & \cdots & Q_s(z_n) \end{vmatrix} \tag{4.12}$$

after deleting the row with index $\lambda$, $\lambda = 0, 1, \cdots, n-1$. Expand by its last row in the form

$$\Delta U_s^{(\lambda)} = \sum_{r=1}^{n} A_r^{(\lambda)} Q_s(z_r) . \tag{4.13}$$

Moreover, let $\underset{\sim}{\Lambda}$ be an $n \times n$ matrix whose $(i, j)$th element is $A_j^{(n-i)}$; similarly let the elements of the $n \times n$ matrices $\underset{\sim}{U}_{s,n}$ and $\underset{\sim}{Q}_{s,n}$ be $U_{s+j-1}^{(n-i)}$ and $Q_{s+j-1}(z_i)$ respectively. Then evidently

$$\underset{\sim}{\Lambda}\, \underset{\sim}{Q}_{s,n} = \Delta \underset{\sim}{U}_{s,n} . \tag{4.14}$$

But by premultiplying $\underset{\sim}{\Lambda}$ by the $n \times n$ alternant $[z_i^{i-1}]$, it is found that

$$|\underset{\sim}{\Lambda}| = \Delta^{n-1} , \tag{4.15}$$

so that $Q_s(\underset{\sim}{z})$, the denominator of the $n$ th order c.f. associated with $(f_l, f_m : C_n)$ (see (4.1)) is given by

$$Q_s(\underset{\sim}{z}) = | U_s^{(n-1)}, U_{s+1}^{(n-2)}, \ldots, U_{s+n-1}^{(0)} | . \tag{4.16}$$

Similarly if $[\underset{\sim}{0}]$ is a row matrix of $n$ zero components and we multiply the bordered matrix of $\underset{\sim}{Q}_{s,n}$ (the determinant of which appears in (4.11e), ii) by

$$\begin{bmatrix} 1 & \underset{\sim}{0} \\ \underset{\sim}{0}' & \underset{\sim}{\Lambda} \end{bmatrix}$$

it will be found that the numerator of the $n$ th order c.f. is

$$\chi_r(\underset{\sim}{z}, f_l, f_m) \equiv \tag{4.17}$$

$$(-1)^n \begin{vmatrix} -(f_l, f_m) & W_r(f_l) & W_{r+1}(f_l) & \cdots & W_{r+n-1}(f_l) \\ U^{(n-1)}(f_m) & U_r^{(n-1)} & U_{r+1}^{(n-1)} & \cdots & U_{r+n-1}^{(n-1)} \\ U^{(n-2)}(f_m) & U_r^{(n-2)} & U_{r+1}^{(n-2)} & \cdots & U_{r+n-1}^{(n-2)} \\ \cdot & \cdot & \cdot & \cdots & \cdot \\ U^{(0)}(f_m) & U_r^{(0)} & U_{r+1}^{(0)} & \cdots & U_{r+n-1}^{(0)} \end{vmatrix}$$

where $U^{(\lambda)}(f_m)$ is defined by the scheme in (4.12) except that the last row is replaced by $f_m(-z_1), f_m(-z_2), \cdots, f_m(-z_n)$; the $W_r$ terms are defined in (4.11e).

It will be noticed that the generalized convergent is the ratio of a bordered determinant to its principal minor (similar to $K_{s-1}(\gamma_1, \beta_1, \alpha_1)/ K_s(\gamma_0, \beta_0, \alpha_0)$), this minor relating solely to the denominator convergents $Q_s(z)$ of the c.f. (4.1), (ii). Moreover the two borders are quite differently structured; the upper border, excepting the first term, relates to linear functions of the basic c.f. numerator convergents $\{P_s(z)\}$, whereas the first column relates solely to the function $f_m(\cdot)$, the polynomial which may (unlike $f_l(\cdot)$) have degree as high as $n$. Also note that $(f_l, f_m)$ will be zero if $l + m < n$; otherwise the generalized convergent will have a basic first term $(-1)^{n+1}(f_l, f_m)$. It is also of interest to point out that

$$W_s(f_l) = \sum_{r=1}^{n} (-1)^{n-r} b_r^{(l)} V_s^{(n-r)} \tag{4.18}$$

where

$$f_l(t) = \sum_{r=1}^{n} b_r^{(l)} t^{n-r},$$

and $V_s^{(\lambda)}$, $\lambda = 0, 1, \ldots, n-1$ is similar in form to $U_s^{(n-1)}$ except that the last row of the numerator determinant is

$$z_1^{\lambda} P_s(z_1), z_2^{\lambda} P_s(z_2), \ldots, z_n^{\lambda} P_s(z_n).$$

Thus $V_s^{(\lambda)}$ is the $n$th divided difference of $z^{\lambda} P_s(z)$ with respect to the arguments $z_1, z_2, \ldots, z_n$; similarly $U_s^{(n-1)}$ is the $n$th divided difference of $Q_s(z)$.

## 4.3 Simultaneous Recurrence Relations

### 4.3.1 J-fraction

From (4.12) it follows from the recurrence for $P_s(z)$, $Q_s(z)$, that

$$\begin{cases} U_s^{(\lambda)} = a_s U_{s-1}^{(\lambda)} - b_{s-1} U_{s-2}^{(\lambda)} - U_{s-1}^{(\lambda-1)} + p_{n-\lambda} U_{s-1}^{(n-1)}, \\ \qquad\qquad (\lambda = 1, 2, \cdots, n-1) \\ U_s^{(0)} = a_s U_{s-1}^{(0)} - b_{s-1} U_{s-2}^{(0)} + p_n U_{s-1}^{(n-1)}, \qquad (s = 2, 3, \cdots) \end{cases}$$

with

$$\begin{cases} U_0^{(\lambda)} = 0, \ \lambda \neq 0, \ U_0^{(0)} = (-1)^{n-1}; \\ U_1^{(\lambda)} = 0, \ \lambda \neq 0, \ \lambda \neq 1; \ U_1^{(0)} = (-1)^{n-1} a_1, \ U_1^{(1)} = (-1)^n, \end{cases} \tag{4.19}$$

where $C_n(t) = \sum_{r=0}^{n} p_r t^{n-r}$. $\qquad (p_0 = 1)$

Similarly,

$$\begin{cases} V_s^{(\lambda)} = a_s V_{s-1}^{(\lambda)} - b_{s-1} V_{s-2}^{(\lambda)} + V_{s-1}^{(\lambda+1)}, & (\lambda = 0, 1, \cdots, n-2); \\ V_s^{(n-1)} = a_s V_{s-1}^{(n-1)} - b_{s-1} V_{s-2}^{(n-1)} + \sum_{r=1}^{n} (-1)^{r-1} p_r V_{s-1}^{(n-r)}, & \\ & (s = 2, 3, \cdots) \end{cases}$$

with

$$\begin{cases} V_0^{(\lambda)} = 0, \quad V_1^{(\lambda)} = 0, \quad \lambda \neq n-1; \\ V_0^{(n-1)} = 0, \quad V_1^{(n-1)} = b_0. \end{cases} \tag{4.20}$$

This very interesting set of *interlaced recurrences* (no $U^{(\lambda)}$ or $V^{(\lambda)}$ has a self-contained recurrence) mimics the basic c.f recurrence contained in

$$\frac{b_0}{z + a_1 -} \; \frac{b_1}{z + a_2 -} \cdots$$

as far as the first three terms are concerned in every case. Thus the basic order involved in the determination of $U_s^{(\lambda)}$, $V_s^{(\lambda)}$ is two, and the generalized convergents to $(f_l, f_m : C_n)$ are then set up by evaluating two $n$th order determinants. For 2nd order c.f.s we have to compute $U_s^{(0)}$, $U_s^{(1)}$, $V_s^{(0)}$, $V_s^{(1)}$ followed by two second order determinants; for 3rd order we need $U_s^{(\lambda)}$, $V_s^{(\lambda)}$ $(\lambda = 0, 1, 2)$ followed by the evaluation of two third order determinants. In the general case there will be $2n$ variables to be used in $n \times n$ determinants, a considerable advance on the single recurrence relation approach described in Sections 3.6 and 3.9; from the latter recall that for the 3rd order c.f.s the recurrences are of order fifteen, and that the order increases at a phenomenal rate.

### 4.3.2 Stieltjes fractions

It is possible for this fraction to have much simpler partial numerators than is the case for its even part. Assume then that with $c_1 = b_0$

$$F(z) = \frac{c_1}{z +} \; \frac{c_2}{1 +} \; \frac{c_3}{z +} \; \frac{c_4}{1 +} \cdots \qquad \left( = \frac{\chi_s(z)}{\omega_s(z)} \right) \tag{4.21}$$

is available. The even convergents of this relate to $P_s(z) = \chi_{2s}(z)$ and $Q_s(z) = \omega_{2s}(z)$. Then set $Q_s(z) = \omega_{2s}(z)$ in (4.12) and use $\omega_{2s}(z) = \omega_{2s-1}(z) + c_{2s}\omega_{2s-2}(z)$; similarly set $P_s(z) = \chi_{2s}(z)$ in (4.11e). We define $X_s^{(\lambda)}$ as $U_s^{(\lambda)}$ except that the last row of (4.12) is $\omega_s(z_1), \omega_s(z_2), \cdots, \omega_s(z_n)$.

Similarly $Y_s^{(\lambda)}$ is $V_s^{(\lambda)}$ with $\chi_s$ replacing $P_s$, in the last row. Then

$$\begin{cases} X_{2s}^{(\lambda)} = X_{2s-1}^{(\lambda)} + c_{2s} X_{2s-2}^{(\lambda)}, & (s = 1, 2, \cdots) \\ Y_{2s}^{(\lambda)} = Y_{2s-1}^{(\lambda)} + c_{2s} Y_{2s-2}^{(\lambda)}. & (\lambda = 0, 1, \cdots, n-1) \end{cases} \tag{4.22}$$

For the odd terms,

$$\begin{cases} X_{2s+1}^{(\lambda)} = -X_{2s}^{(\lambda-1)} + p_{n-\lambda} X_{2s}^{(n-1)} + c_{2s+1} X_{2s-1}^{(\lambda)}, & (\lambda = 1, 2, \cdots, n-1) \\ X_{2s+1}^{(0)} = p_n X_{2s}^{(n-1)} + c_{2s+1} X_{2s-1}^{(0)}, \end{cases} \tag{4.23}$$

$$\begin{cases} Y_{2s+1}^{(\lambda)} = Y_{2s}^{(\lambda+1)} + c_{2s+1} Y_{2s-1}^{(\lambda)}, & (\lambda = 0, 1, \cdots, n-2) \\ Y_{2s+1}^{(n-1)} = \sum_{r=1}^{n} (-1)^{r-1} p_r Y_{2s}^{(n-r)} + c_{2s+1} Y_{2s-1}^{(n-1)}. & (s = 1, 2, \cdots) \end{cases} \tag{4.24}$$

The initial Values are:

$$\begin{cases} X_0^{(0)} = (-1)^{n-1}, \quad X_1^{(0)} = 0; \\ X_0^{(1)} = 0, \quad X_1^{(1)} = (-1)^n; \\ X_0^{(\lambda)} = X_1^{(\lambda)} = 0, \qquad (\lambda = 2, 3, \cdots, n-1); \end{cases}$$

$$\begin{cases} Y_0^{(\lambda)} = 0, \quad Y_1^{(\lambda)} = 0, \qquad (\lambda = 0, 1, \cdots, n-2); \\ Y_0^{(n-1)} = 0, \quad Y_1^{(n-1)} = c_1. \end{cases}$$

## 4.4 Decreasing Sequences

The convergents discussed so far consist of increasing sequences when $f_l(t) = f_m(t)$, in view of Parseval's formula. Decreasing sequences can sometimes be constructed, but the form will depend on the polynomials $f_l$, $f_m$, $C_n$. For example, see the 2nd order example in Section 3.4 and the result in (3.41). Further examples are given in the sequel, and to illustrate the general approach consider

$$I(\pi, k) = \int_0^\infty \frac{d\psi(t)}{t\,\pi^2(t) + k} \tag{4.25}$$

where $k > 0$, $\pi(t)$ is a real polynomial of degree $m$, and $\psi(t)$ is the unique solution of a Stieltjes' moment problem. Then

$$kI = c_1 - \int_0^\infty \frac{\pi^2(t)\{td\,\psi(t)\}}{t\,\pi^2(t)+k} \tag{4.26}$$

using the c.f. (4.21) for the basic $F(z)$. Clearly we can set up approximants in the form of increasing sequences to the integral in (4.26) using (4.1) with $f_l(t) = f_m(t) = \pi(t) = C_n(t) = t\,\pi^2(t) + k$ and $td\,\psi(t)$ for $d\,\psi(t)$. *We want to show that approximants to I can be set up in terms of the odd convergents of (4.21) directly.* In fact,

$$I(\pi, k) = \underset{r\to\infty}{l.\ d.\ s.}\ \chi_{2r+1}(\underset{\sim}{z})/\,\omega_{2r+1}(\underset{\sim}{z}) \tag{4.27}$$

where

$$\omega_{2r+1}(\underset{\sim}{z}) = |\,X_{2r+1}^{(n-1)},\ X_{2r+3}^{(n-2)},\ \cdots,\ X_{2r+2n-1}^{(0)}\,|\,,$$

and

$$\chi_{2r+1}(z) = \frac{(-1)^n}{k}\begin{vmatrix} 0 & W_{2r+1}(\pi) & W_{2r+3}(\pi) & \cdots & W_{2r+2n-1}(\pi) \\ X^{(n-1)}(\hat{\pi}) & X_{2r+1}^{(n-1)} & X_{2r+3}^{(n-1)} & \cdots & X_{2r+2n-1}^{(n-1)} \\ X^{(n-2)}(\hat{\pi}) & X_{2r+1}^{(n-2)} & X_{2r+3}^{(n-2)} & \cdots & X_{2r+2n-1}^{(n-2)} \\ \cdot & \cdot & \cdot & \cdots & \cdot \\ X^{(0)}(\hat{\pi}) & X_{2r+1}^{(0)} & X_{2r+3}^{(0)} & \cdots & X_{2r+2n-1}^{(0)} \end{vmatrix}$$

with

(i) $W_{2r+1}(\pi)$ given by (4.11e) with $P_s$ replaced by $\chi_{2s+1}$; $W_{2r+1}(\pi)$ is therefore a linear function of $Y_{2r+1}^{(0)}, \cdots, Y_{2r+1}^{(n-1)}$.

(ii) $X_{2s+1}^{(\lambda)}$ given by (4.12) with the last row having $\omega_{2s+1}(z)$ for $Q_s(z)$.

(iii) $X^{(\lambda)}(\hat{\pi})$ given by (4.12) with the last row having $\hat{\pi} = -z\,\pi(z)$ for $Q_s(z)$.

Briefly, for the odd part (4.21) (see (1.22))

$$\int_0^\infty \frac{td\,\psi(t)}{t+z} = \frac{c_1c_2}{z+c_2+c_3-}\ \frac{c_3c_4}{z+c_4+c_5-}\ \cdots, \tag{4.28}$$

the $s$ th convergent is $\hat{P}_s(z)/\hat{Q}_x(z)$, where

$$z\,\hat{P}_s(z) = c_1\omega_{2s+1}(z) - z\,\chi_{2s+1}(z)\,,$$

$$z\,\hat{Q}_s(z) = \omega_{2s+1}(z)$$

in terms of the odd convergents of (4.21). Now for $W_s(\pi)$ in (4.11e), we have

$$\gamma_r^{(s)} = \pi(-z_r)\{c_1\omega_{2s+1}(z_r)/z_r - \chi_{2s+1}(z_r)\}. \tag{4.29}$$

Hence

$$W_s(\pi) = c_1\hat{W}_{2s+1}(\pi) - W^*_{2s+1}(\pi) \tag{4.30}$$

where $W^*_{2s+1}(\pi)$ is the same as $W_s(\pi)$ with the last row replaced by $\chi_{2s+1}(z_1)\pi(-z_1)$, $\chi_{2s+1}(z_2)\pi(-z_2)$, $\cdots$. As for $\hat{W}_{2s+1}(\pi)$ this has the last row of $W_s(\pi)$ replaced by $\omega_{2s+1}(z_1)\pi(-z_1)/z_1$, $\omega_{2s+1}(z_2)\pi(-z_2)/z_2$, $\cdots$. Expand this determinant by its last row, so that

$$\hat{W}_{2s+1}(\pi) = \sum_{r=1}^{n} \omega_{2s+1}(z_r)\pi(-z_r)A_n^{(r)}/z_r\,, \tag{4.31}$$

and now consider (4.11b) where $Q$ is replaced by $\hat{Q}$. All terms in the first row except the first can be eliminated by linear operations on rows 2 through $n+1$. In this process, noting that $(f_l, f_m) = 0$, the first term in the first row becomes

$$-\sum_{r=1}^{n} \pi^2(-z_r)A_n^{(r)} \qquad (n = 2m+1)$$

where $-z_r\pi^2(-z_r) + k = 0$, $r = 1, 2, \cdots, 2m+1$. Hence (4.11b) becomes

$$\frac{(-1)^n}{\Delta}\begin{vmatrix} (-1)^n c_1 & -W_{2r+1}(\pi) & \cdots & -W_{2r+n-1}(\pi) \\ \pi(-z_1) & \omega_{2r+1}(z_1)/z_1 & \cdots & \omega_{2r+n-1}(z_1)/z_1 \\ \cdot & \cdot & \cdots & \cdot \\ \cdot & \cdot & \cdots & \cdot \\ \pi(-z_n) & \omega_{2r+1}(z_n)/z_n & \cdots & \omega_{2r+n-1}(z_n)/z_n \end{vmatrix}$$

which used in (4.26) after simplification leads to (4.27).

## 4.5. Some Special Cases

### 4.5.1 Quadratic Denominator

In the notation of (4.1) and Section 4.3.1, and assuming uniqueness for the Stieltjes' moment problem, we have with $(t+z_1)(t+z_2) > 0$ for $t \in (0, \infty)$, and

$$F_2(i) = \int_0^\infty \frac{t^i\,d\psi(t)}{(t+z_1)(t+z_2)}, \qquad (i = 0, 1, 2)$$

$$\omega_r(\underset{\sim}{z}) = \begin{bmatrix} U_r^{(1)} & U_{r+1}^{(1)} \\ U_r^{(0)} & U_{r+1}^{(0)} \end{bmatrix}.$$

(a) $\underline{f_1 = f_m = 1}$.

$$F_2(0) = \underset{r\to\infty}{l.\ i.\ s.} |\, U_r^{(1)}, V_{r+1}^{(0)} \,| / \omega_r(\underset{\sim}{z}). \tag{4.32}$$

If $z_1 = x + iy$, $z_2 = x - iy$, $y \neq 0$, then

$$y^2 F_2(0) = \underset{r\to\infty}{l.\ d.\ s.} \begin{bmatrix} xV_r^{(0)} - V_r^{(1)} & xV_{r+1}^{(0)} - V_{r+1}^{(1)} \\ xU_r^{(1)} - U_r^{(0)} & xU_{r+1}^{(1)} - U_{r+1}^{(0)} \end{bmatrix} / \omega_r(\underset{\sim}{z}). \tag{4.33}$$

Note that in this case

$$F_2(0) = \underset{r\to\infty}{l.\ d.\ s.} \left( \frac{|\, U_r^{(1)}, V_{r+1}^{(0)} \,|}{|\, U_r^{(1)}, U_{r+1}^{(0)} \,|} + \frac{\prod_{s=0}^{r} b_s}{y^2 |\, U_r^{(1)}, U_{r+1}^{(0)} \,|} \right).$$

(b) $\underline{f_l = 1, f_m = t;\ f_l = t, f_m = 1}$.

$$\begin{aligned} F_2(1) &= \lim_{r\to\infty} |\, V_r^{(0)}, U_{r+1}^{(0)} \,| / \omega_r(\underset{\sim}{z}) \\ &= \lim_{r\to\infty} |\, V_r^{(1)}, U_{r+1}^{(1)} \,| / \omega_r(\underset{\sim}{z}). \end{aligned} \tag{4.34}$$

(c) $\underline{f_l = 1, f_m = t^2;\ f_l = f_m = t;\ f_l = t^2, f_m = 1}$.

$$\begin{aligned} F_2(2) &= c_1 + \lim_{r\to\infty} \{ p_1 |\, V_r^{(0)}, U_{r+1}^{(0)} \,| - p_2 |\, V_r^{(0)}, U_{r+1}^{(1)} \,| \} / \omega_r(\underset{\sim}{z}) \\ &= \underset{r\to\infty}{l.\ d.\ s.} \{ c_1 - |\, V_r^{(1)}, U_{r+1}^{(0)} \,| / \omega_r(\underset{\sim}{z}) \} \qquad (4.35) \\ &= \lim_{r\to\infty} \{ c_1 - p_1 |\, V_r^{(1)}, U_{r+1}^{(1)} \,| + p_2 |\, V_r^{(0)}, U_{r+1}^{(1)} \,| \} / \omega_r(\underset{\sim}{z}). \end{aligned}$$

(The $U$'s and $V$'s follow (4.19) and (4.20) with $n = 2$ and $C_2(t) = t^2 + p_1 t + p_2$. Moreover, $a_s$, $b_s$ refer to the J. f. for $F(z)$. For $f_l = t^2$, $f_m = 1$ use $t^2 = C_2(t) - (p_1 t + p_2)$.)

### 4.5.2 Cubic Denominators

Again, let

$$F_3(i) = \int_0^\infty \frac{t^i\, d\psi(t)}{(t+z_1)(t+z_2)(t+z_3)}, \qquad (i = 0, 1, 2, 4)$$

with $(t+z_1)(t+z_2)(t+z_3)>0$ for $t\in(0,\infty)$, and $\omega_r(\underset{\sim}{z}) = |U_r^{(2)}, U_{r+1}^{(1)}, U_{r+2}^{(0)}|$:

(d) $\underline{f_l = f_m = 1.}$

$$F_3(0) = \underset{r\to\infty}{l.\ i.\ s.} |V_r^{(0)}, U_{r+1}^{(2)}, U_{r+2}^{(1)}| / \omega_r(\underset{\sim}{z}) \ . \tag{4.36}$$

(e) $\underline{f_l = 1, f_m = t;\ f_l = t, f_m = 1.}$

$$\begin{aligned} F_3(1) &= -\lim_{r\to\infty} |V_r^{(0)}, U_{r+1}^{(2)}, U_{r+2}^{(0)}| / \omega_r(\underset{\sim}{z}) \\ &= -\lim_{r\to\infty} |V_r^{(1)}, U_{r+1}^{(2)}, U_{r+2}^{(1)}| / \omega_r(\underset{\sim}{z}) \end{aligned} \tag{4.37}$$

(f) $\underline{f_l = t^2, f_m = 1;\ f_l = f_m = t;\ f_l = 1, f_m = t^2.}$

$$\begin{aligned} F_3(2) &= \lim_{r\to\infty} |V_r^{(2)}, U_{r+1}^{(2)}, U_{r+2}^{(1)}| / \omega_r(\underset{\sim}{z}) \\ &= \underset{r\to\infty}{l.\ i.\ s.} |V_r^{(1)}, U_{r+1}^{(2)}, U_{r+2}^{(0)}| / \omega_r(\underset{\sim}{z}) \\ &= \lim_{r\to\infty} |V_r^{(0)}, U_{r+1}^{(1)}, U_{r+2}^{(0)}| / \omega_r(\underset{\sim}{z}) \ . \end{aligned} \tag{4.38}$$

(g) $\underline{f_l = t^2, f_m = t^2;\ f_l = t, f_m = t^3.}$

$$\begin{aligned} F_3(4) &= (a_1 - p_1)b_0 + \underset{r\to\infty}{l.\ i.\ s.} |V_r^{(2)}, U_{r+1}^{(1)}, U_{r+2}^{(0)}| / \omega_r(\underset{\sim}{z}) \\ &= (a_1 - p_1)b_0 - \lim_{r\to\infty} (p_1\theta_1 - p_2\theta_2 + p_3\theta_3) / \omega_r(\underset{\sim}{z}) \end{aligned} \tag{4.39}$$

where

$$\begin{aligned} \theta_1 &= |V_r^{(1)}, U_{r+1}^{(1)}, U_{r+2}^{(0)}| \\ \theta_2 &= |V_r^{(1)}, U_{r+1}^{(2)}, U_{r+2}^{(0)}| \\ \theta_3 &= |V_r^{(1)}, U_{r+1}^{(2)}, U_{r+2}^{(1)}| \ . \end{aligned}$$

(The $U$'s and $V$'s follow (4.19) and (4.20) with $n=3$ and $C_3(t) = t^3 + p_1t^2 + p_2t + p_3$. As before, the $a$'s and $b$'s refer to the J.f. for $F(z)$.)

**Example 4.1**

$$F(z) = \int_0^\infty \frac{e^{-t}\,dt}{t+z}$$

$$= \frac{1}{z+1-}\;\frac{1^2}{z+3-}\;\frac{2^2}{z+5-}\;\cdots\;. \quad \text{(c.f. Example 3.1)}$$

$$I = \int_0^\infty \frac{e^{-t}\,dt}{t^2+1} = 0.62144962$$

$$= \underset{r\to\infty}{l.\ i.\ s.} \,|\, V_r^{(0)}, U_{r+1}^{(1)} \,| \div |\, U_r^{(0)}, U_{r+1}^{(1)} \,|$$

$$= \underset{r\to\infty}{l.\ d.\ s.} \,|\, V_r^{(1)}, U_{r+1}^{(0)} \,| \div |\, U_r^{(1)}, U_{r+1}^{(0)} \,|$$

Recursive Scheme.

Define

$$\psi(\theta_s) = \theta_s - (2s-1)\theta_{s-1} + (s-1)^2\theta_{s-2}.$$

Then

$$\psi(U_s^{(0)}) = U_{s-1}^{(1)}$$
$$\psi(U_s^{(1)}) = -U_{s-1}^{(0)}$$
$$\psi(V_s^{(0)}) = V_{s-1}^{(1)}$$
$$\psi(V_s^{(1)}) = -V_{s-1}^{(0)}$$

and

| $s$ | $U_s^{(0)}$ | $U_s^{(1)}$ | $V_s^{(0)}$ | $V_s^{(1)}$ | Inc. Sequence | Dec. Sequence |
|---|---|---|---|---|---|---|
| 0 | -1 | 0 | 0 | 0 | | |
| 1 | -1 | 1 | 0 | 1 | 0.3333 | 0.6667 |
| 2 | -1 | 4 | 1 | 3 | 0.5172 | 0.6552 |
| 3 | 3 | 17 | 8 | 10 | 0.5886 | 0.6530 |
| 4 | 47 | 80 | 57 | 35 | 0.6120 | 0.6448 |
| 5 | 455 | 401 | 420 | 98 | 0.6186 | 0.6364 |
| 6 | 4231 | 1956 | 3293 | -217 | 0.6199 | 0.6301 |
| 7 | 40579 | 6761 | 27472 | -9642 | 0.6200 | 0.6261 |
| 8 | 408127 | -35008 | 241081 | -161469 | - | - |

Example 4.2

$$F(z) = \frac{1}{\surd(2\pi)} \int_{-\infty}^{\infty} \frac{e^{-t^2/2}dt}{t+z}$$

$$= \frac{1}{z-} \; \frac{1}{z-} \; \frac{2}{z-} \; \frac{3}{z-} \cdots \quad (\mathrm{Im}(z) \neq 0)$$

$$I = \frac{1}{\surd(2\pi)} \int_{-\infty}^{\infty} \frac{e^{-t^2/2}dt}{t^2-t+1} = 0.762826$$

$$= \frac{4}{3} - \frac{4}{3} \frac{1}{\surd(2\pi)} \int_{-\infty}^{\infty} \frac{(t-1/2)^2 e^{-t^2/2}dt}{t^2-t+1}$$

$$= \underset{r\to\infty}{l.\ i.\ s.} \mid U_r^{(1)}, V_{r+1}^{(0)} \mid \div \mid U_r^{(1)}, U_{r+1}^{(0)} \mid$$

$$= \underset{r\to\infty}{l.\ d.\ s.} \frac{2}{3} \begin{vmatrix} V_r^{(1)} + V_r^{(0)}/2 & V_{r+1}^{(1)} + V_{r+1}^{(0)}/2 \\ U_r^{(1)} + 2U_r^{(0)} & U_{r+1}^{(1)} + 2U_{r+1}^{(0)} \end{vmatrix} \div \mid U_r^{(1)}, U_{r+1}^{(0)} \mid$$

| | 1 | 2 | 3 | 4 | 5 | 6 | 10 | 11 |
|---|---|---|---|---|---|---|---|---|
| Inc. Seq. | 0.5000 | 0.5714 | 0.6471 | 0.6897 | 0.7052 | 0.7286 | 0.75239 | 0.75243 |
| Dec. Seq. | 1.1667 | 0.9524 | 0.8824 | 0.8473 | 0.8129 | 0.8057 | 0.7764 | 0.7710 |

## 4.6 Some Special Functions and 2nd Order c.f.s

### 4.6.1 Complex Error Function

For this (see Gautschi, 1964)

$$w(z) = e^{-z^2} erfc(-iz) \tag{4.40}$$

$$= \frac{i}{\pi} \int_{-\infty}^{\infty} \frac{e^{-t^2}dt}{z-t} \quad (\mathrm{Im}(z) > 0)$$

and

$$\pi\,\mathrm{Re}\{w(x+iy)\} = y \int_{-\infty}^{\infty} \frac{e^{-t^2}dt}{(t-x)^2+y^2} \quad (x \text{ real},\ y>0) \tag{4.40a}$$

$$\pi\,\mathrm{Im}\{w(x+iy)\} = \int_{-\infty}^{\infty} \frac{(x-t)e^{-t^2}dt}{(t-x)^2+y^2} \tag{4.40b}$$

The basic c.f. is

$$F(z) = \frac{1}{\sqrt{\pi}} \int_{-\infty}^{\infty} \frac{e^{-t^2}dt}{t+z} = \frac{1}{z-}\ \frac{\frac{1}{2}}{z-}\ \frac{\frac{2}{2}}{z-}\ \frac{\frac{3}{2}}{z-} \cdots .$$

(Note that

$$\mathrm{Re}\left\{\frac{i\sqrt{\pi}w(z)}{z}\right\} = \frac{1}{\sqrt{\pi}} \int_0^{\infty} \frac{e^{-t}t^{-1/2}(t-x^2+y^2)dt}{(t-x^2+y^2)^2+4x^2y^2}$$

$$= \frac{\sqrt{\pi}}{x^2+y^2}(yw_{re} - xw_{im}),$$

$$\mathrm{Im}\left\{\frac{i\sqrt{\pi}w(z)}{z}\right\} = \frac{2xy}{\sqrt{\pi}} \int_0^{\infty} \frac{e^{-t}t^{-1/2}dt}{(t-x^2+y^2)^2+4x^2y^2}$$

$$= \frac{\sqrt{\pi}}{x^2+y^2}(yw_{im} + xw_{re}),$$

when $w(z) = w_{re} + iw_{im}$.)

Use (4.19), (4.20), (4.32), and (4.33) to derive approximants to $(\sqrt{\pi}/2)\,\mathrm{Re}\{w(2+2i)\} = 0.131120$ as follows:

| $s$ | $U_s^{(0)}$ | $U_s^{(1)}$ | $V_s^{(0)}$ | $V_s^{(1)}$ | Inc. Seq. | Dec. Seq. |
|---|---|---|---|---|---|---|
| 0 | -1 | 0 | 0 | 0 | | |
| 1 | 0 | 1 | 0 | 1 | 0.11 | 0.132 |
| 2 | 8.5 | -4 | 1 | -4 | 0.1305 | 0.132 |
| 3 | -32 | 6.5 | -4 | 7 | 0.131038 | 0.1313 |
| 4 | 39.25 | 12 | 5.5 | 10 | 0.131067 | 0.1311 |
| 5 | 160 | -100.25 | 18 | -98 | 0.131113 | 0.1311 |
| 6 | -900.125 | 211 | -111.75 | 223 | 0.1311181 | 0.13112 |
| 7 | 1208 | 356.875 | 169 | 296 | 0.1311185 | 0.13112 |
| 8 | 6005.4375 | -3374 | 687.125 | -3316.5 | -- | -- |

Using this table, approximants to $\sqrt{\pi}\,\mathrm{Im}\{w(2+2i)\} = 0.2325105$ are:

| $s$ | 1 | 2 | 3 | 4 | 5 | 6 | 7 |
|---|---|---|---|---|---|---|---|
| Approx. | 0.235294 | 0.233677 | 0.232349 | 0.232501 | 0.232519 | 0.232508 | 0.232510 |

(Note: A single 4th order recursive scheme is given in example 3.8 for a similar case.)

Even though the integrand in (4.40b) changes sign, it is still possible to set up increasing and decreasing sequences. For suppose $z_1 = u + iv$ and

$z_2 = u - iv$ $(v > 0)$ then for real $E$ and $F$ (not both zero) we can determine real $a$ and $b$ such that

$$Et + F = (at + b)^2 - a^2 C_2(t). \tag{4.41a}$$

For

$$2a^2 v = R + Eu - F, \qquad 2ab = E + 2a^2 u,$$
$$2b^2 = 4abu - 2a^2u^2 + R + F - Eu,$$

where

$$R = \sqrt{\{E^2v^2 + (F - Eu)^2\}},$$

and we choose

$$a = \{\sqrt{(R + Eu - F)}\} / \sqrt{(2v)},$$
$$b = (E + 2a^2 u) / (2a).$$

Similarly, for given $E^*$, $F^*$ (real and not both zero) we can determine $a_1, b_1$ so that

$$(a_1 t + b_1)^2 = a_1^2 C_2(t) - E^* t - F^*. \tag{4.41b}$$

$$(z_1 = u + iv,\ z_2 = u - iv)$$

For the imaginary part of the error function (4.40b) we find (with $y > 0$)

$$a = 1/\sqrt{(2y)}, \qquad b = -(y + x)/\sqrt{(2y)},$$

leading to

$$\pi \operatorname{Im}\{w(x + iy)\} = \tag{4.42a}$$
$$\underset{r\to\infty}{l.i.s.} \frac{-(x+y)^2 |V_r^{(0)}, U_{r+1}^{(1)}| - 2(x+y)|V_r^{(0)}, U_{r+1}^{(0)}| - |V_r^{(1)}, U_{r+1}^{(0)}|}{2y\, |U_r^{(1)}, U_{r+1}^{(0)}|}.$$

Similarly

$$a_1 = 1/\sqrt{(2y)}, \qquad b_1 = (y - x)/\sqrt{(2y)};$$

leading to

$$\pi \operatorname{Im}\{w(x + iy)\} = \tag{4.42b}$$
$$\underset{r\to\infty}{l.d.s.} \frac{(y-x)^2 |V_r^{(0)}, U_{r+1}^{(1)}| - 2(y-x)|V_r^{(0)}, U_{r+1}^{(0)}| + |V_r^{(1)}, U_{r+1}^{(0)}|}{2y\, |U_r^{(1)}, U_{r+1}^{(0)}|}.$$

Numerical.

Defining $A_r$, $B_r$, $C_r$ to be the three determinantal ratios in (4.42) ($A_r = |V_r^{(0)}, U_{r+1}^{(1)}| / |U_r^{(1)}, U_{r+1}^{(0)}|$ for example) we find for $x = 2$ and $y = 3$.

| $r$ | $A_r$ | $B_r$ | $C_r$ |
|---|---|---|---|
| 1 | 13.5 | 0 | -1 |
| 2 | 181 | 2 | -14.5 |
| 3 | 2598.5 | 29 | -208.75 |
| 4 | 38842.125 | 433.5 | -3121.125 |
| 5 | 603814.4375 | 6745.25 | -48522.4375 |

| $r$ | 1 | 2 | 3 | 4 | 5 |
|---|---|---|---|---|---|
| Inc. Seq. | 0.142 | 0.1434 | 0.14369 | 0.143767 | 0.1437672 |
| Dec. Seq. | 0.154 | 0.1443 | 0.14378 | 0.143779 | 0.1437692 |

($\sqrt{\pi}\,\mathrm{Im}\{w(2+i3)\} = 0.1437690$).

### 4.6.2 Sine and Cosine Integrals

These are defined by (see Gautschi, 1964)

$$Si(t) = \int_0^t (\sin x / x)dx , \quad Ci(t) = -\int_t^\infty (\cos x / x)dx , \tag{4.43}$$

which are equivalent to

$$\begin{aligned} \pi/2 - Si(t) &= \int_0^\infty e^{-x} A_2(x)dx / C(x) \\ Ci(t) &= \int_0^\infty e^{-x} A_1(x)dx / C(x), \qquad (\mathrm{Re}(t) > 0) \end{aligned} \tag{4.44}$$

where

$$\begin{cases} A_1(x) = t \sin t - x \cos t \\ A_2(x) = t \cos t + x \sin t \\ C(x) = x^2 + t^2 . \end{cases}$$

The basic c.f. is that for $F(t)$ in Example 4.1.

In the notation of 4.6.1,

$$a^2(t^2 + x^2) + t \cos t + x \sin t = (ax + b)^2$$

where

$$a = \sqrt{\left(\frac{1 - \cos t}{2t}\right)}, \quad b = \sqrt{\left(\frac{t + t \cos t}{2}\right)}; \quad (t > 0).$$

Similarly

$$a_1^2 (t^2 + x^2) - t \cos t - x \sin t = (a_1 x - b_1)^2$$

where

$$a_1 = \sqrt{\left(\frac{1 + \cos t}{2t}\right)}, \quad b_1 = \sqrt{\left(\frac{t - t \cos t}{2}\right)}.$$

Then

$$Si(t) = \underset{r \to \infty}{l.\ i.\ s.}\, (t_r) = \underset{r \to \infty}{l.\ d.\ s.}\, (t_r^*), \qquad (t > 0) \tag{4.45}$$

where

$$t_r \equiv \frac{\pi}{2} - \left\{ \frac{(1 + \cos t)}{2t} A_r + (\sin t) B_r + \frac{t(1 - \cos t)}{2} C_r \right\},$$

$$t_r^* = \frac{\pi}{2} + \frac{(1 - \cos t)}{2t} A_r - (\sin t) B_r + \frac{t(1 + \cos t)}{2} C_r .$$

Again for $Ci(t)$ we have the increasing sequence $(\xi_r)$ and decreasing sequence $(\xi_r^*)$,

$$\xi_r = -\left\{ \frac{(1 - \sin t)}{2t} A_r + (\cos t) B_r + \frac{t(1 + \sin t)}{2} C_r \right\}, \tag{4.46}$$

$$\xi_r^* = \left\{ \frac{(1 + \sin t)}{2t} A_r - (\cos t) B_r + \frac{t(1 - \sin t)}{2} C_r \right\}.$$

<u>Numerical $t = 1$</u>

| $r$ | $A_r$ | $B_r$ | $C_r$ |
|---|---|---|---|
| 2 | 2 | 1 | -1 |
| 3 | 19 | 11 | -15 |
| 4 | 11319 | 6195 | -10743 |
| 5 | 513373 | 278705 | -498973 |
| 6 | 31989659 | 17392615 | -31471259 |

| $r$ | $t_r$ | $t_r^*$ | $\xi_r$ | $\xi_r^*$ |
|---|---|---|---|---|
| 2 | 0.8535 | 1.1868 | 0.0740 | 0.4073 |
| 3 | 0.8659 | 1.0039 | 0.2194 | 0.3573 |
| 4 | 0.9179 | 0.9507 | 0.3217 | 0.3545 |
| 5 | 0.9321 | 0.94995 | 0.3324 | 0.3503 |
| 6 | 0.9397 | 0.94991 | 0.3357 | 0.3459 |

$$0.9397 < Si(1) < 0.94991$$
$$0.336 < Ci(1) < 0.346$$

### 4.6.3 Fresnel Integrals

By definition (see Gautschi, 1964)

$$S\left(\sqrt{\frac{2t}{\pi}}\right) = \frac{1}{\sqrt{(2\pi)}} \int_0^t \frac{\sin x}{\sqrt{x}}\, dx \tag{4.47}$$
$$C\left(\sqrt{\frac{2t}{\pi}}\right) = \frac{1}{\sqrt{(2\pi)}} \int_0^t \frac{\cos x}{\sqrt{x}}\, dx$$

and we assume $t > 0$. These are equivalent to

$$C\left(\sqrt{\frac{2t}{\pi}}\right) - \frac{1}{2} = \frac{1}{\pi}\left(\sqrt{\frac{t}{2}}\right) \int_0^\infty \frac{e^{-x} x^{-1/2}(t \sin t - x \cos t)dx}{t^2 + x^2} \tag{4.48}$$
$$\frac{1}{2} - S\left(\sqrt{\frac{2t}{\pi}}\right) = \frac{1}{\pi}\left(\sqrt{\frac{t}{2}}\right) \int_0^\infty \frac{e^{-x} x^{-1/2}(t \cos t + x \sin t)dx}{t^2 + x^2}.$$

The basic c.f. is

$$F(z) = \frac{1}{\sqrt{\pi}} \int_0^\infty \frac{e^{-t} t^{-1/2} dt}{t + z} \tag{4.49}$$
$$= \frac{1}{z + \frac{1}{2} -}\; \frac{1\cdot\frac{1}{2}}{z + \frac{5}{2} -}\; \frac{2\cdot\frac{3}{2}}{z + \frac{7}{2} -} \cdots .$$

It is convenient to define

$$C^*(t) = \sqrt{\frac{2\pi}{t}} \left\{C\left(\sqrt{\frac{2t}{\pi}}\right) - \frac{1}{2}\right\} \tag{4.50}$$
$$S^*(t) = \sqrt{\frac{2\pi}{t}} \left\{\frac{1}{2} - S\left(\sqrt{\frac{2t}{\pi}}\right)\right\}.$$

Bounds for $C^*(t)$ are now set up following (4.46) and the use of (4.49) in the terms $A_r, B_r, C_r$. Similarly, bounds for $S^*(t)$ are set. We have

$$t_r = -\left\{\frac{(t-\cos t)}{2t}A_r - (\sin t)B_r + \frac{t(1+\cos t)}{2}C_r\right\}, \qquad (4.51)$$

$$t_r^* = \left\{\frac{(1+\cos t)}{2t}A_r + (\sin t)B_r + \frac{t(1-\cos t)}{2}C_r\right\}.$$

Numerical $t = 5$.

| | $C^*(5)$ | | $S^*(5)$ | |
|---|---|---|---|---|
| 1 | -0.1955 | -0.1916 | 0.0351 | 0.0390 |
| 2 | -0.1925 | -0.1922 | 0.0354 | 0.0385 |
| 3 | -0.19230 | 0.19226 | 0.03816 | 0.03821 |
| 4 | -0.192303 | -0.192293 | 0.038172 | 0.038181 |

## 4.7 Determinantal Identities

### 4.7.1 Special Cases

For conventional c.f.s there is the basic identity (1.17)

$$\omega_{s-1}\chi_s - \omega_s\chi_{s-1} = b_{s-1}b_{s-2}\cdots b_1.$$

With $n$ th order c.f.s, there are many identities involving subsets of $V^{(0)}$, $V^{(1)}, \cdots, V^{(n-1)}$, and $U^{(0)}, U^{(1)}, \cdots, U^{(n-1)}$. The identifies are of more than theoretical interest since the simplest of them provide essential numerical checks, and can indicate the presence of overflow or underflow problems in computation. Examples for 2nd-order c.f.s are in evidence in (4.34) and for 3rd order in (4.37) and (4.38). A more complete listing is

Second-Order ($n = 2$)

$$|V_s^{(0)}, U_{s+1}^{(0)}| = |V_s^{(1)}, U_{s+1}^{(1)}| \qquad (s = 1, 2, \cdots) \qquad (4.52)$$

$$|V_s^{(0)}, V_{s+1}^{(1)}, U_{s+2}^{(1)}| = \alpha_s V_{s+1}^{(0)} \qquad (\alpha_s = \prod_{\lambda=0}^{s} b_\lambda)$$

$$|V_s^{(1)}, V_{s+1}^{(0)}, U_{s+2}^{(0)}| = -\alpha_s V_{s+1}^{(1)}$$

$$|U_s^{(0)}, V_{s+1}^{(1)}, U_{s+2}^{(1)}| = \alpha_s U_{s+1}^{(0)}$$

$$|U_s^{(1)}, V_{s+1}^{(0)}, U_{s+2}^{(0)}| = -\alpha_s U_{s+1}^{(1)}$$

$$|V_s^{(0)}, U_{s+1}^{(0)}, V_{s+2}^{(1)}, U_{s+3}^{(1)}| = \alpha_s \alpha_{s+1}. \qquad (s = 0, 1, \cdots)$$

<u>Third Order ($n = 3$)</u>

$$\begin{cases} | U_s^{(0)}, V_{s+1}^{(1)}, U_{s+2}^{(1)} | = | U_s^{(0)}, V_{s+1}^{(2)}, U_{s+2}^{(2)} | , \\ | U_s^{(1)}, V_{s+1}^{(2)}, U_{s+2}^{(2)} | = - | U_s^{(1)}, V_{s+1}^{(0)}, U_{s+2}^{(0)} | , \\ | U_s^{(2)}, V_{s+1}^{(0)}, U_{s+2}^{(0)} | = | U_s^{(2)}, V_{s+1}^{(1)}, U_{s+2}^{(1)} | , \quad (s = 2, 3, \cdots) \end{cases}$$

$$\begin{cases} | V_s^{(0)}, U_{s+1}^{(0)}, V_{s+2}^{(1)}, U_{s+3}^{(1)} | = -\alpha_s | V_{s+1}^{(2)}, U_{s+2}^{(2)} | , \\ | V_s^{(1)}, U_{s+1}^{(1)}, V_{s+2}^{(2)}, U_{s+3}^{(2)} | = -\alpha_s | V_{s+1}^{(0)}, U_{s+2}^{(0)} | , \\ | V_s^{(2)}, U_{s+1}^{(2)}, V_{s+2}^{(0)}, U_{s+3}^{(0)} | = -\alpha_s | V_{s+1}^{(1)}, U_{s+2}^{(1)} | , \quad (s = 1, 2, \cdots) \end{cases} \tag{4.53}$$

$$\begin{cases} | V_s^{(0)}, V_{s+1}^{(1)}, V_{s+2}^{(2)}, U_{s+3}^{(2)} | = -\alpha_s | V_{s+1}^{(0)}, V_{s+2}^{(1)} | , \\ | V_s^{(0)}, V_{s+1}^{(2)}, V_{s+2}^{(1)}, U_{s+3}^{(1)} | = \alpha_s | V_{s+1}^{(0)}, V_{s+2}^{(2)} | , \quad (s = 1, 2, \cdots) \end{cases}$$

$$| V_s^{(0)}, V_{s+1}^{(1)}, U_{s+2}^{(1)}, V_{s+3}^{(2)}, U_{s+4}^{(2)} | = \alpha_s \alpha_{s+1} V_{s+2}^{(0)} , \quad (s = 0, 1, \cdots)$$

$$| V_s^{(0)}, U_{s+1}^{(0)}, V_{s+2}^{(1)}, U_{s+3}^{(1)}, V_{s+4}^{(2)}, U_{s+5}^{(2)} | = -\alpha_s \alpha_{s+1} \alpha_{s+2} . \quad (s = 0, 1, \cdots)$$

### 4.7.2 A General Case

There is no general formula to cover all possibilities for the combinations of the $U$'s and $V$'s. We content ourselves by indicating one approach to one case. Let

$$\phi_s \begin{pmatrix} \hat{a}_1, \hat{a}_2, \cdots, \hat{a}_n \\ \hat{b}_1, \hat{b}_2, \cdots, \hat{b}_m \end{pmatrix}$$

be a determinant of order $n + m$ with elements $P_{r+s-1}(\hat{a}_1)$, $P_{r+s-1}(\hat{a}_2)$, $\cdots$, $P_{r+s-1}(\hat{a}_n)$, $Q_{r+s-1}(\hat{b}_1)$, $Q_{s+s-1}(\hat{b}_2)$, $\cdots$, $Q_{r+s-1}(\hat{b}_m)$, in the $r$th column, $r = 1, 2, \cdots, n+m$ ($P, Q$ defined in (4.1)).

Apply the recursion

$$\phi_r(z) = (z + a_r)\phi_{r-1}(z) - b_{r-1}\phi_{r-2}(z) \tag{4.54}$$

(satisfied by $P(z)$, $Q(z)$) to

$$\phi_s \begin{pmatrix} z_2, z_3, \cdots, z_n \\ z_1, z_2, \cdots, z_n \end{pmatrix} \tag{4.55}$$

to eliminate $P_{s+2n-2}(z_2)$, $Q_{s+2n-2}(z_2)$, followed by the elimination of $P_{s+2n-3}(z_2)$, $Q_{s+2n-3}(z_2)$, and so on, arriving at

$$\Phi_s \begin{vmatrix} z_2, z_3, \cdots, z_n \\ z_1, z_2, \cdots, z_n \end{vmatrix} = (-1)^{n-1}\alpha_s(z_2 - z_1) \times \prod_{\lambda=3}^{n}(z_\lambda - z_2)^2\Phi_{s+1}\begin{vmatrix} z_3, \cdots, z_n \\ z_1, z_3, \cdots, z_n \end{vmatrix} \tag{4.56}$$

where

$$\alpha_s = \prod_{\lambda=0}^{s} b_\lambda,$$

and it is assumed $n > 2$. Repeating the process, but now dealing with terms involving $z_3$, we have

$$\Phi_s \begin{vmatrix} z_2, \cdots, z_n \\ z_1, \cdots, z_n \end{vmatrix} = (-1)^{(n-1)+(n-2)}\alpha_s\,\alpha_{s+1}(z_2 - z_1)(z_3 - z_1) \times \prod_{\mu=2}^{3}\prod_{\lambda=\mu+1}^{n}(z_\lambda - z_\mu)^2\,\Phi_{s+2}\begin{vmatrix} z_4, \cdots, z_n \\ z_1, z_4, \cdots, z_n \end{vmatrix}. \quad (n > 3) \tag{4.57}$$

The condensation process is continued until the $\Phi$ term becomes

$$\Phi_{s+n-2}\begin{vmatrix} z_n \\ z_1, z_n \end{vmatrix},$$

that is

$$\begin{vmatrix} P_{s+n-2}(z_n) & P_{s+n-1}(z_n) & P_{s+n}(z_n) \\ Q_{s+n-2}(z_1) & Q_{s+n-1}(z_1) & Q_{s+n}(z_1) \\ Q_{s+n-2}(z_n) & Q_{s+n-1}(z_n) & Q_{s+n}(z_n) \end{vmatrix}$$

or $-(z_n - z_1)\alpha_{s+n-2}Q_{s+n-1}(z_1)$. Hence we have

$$\Phi_s \begin{vmatrix} z_2, z_3, \cdots, z_n \\ z_1, z_2, \cdots, z_n \end{vmatrix} = (-1)^{n(n-1)/2}\prod_{\lambda=s}^{s+n-2}\alpha_\lambda \prod_{k=2}^{n-1}\prod_{j=k+1}^{n}(z_j - z_k)Q_{s+n-1}(z_1). \tag{4.58}$$

Similarly we prove the identity,

$$\Phi_s \begin{vmatrix} z_1, z_2, \cdots, z_n \\ z_2, z_3, \cdots, z_n \end{vmatrix} = \tag{4.59}$$

$$(-1)^{n(n+1)/2-1}\Delta_n \prod_{\lambda=s}^{s+n-2} \lambda \prod_{k=2}^{n-1} \prod_{j=k+1}^{n} (z_j - z_k) P_{s+n-1}(z_1)$$

where $\Delta_n = |z_1^0, z_2^1, \cdots, z_n^{n-1}|$.

Further elaboration of these results is given in Shenton (1957c). That there are many aspects of the problem, defying a unified result, may be gathered from the tentative assessment of the number of distinct identities for various orders $n$ of generalized c.f.s:

| $n$ | 2 | 3 | 4 | 5 | 6 | 7 | 8 |
|---|---|---|---|---|---|---|---|
| No. of identities | 5 | 27 | 119 | 495 | 2015 | 8127 | 32639 |

## 4.8 Further Generalization

As was pointed out in Section 1.10, the Padé table consists formally of c.f.s for a series

$$f(z) = a_0 + a_1 z + \cdots, \tag{4.60}$$

its successive truncations from the left, and the same set for the inverse series ($a_0 \neq 0$).

From the viewpoint of integral representations, the corresponding complex of c.f.s arises for the most part from

$$\int_a^b \frac{t^s \, d\psi(t)}{1 + zt}$$

with $s = 0, 1, \cdots$, and its truncation. However, the analogue for generalized c.f.s is not obvious since the truncation of the series development of $(f_l, f_m : C_n)$ is not unique, depending on which parameters in $C_n$ are used in the asymptotic development (note that terms in a c.f. developed from a series are sensitively poised with respect to the series coefficient). A modification of any coefficient in the series may destroy the closed form structure of the c.f. partial numerators (and, or, denominators). Similarly, new terms cannot be added in general to the left of a series (such as $a_{-1}z^{-1}$ in (4.60)). For a Stieltjes c.f. corresponding to (4.57), Wall (1929) has shown that there will be a Stieltjes c.f. for the series $a_{-1}z^{-1} + f(z)$ if and only if the sum of the even partial denominators

($k_{2s}$ in $F(n)$ in Section 1.3) converges. This type of property should be kept in mind when summing divergent series for which there is anomolous behavior in the first few coefficients which can be alleviated by truncation.

Modifications of the basic c.f. itself lend themselves directly to the generalized case. For example, in the notation of (1.21),

$$\frac{c_1}{F(n)} - n = \frac{c_2}{1+} \frac{c_2}{n+} \cdots, \tag{4.61}$$

so that if

$$\phi_1(n) = \frac{c_1}{nF(n)} - 1$$

then

$$\phi_1(n) = \frac{c_2}{n+} \frac{c_3}{1+} \frac{c_4}{n+} \cdots.$$

Moreover for the unique Stieltjes moment problem

$$\phi_1(n) = \int_0^\infty \frac{d\psi^*(t)}{t+n}$$

where $\psi^*(\cdot)$ has the usual properties. Hence there will be the basic 2nd order c.f. for

$$\frac{\phi_1(z_2) - \phi_1(z_1)}{z_1 - z_2} = \left\{\frac{c_1}{z_2F(z_2)} - \frac{c_1}{z_1F(z_1)}\right\} \Big/ (z_1 - z_2) \tag{4.62}$$

including the confluent case.

The truncation process could be continued further.

## 4.9 Some Further Formal Results Based on the Formulas (4.11)

### 4.9.1 An Extensional (refer to Aitken, 1946)

Let $n = 1$, $f_l(t) = 1$, $f_m(t) = t + x_2$ and $C(t) = t + x_1$. Then

$$\int_0^\infty \frac{d\psi(t)}{t+x_1} = \lim_{s\to\infty} \left|\frac{|P_s(x_1), Q_{s+1}(x_2)|^+}{|Q_s(x_1), Q_{s+1}(x_2)|}\right|. \tag{4.63}$$

$$(x_1, x_2 > 0,\ x_1 \neq x_2)$$

(The notations $|\cdot,\cdot,\cdot|^{+}$ and $^{+}|\cdot,\cdot,\cdot|$ are defined in Section 4.1.)

Note also that the choice $n = 2$, $f_l(t) = t + x_1$, $f_m(t) = t + x_2$, leads to

$$\int_0^\infty \frac{(t + x_1)(t + x_2)d\psi(t)}{(t + x_1)(t + x_2)} = b_0$$

$$= a_0 + \lim_{s\to\infty} \frac{(x_1 - x_2)(b_0 b_1 \cdots b_s)}{|Q_s(x_1), Q_{s+1}(x_2)|}.$$

Thus

$$\lim_{s\to\infty}\left\{\frac{B_s}{|Q_s(x_1), Q_{s+1}(x_2)|}\right\} = 0, \tag{4.64}$$

$$(x_1, x_2 > 0;\ x_1 = x + iy,\ x_2 = x - iy,\ y > 0)$$

where $B_s = b_s b_{s-1} \cdots b_0$.

### 4.9.2 Product of Stieltjes Transforms

A form for

$$P(x_1, x_2) = \int_0^\infty \frac{d\psi(t)}{t + x_1} \cdot \int_0^\infty \frac{d\psi(t)}{t + x_2}. \qquad (x_1, x_2 > 0)$$

Expand by Laplace the determinant

$$\begin{vmatrix} P_s(x_1) & P_{s+1}(x_1) & 0 & 0 \\ P_s(x_2) & P_{s+1}(x_2) & P_s(x_2) & P_{s+1}(x_2) \\ Q_s(x_1) & Q_{s+1}(x_1) & Q_s(x_1) & Q_{s+1}(x_1) \\ Q_s(x_2) & Q_{s+1}(x_2) & Q_s(x_2) & Q_{s+1}(x_2) \end{vmatrix} = 0.$$

We have

$$|P_s(x_1), P_{s+1}(x_2)|\ |Q_s(x_1), Q_{s+1}(x_2)|$$

$$= B_s^2 - |P_s(x_1), Q_{s+1}(x_2)|^{+}\ |P_s(x_2), Q_{s+1}(x_1)|^{+}.$$

$$(B_s = b_s b_{s-1} \cdots b_0)$$

Hence

$$\lim_{s\to\infty}\frac{|P_s(x_1), P_{s+1}(x_2)|}{|Q_s(x_1), Q_{s+1}(x_2)|} = \lim_{s\to\infty}\left\{\frac{B_s}{|Q_s(x_1), Q_{s+1}(x_2)|}\right\}^2 - \lim_{s\to\infty}\frac{|P_s(x_1), Q_{s+1}(x_2)|^{+}|P_s(x_2), Q_{s+1}(x_1)|^{+}}{|Q_s(x_1), Q_{s+1}(x_2)|^2}$$

or from (4.63)

$$\int_0^\infty \frac{d\psi(t)}{t+x_1}\int_0^\infty \frac{d\psi(t)}{t+x_2} = \lim_{s\to\infty}\left\{\frac{|P_s(x_1), P_{s+1}(x_2)|}{|Q_s(x_1), Q_{s+1}(x_2)|} - \frac{B_s^2}{|Q_s(x_1), Q_{s+1}(x_2)|^2}\right\}. \tag{4.65}$$

In effect then, the two-part product of Stieltjes integrals may be approximated as a single 2nd order c.f., namely

$$F(x_1)F(x_2) = \lim_{s\to\infty}\frac{|P_s(x_1), P_{s+1}(x_2)|}{|Q_s(x_1), Q_{s+1}(x_2)|}, \tag{4.66a}$$

which may be compared with the one-part form

$$F(x) = \lim_{s\to\infty}\left\{\frac{P_s(x)}{Q_s(x)}\right\}.$$

To derive a formula in the three-part case we need one basic formula.

Note that (4.66a) may be expressed as

$$\lim_{s\to\infty}\frac{|V_s^{(0)}, V_s^{(1)}|}{|U_s^{(0)}, U_s^{(1)}|}, \tag{4.66b}$$

the components following the recurrences in Section 4.3.1 with $C_2(t) = (t+x_1)(t+x_2)$.

### 4.9.3 A Simple Identity

From

$$\begin{vmatrix} P_s(x_1) & P_{s+1}(x_1) & P_{s+2}(x_1) \\ Q_s(x_1) & Q_{s+1}(x_1) & Q_{s+2}(x_1) \\ P_s(x_2) & P_{s+1}(x_2) & P_{s+2}(x_2) \end{vmatrix} = \begin{vmatrix} P_s(x_1) & P_{s+1}(x_1) & 0 \\ Q_s(x_1) & Q_{s+1}(x_1) & 0 \\ P_s(x_2) & P_{s+1}(x_2) & (x_2-x_1)P_{s+1}(x_2) \end{vmatrix},$$

and

$$P_s(\theta)Q_{s+1}(\theta) - P_{s+1}(\theta)Q_s(\theta) = -b_s b_{s-1} \cdots b_0, \tag{4.67}$$

we have

$$|P_s(x_1), Q_{s+1}(x_1), P_{s+2}(x_2)|^+ = -(x_2 - x_1)P_{s+1}(x_2)(b_s b_{s-1} \cdots b_0), \tag{4.68}$$

and

$$|P_s(x_3), Q_{s+1}(x_2), Q_{s+2}(x_3)|^+ = (x_2 - x_3)Q_{s+1}(x_2)(b_s b_{s-1} \cdots b_0). \tag{4.69}$$

### 4.9.4 A Product of Three Stieltjes Transforms

Now consider the Laplacian expansion of

$$X(x_1, x_2, x_3) = \tag{4.70}$$

$$\begin{vmatrix} P_s(x_1) & P_{s+1}(x_1) & P_{s+2}(x_1) & 0 & 0 & 0 \\ P_s(x_2) & P_{s+1}(x_2) & P_{s+2}(x_2) & 0 & 0 & 0 \\ P_s(x_3) & P_{s+1}(x_3) & P_{s+2}(x_3) & P_s(x_3) & P_{s+1}(x_3) & P_{s+2}(x_3) \\ Q_s(x_1) & Q_{s+1}(x_1) & Q_{s+2}(x_1) & Q_s(x_1) & Q_{s+1}(x_1) & Q_{s+2}(x_1) \\ Q_s(x_2) & Q_{s+1}(x_2) & Q_{s+2}(x_2) & Q_s(x_2) & Q_{s+1}(x_2) & Q_{s+2}(x_2) \\ Q_s(x_3) & Q_{s+1}(x_3) & Q_{s+2}(x_3) & Q_s(x_3) & Q_{s+1}(x_3) & Q_{s+2}(x_3) \end{vmatrix}$$

using its first three columns (we assume $x_1, x_2, x_3$ are distinct and in general subscribe to the conditions involved in the validity of third order c.f.s., i.e. $(t + x_1)(t + x_2)(t + x_3) > 0$ for $t \in (0, \infty)$.)

$$X(x_1, x_2, x_3) = 0$$

leads to

$$\begin{aligned} &|P_s(x_1), P_{s+1}(x_2), P_{s+2}(x_3)|\,|Q_s(x_1), Q_{s+1}(x_2), Q_{s+2}(x_2)| - \\ &|P_s(x_1), P_{s+1}(x_2), Q_{s+2}(x_1)|^+\,|P_s(x_3), Q_{s+1}(x_2), Q_{s+2}(x_3)|^+ + \\ &|P_s(x_1), P_{s+1}(x_2), Q_{s+2}(x_2)|^+\,|P_s(x_3), Q_{s+1}(x_1), Q_{s+2}(x_3)|^+ - \\ &|P_s(x_1), P_{s+1}(x_2), Q_{s+2}(x_3)|^+\,|P_s(x_3), Q_{s+1}(x_1), Q_{s+2}(x_2)|^+ = 0. \end{aligned}$$

Thus since

$$\Delta_s(x_1, x_2, x_3) = |Q_s(x_1), Q_{s+1}(x_2), Q_{s+2}(x_3)| > 0,$$

we have

$$\frac{|P_s(x_1), P_{s+1}(x_2), P_{s+3}(x_3)|}{\Delta_s(x_1, x_2, x_3)} = \tag{4.71}$$

$$- \Delta_0(x_1, x_2, x_3) B_s^2 \left\{ \frac{P_{s+1}(x_2) Q_{s+1}(x_2)}{(x_3 - x_1)\Delta_s} - \frac{P_{s+1}(x_1) Q_{s+1}(x_1)}{(x_2 - x_3)\Delta_s} \right\}$$

$$+ \frac{|P_s(x_1), P_{s+2}(x_2), Q_{s+2}(x_3)|^+}{|Q_s(x_1), Q_{s+1}(x_2), Q_{s+2}(x_3)|} \frac{|P_s(x_3), Q_{s+1}(x_1), Q_{s+2}(x_2)|^+}{|Q_s(x_1), Q_{s+1}(x_2), Q_{s+2}(x_3)|}.$$

We now need to explore expressions for the second product ratios in (4.65). Expand by Laplace,

$$\begin{vmatrix} P_s(x_1) & P_{s+1}(x_1) & P_{s+2}(x_1) & P_s(x_1) & P_{s+1}(x_1) & P_{s+2}(x_1) \\ P_s(x_2) & P_{s+1}(x_2) & P_{s+2}(x_2) & 0 & 0 & 0 \\ Q_s(x_3) & Q_{s+1}(x_3) & Q_{s+2}(x_3) & 0 & 0 & 0 \\ Q_s(x_1) & Q_{s+1}(x_1) & Q_{s+2}(x_1) & Q_s(x_1) & Q_{s+1}(x_1) & Q_{s+2}(x_1) \\ Q_s(x_2) & Q_{s+1}(x_2) & Q_{s+2}(x_2) & Q_s(x_2) & Q_{s+1}(x_2) & Q_{s+2}(x_2) \\ Q_s(x_3) & Q_{s+1}(x_3) & Q_{s+2}(x_3) & Q_s(x_3) & Q_{s+1}(x_3) & Q_{s+2}(x_3) \end{vmatrix} = 0$$

yielding,

$$\begin{aligned} &|P_s(x_1), P_{s+1}(x_2), Q_{s+2}(x_3)|^+ \; |Q_s(x_1), Q_{s+1}(x_2), Q_{s+3}(x_2)| - \\ &|P_s(x_2), Q_{s+1}(x_3), Q_{s+2}(x_1)|^+ \; |P_s(x_1), Q_{s+1}(x_2), Q_{s+2}(x_3)|^+ + \\ &|P_s(x_2), Q_{s+1}(x_3), Q_{s+2}(x_2)|^+ \; |P_s(x_1), Q_{s+1}(x_1), Q_{s+2}(x_3)|^+ = 0. \end{aligned} \tag{4.72}$$

Thus,

$$\begin{aligned} &\frac{|P_s(x_1), P_{s+1}(x_2), Q_{s+2}(x_3)|^+}{|Q_s(x_1), Q_{s+1}(x_2), Q_{s+2}(x_3)|} \\ &= \frac{|P_s(x_2), Q_{s+1}(x_3), Q_{s+2}(x_1)|^+}{|Q_s(x_1), Q_{s+1}(x_2), Q_{s+2}(x_3)|} \frac{|P_s(x_1), Q_{s+1}(x_2), Q_{s+2}(x_3)|^+}{|Q_s(x_1), Q_{s+1}(x_2), Q_{s+2}(x_3)|} \\ &\qquad + \frac{(x_3 - x_2)(x_3 - x_1) B_s^2 Q_{s+1}^2(x_3)}{|Q_s(x_1), Q_{s+1}(x_2), Q_{s+2}(x_3)|^2}. \end{aligned} \tag{4.73}$$

But by the choice $n = 3$, $f_l(t) = (t + x_1)(t + x_2)$, and $f_m(t) = (t + x_2)(t + x_3)$, we find

$$\frac{B_s Q_{s+1}(x_2)}{|Q_s(x_1), Q_{s+1}(x_2), Q_{s+2}(x_3)|} \to 0 \quad \text{as} \quad s \to \infty. \tag{4.74}$$

Hence, ignoring components which tend to zero as $s \to \infty$, we have

$$\lim_{s\to\infty}\left\{\frac{|P_s(x_1), P_{s+1}(x_2), P_{s+2}(x_3)|}{|Q_s(x_1), Q_{s+1}(x_2), Q_{s+3}(x_3)|}\right\}$$

$$= \lim_{s\to\infty}\left\{\frac{|P_s(x_2), Q_{s+1}(x_3), Q_{s+2}(x_1)|^+}{|Q_s(x_1), Q_{s+1}(x_2), Q_{s+2}(x_3)|}\;\frac{|P_s(x_1), Q_{s+1}(x_2), Q_{s+2}(x_3)|^+}{|Q_s(x_1), Q_{s+1}(x_2), Q_{s+2}(x_3)|}\right.$$

$$\left.\times\frac{|P_s(x_3), Q_{s+1}(x_1), Q_{s+2}(x_2)|^+}{|Q_s(x_1), Q_{s+1}(x_2), Q_{s+2}(x_3)|}\right\}$$

$$= \int_0^\infty \frac{d\psi(t)}{t+x_1}\int_0^\infty \frac{d\psi(t)}{t+x_2}\int_0^\infty \frac{d\psi(t)}{t+x_3}. \tag{4.75}$$

In this formula $(t+x_1)(t+x_2)(t+x_3)>0$ for $t \in (0,\infty)$. Note the alternative form relating to the sequence

$$|V_s^{(0)}, V_{s+1}^{(1)}, V_{s+2}^{(2)}| \;\div\; |U_s^{(0)}, U_{s+1}^{(1)}, U_{s+2}^{(2)}|.$$

### 4.9.5 Further Extensionals

Results similar to (4.63) have been found. Under the usual conditions assumed, we have for example:

$$F(x_1) = \int_0^\infty \frac{d\psi(t)}{t+x_1}$$

$$= \lim_{s\to\infty}\frac{|P_s(x_1), Q_{s+1}(x_2), \cdots Q_{s+n-1}(x_n)|^+}{|Q_s(x_1), Q_{s+1}(x_2), \cdots Q_{s+n-1}(x_n)|}, \quad (n \geqslant 2) \tag{4.76}$$

$$F(x_1)F(x_2) = \lim_{s\to\infty}\frac{|P_s(x_1), P_{s+1}(x_2), Q_{s+2}(x_3), Q_{s+3}(x_4)|^+}{|Q_s(x_1), Q_{s+1}(x_2), Q_{s+2}(x_3), Q_{s+3}(x_4)|}, \tag{4.77}$$

$$\frac{F(x_1)F(x_2)}{F(x_3)F(x_4)} = \lim_{s\to\infty}\frac{|P_s(x_1), P_{s+1}(x_2), Q_{s+2}(x_3), Q_{s+3}(x_4)|^+}{|Q_s(x_1), Q_{s+1}(x_2), P_{s+2}(x_3), P_{s+3}(x_4)|^+}, \tag{4.78}$$

$$\lim_{s\to\infty}\frac{|P_s(x_1), Q_{s+1}(x_1), Q_{s+2}(x_2), \cdots Q_{s+n-1}(x_{n-1})|^+}{|Q_s(x_1), Q_{s+1}(x_2), Q_{s+2}(x_3), \cdots Q_{s+n-1}(x_n)|} = 0. \quad (n>2) \tag{4.79}$$

Further extensions such as the product of 4 components in (4.75) are possible. However the present approach using an initial zero determinant followed by Laplacian expansion becomes complicated.

### 4.9.6 Numerical Examples

For the function

$$F(x) = \frac{1}{2\pi}\int_0^4 \frac{\sqrt{(4/t-1)}\,dt}{t+x}, \qquad (x>0)$$

$F(1)F(2) = (\sqrt{15} - \sqrt{3} - \sqrt{5} + 1)/4$. Defining $\Theta_s(x_1, x_2)$ to be the determinantal ratio in (4.66) we have:

| $s$ | $\Theta_s(1,2)$ | |
|---|---|---|
| 2 | 13/62 | 0.21 |
| 3 | 133/595 | 0.2235 |
| 4 | 1309/5797 | 0.22581 |
| 5 | 12804/56616 | 0.226155 |
| 6 | 125124/553139 | 0.226207 |
| 7 | 1222571/5404469 | 0.2262148 |
| 8 | 11945339/52805026 | 0.22621595 |
| 9 | 116713619/515938577 | 0.226216112 |

(Comment: This is evidently an increasing sequence.)

For a triple product,

$$F(1)F\left\{\exp\left(\frac{2i\pi}{3}\right)\right\}F\left\{\exp\left(\frac{-2i\pi}{3}\right)\right\}$$
$$= \frac{1}{8}(\sqrt{5}-1)\{1+\sqrt{13}-\sqrt{(2\sqrt{13}-2)}\},$$

the first few approximants being

$$\begin{aligned}
1/6 &= 0.167\\
15/72 &= 0.208\\
119/319 &= 0.373\\
888/2511 &= 0.3536\\
6921/19350 &= 0.35767\\
53976/150104 &= 0.359591\\
418712/1166490 &= 0.358950.
\end{aligned}$$

For a confluent case,

$$F^2(x) = \lim_{s\to\infty}\left\{\frac{P_s^2(x)+b_sP_{s-1}^2(x)+b_sb_{s-1}P_{s-2}^2(x)+\cdots}{Q_s^2(x)+b_sQ_{s-1}^2(x)+b_sb_{s-1}Q_{s-2}^2(x)+\cdots}\right\}$$

and $F^2(4) = (3-2\sqrt{2})/4$ with approximants:

| $s$ | Approximants | | |
|---|---|---|---|
| 1 | 1/26 | | |
| 2 | $\dfrac{6^2+1^2}{29^2+26}$ | = | $\dfrac{37}{867}$ |
| 3 | $\dfrac{35^2+37}{169^2+867}$ | = | $\dfrac{1262}{29428}$ |
| 4 | $\dfrac{204^2+1262}{985^2+29478}$ | = | $\dfrac{42878}{999653}$ |
| 5 | $\dfrac{1189^2+42878}{5741^2+999653}$ | = | $\dfrac{1456599}{33958734}$ |
| 6 | $\dfrac{6930^2+1456599}{33461^2+33958734}$ | = | $\dfrac{49481499}{1153597255}$ |
| 7 | $\dfrac{40391^2+49481499}{195025^2+1153597255}$ | = | $\dfrac{1680914380}{39188347880}$ |

$(P_0 = 0, Q_0 = 1;\ P_1 = 1, Q_1 = 5;\ P_2 = 6, Q_2 = 29;\ P_3 = 35, Q_3 = 169)$.

The sequence is increasing. For if

$$C_s = |P_s(x_1), P_{s+1}(x_2)| \div |Q_r(x_1), Q_{s+1}(x_2)|$$

then

$$C_{s+1} - C_s = \frac{B_s\left\{\dfrac{P_{s+1}(x_2)Q_{s+1}(x_2) - P_{s+1}(x_1)Q_{s+1}(x_1)}{(x_2-x_1)}\right\}}{\left\{\dfrac{|Q_s(x_1), Q_{s+1}(x_2)|}{(x_2-x_1)}\ \dfrac{|Q_{s+1}(x_1), Q_{s+2}(x_2)|}{(x_2-x_1)}\right\}}$$

which is clearly greater than 0 in the confluent case.

## Further Examples

### Further Example 4.1

$$F(z) = \frac{1}{2\pi}\int_0^4 \frac{\sqrt{(4/t-1)}dt}{t+z} = \frac{1}{z+}\;\frac{1}{1+}\;\frac{1}{z+}\;\frac{1}{1+}\;\cdots$$

$$I = \frac{1}{2\pi}\int_0^4 \frac{\sqrt{(4/t-1)}dt}{t^3+64} = \sqrt{2}/96 = 0.0147313913$$

Use (4.36) and (4.27) to derive:

<u>Increasing Sequence</u>

| | | | |
|---|---|---|---|
| 1/69 | 83/5646 | 6747/458014 | 547749/37183095 |
| 0.014492 | 0.014701 | 0.014731 | 0.01473113 |

<u>Decreasing Sequence</u>

| | | | |
|---|---|---|---|
| 1/64 | 74/4992 | 5998/406784 | 486354/33014400 |
| 0.015625 | 0.014824 | 0.014744 | 0.01473157 |

### Further Example 4.2

Use

$$\ln(1+1/z) = \frac{1}{z+}\;\frac{1^2}{2+}\;\frac{1^2}{3z+}\;\frac{2^2}{4+}\;\frac{2^2}{5z+}\;\frac{3^2}{6+}\;\frac{3^2}{7z+}\;\cdots$$

$$= \frac{1}{z+}\;\frac{1^2/2}{1+}\;\frac{1^2/(2\cdot3)}{z+}\;\frac{2^2/(3\cdot4)}{1+}\;\frac{2^2/(4\cdot5)}{z+}\;\frac{3^2/(5\cdot6)}{1+}\;\frac{3^2/(6\cdot7)}{z+}\;\cdots$$

$$= \frac{1}{z+\frac{1}{2}-}\;\frac{1^4/(1\cdot2^2\cdot3)}{z+\frac{1}{2}-}\;\frac{2^4/(3\cdot4^2\cdot5)}{z+\frac{1}{2}-}\;\frac{3^4/(5\cdot6^2\cdot7)}{z+\frac{1}{2}-}\;\cdots.$$

(See Example 3.4).

to approximate

$$\int_0^1 \frac{dt}{t^3+1} = \frac{1}{3}\{\ln 2 + \frac{\pi}{\sqrt{3}}\}.$$

From (4.36) and (4.27) derive the approximants:

$$k^*_{10}/k_{10} = 1865939414976/2232924613920 = 0.835648913$$

$$k^*_{11}/k_{11} = 159988981829040/191454782500800 = 0.835648917$$

$$k^*_{12}/k_{12} = 88249150080/105605553600 = 0.835648761$$

$$k^*_{13}/k_{13} = 2674144739158740/3200081746654800 = 0.835648884$$

Note the Gauss c.f. form

$$\frac{1}{1+}\;\frac{\frac{1}{1\cdot 4}}{1+}\;\frac{\frac{3^2}{4\cdot 7}}{1+}\;\frac{\frac{4^2}{7\cdot 10}}{1+}$$

$$= \frac{1}{1+}\;\frac{1}{4+}\;\frac{3^2}{7+}\;\frac{4^2}{10+}\;\frac{6^2}{13+}\;\frac{7^2}{16+}\;\frac{9^2}{19+}\;\frac{10^2}{22+}\;\cdots .$$

**Further Example 4.3**

$$I = \frac{1}{2\pi}\int_0^4 \frac{\sqrt{(4/t-1)}\,dt}{(t+4)(t^2+4)} = \frac{\sqrt{2}}{40}\left\{1 + \sqrt{(\sqrt{5}-1)} - \frac{1}{\sqrt{(\sqrt{5}-1)}}\right\}$$

**Table 4.1 Values for $I$**

| $s$ | $U_s^{(0)}$ | $U_s^{(1)}$ | $U_s^{(2)}$ | $V_s^{(0)}$ | $V_s^{(1)}$ | $V_s^{(2)}$ |
|---|---|---|---|---|---|---|
| 0 | 1 | 0 | 0 | 0 | 0 | 0 |
| 1 | 0 | -1 | 0 | 0 | 0 | 1 |
| 2 | 1 | -1 | 0 | 0 | 0 | 1 |
| 3 | 0 | -2 | 1 | 0 | 1 | 5 |
| 4 | 1 | -3 | 1 | 0 | 1 | 6 |
| 5 | 16 | 1 | 8 | 1 | 7 | 25 |
| 6 | 17 | -2 | 9 | 1 | 8 | 31 |
| 7 | 160 | 20 | 46 | 9 | 38 | 133 |
| 8 | 177 | 18 | 55 | 10 | 46 | 164 |
| 9 | 1040 | 63 | 248 | 55 | 202 | 765 |
| 10 | 1217 | 81 | 303 | 65 | 248 | 929 |
| 11 | 5888 | 58 | 1379 | 303 | 1131 | 4529 |
| 12 | 7105 | 139 | 1682 | 368 | 1379 | 5458 |
| 13 | 32800 | -319 | 7968 | 1682 | 6589 | 26733 |
| 14 | 39905 | -180 | 9650 | 2050 | 7968 | 32191 |
| 15 | 187200 | -1624 | 46748 | 9650 | 38780 | 156425 |
| 16 | 227105 | -1804 | 56398 | 11700 | 46748 | 188616 |
| 17 | 1089568 | -3137 | 274144 | 56348 | 227396 | 911097 |
| 18 | 1316673 | -4941 | 330542 | 68098 | 274144 | 1099713 |
| 19 | 6378240 | 2358 | 1601253 | 330542 | 1327109 | 5302941 |
| 20 | 7694913 | -2583 | 1931795 | 398640 | 1601253 | 6402654 |
| 21 | 27286960 | 34625 | 9331016 | 1931795 | 7729763 | 30886785 |

Table 4.2 Approximants to $I$ (True value; 0.042862417)

| $s$ | Increasing Sequence | | Decreasing Sequence | |
|---|---|---|---|---|
| 0 | 0 | 0 | $\frac{1}{16}$ | 0.0625 |
| 1 | $\frac{1}{33}$ | 0.0303 | $\frac{42}{928}$ | 0.045259 |
| 2 | $\frac{67}{1586}$ | 0.042245 | $\frac{1914}{44608}$ | 0.042907 |
| 3 | $\frac{3331}{77746}$ | 0.042845 | $\frac{93002}{2168544}$ | 0.042887 |
| 4 | $\frac{162133}{3783907}$ | 0.042848 | $\frac{4526755}{105596048}$ | 0.04286860 |
| 5 | $\frac{7894520}{184189955}$ | 0.04286075 | $\frac{220339700}{5140591040}$ | 0.042862717 |
| 6 | $\frac{384334220}{8966700900}$ | 0.042862389 | $\frac{10726331700}{250250205440}$ | 0.042862429 |
| 7 | $\frac{18709953420}{436512011620}$ | 0.042862402 | $\frac{522173026420}{12182535600192}$ | 0.0428624257 |
| 8 | $\frac{910827056992}{21250018162437}$ | 0.04286241311 | $\frac{25420117402917}{593063078890896}$ | 0.0428624177 |

(The increasing sequence is derived from

$$|V_r^{(0)},\ U_{r+1}^{(2)},\ U_{r+2}^{(1)}| \div |U_r^{(0)},\ U_{r+1}^{(2)},\ U_{r+2}^{(1)}|.$$

The decreasing sequence uses $C_3(t) = t(t+2)^2 + 16$ and the method of Section 4.4 (in particular, (4.27)).

**Further Example 4.4**

The polygamma function

$$\psi_m(z) = \frac{d^{m+1}\ln\Gamma(z)}{dz^{m+1}}, \qquad (m = 1, 2, \cdots)$$

along with a related c.f. is given in Bowman and Shenton (1988, p216). We have

$$(-1)^{m+1}\psi_m(z) = \frac{(m-1)!}{z^m} + \frac{m!}{2z^{m+!}} + \left(\frac{2\pi}{z}\right)^m g_m(z), \quad (\mathrm{Re}(z) > 0)$$

where

$$\begin{cases} g_m(z) = \displaystyle\int_0^\infty \frac{x^{m/2}\theta_m(y)\,dx}{(x+z^2)(y-1)^{m+1}}, & (y = e^{2\pi\sqrt{x}}) \\ \theta_m(y) = y(1-y)\dfrac{d\theta_{m-1}}{dy} + my\theta_{m-1}. & (m = 0, 1, \cdots) \end{cases}$$

For $m = 0$, the expression $(m-1)!/z^m$ is replaced by $-\ln z$ and 0! by unity.

For example,

$$\psi_3(z) = \frac{2}{z^3} + \frac{3}{z^4} + \left(\frac{2\pi}{z}\right)^3 \int_0^\infty \frac{(y + 4y^2 + y^3)\sqrt{x^3}dx}{(x + z^2)(y-1)^4}.$$

Now we may carry out the transformation $x = t^2$, and find

$$g_m(z) = \int_{-\infty}^{\infty} \frac{t^{m+1}\theta_m(e^{2\pi t})dt}{(t^2 + z^2)(e^{2\pi t} - 1)^{m+1}} = \int_{-\infty}^{\infty} \frac{t^{m+1}\phi_m(e^{2\pi t})dt}{(t^2 + z^2)(e^{\pi t} - e^{-\pi t})^{m+1}}$$

in which

$$\phi_m(e^{2\pi t}) = \phi_m(e^{-2\pi t}).$$

Hence

$$g_m(z) = \frac{i}{z}\int_{-\infty}^{\infty} \frac{d\,\sigma(t)}{t + iz} \qquad (\mathrm{Re}(z) > 0)$$

where $\sigma(t)$ is a distribution function on $(-\infty, \infty)$, i.e.

$$(-1)^{m+1}\psi_m(z) = \frac{(m-1)!}{z^m} + \frac{m!}{2z^{m+1}} + \frac{i}{z^{m+1}}\int_{-\infty}^{\infty} \frac{\Theta_m(t)dt}{t + iz},$$

where

$$i\int_{-\infty}^{\infty} \frac{\Theta_m(t)dt}{t + iz} = \frac{c_0^{(m)}}{z+}\ \frac{c_1^{(m)}}{z+}\ \cdots,$$

or

$$\int_{-\infty}^{\infty} \frac{\Theta_m(t)dt}{t + w} = \frac{c_0^{(m)}}{w-}\ \frac{c_1^{(m)}}{w-}\ \cdots. \qquad \left(\begin{matrix}\mathrm{Im}(w) > 0,\\ \mathrm{Re}(w) \neq 0\end{matrix}\right)$$

(For further details, see Shenton and Bowman (1971) and Bowman and Shenton (1983b).)

Taking $z = re^{i\theta}$, $-\pi/2 < \theta < \pi/2$, $\theta \neq 0$, with

$$\Phi_m(z) = (-1)^{m+1}\psi_m(z) - \frac{(m-1)!}{z^m} - \frac{m!}{2z^{m+1}},$$

we find for $m \geqslant 1$, the real and imaginary parts

$$\mathrm{Re}\{\Phi_m(z)\} = \frac{1}{r^{m+1}} \int_{-\infty}^{\infty} \frac{[t\sin\{(m+1)\theta\} + r\cos\{(m+2)\theta\}]\Theta_m(t)dt}{t^2 - 2tr\sin(\theta) + r^2},$$

$$\mathrm{Im}\{\Phi_m(z)\} = \frac{1}{r^{m+1}} \int_{-\infty}^{\infty} \frac{[t\cos\{(m+1)\theta\} - r\sin\{(m+2)\theta\}]\Theta_m(t)dt}{t^2 - 2tr\sin(\theta) + r^2}.$$

To produce monotonic approximating sequences, we follow the idea in Section 4.3.2 on Stieltjes fractions. If

$$t\sin\{(m+1)\theta\} + r\cos\{(m+2)\theta\} = (C_1t + D_1)^2 - C_1^2(t^2 - 2tr\sin(\theta) + r^2)$$

then

$$C_1 = \sqrt{\left\{\frac{1-\cos\{(m+1)\theta\}}{2r\cos(\theta)}\right\}}, \qquad D_1 = \frac{\sin\{(m+1)\theta\} - \sin(\theta)}{2C_1\cos(\theta)}.$$

$$(-\pi/2 < \theta < \pi/2,\ \theta \neq 0,\ m = 1, 2, \cdots)$$

Similarly, if

$$t\sin\{(m+1)\theta\} + r\cos\{(m+2)\theta\} = C_2^2(t^2 - 2tr\sin(\theta) + r^2) - (C_2t + D_2)^2,$$

then

$$C_2 = \sqrt{\left\{\frac{1+\cos\{(m+1)\theta\}}{2r\cos(\theta)}\right\}}, \qquad D_2 = -\frac{\sin\{(m+2)\theta\} + \sin(\theta)}{2C_2\cos(\theta)}.$$

We now have

$$\mathrm{Re}\{\Phi_m(z)\} = \frac{1}{r^{m+1}}\, \underset{t\to\infty}{l.\ i.\ s}\{P_t^{(m)}(r;\theta)\} = \frac{1}{r^{m+1}}\, \underset{t\to\infty}{l.\ d.\ s}\{Q_t^{(m)}(r;\theta)\}$$

where

(i) $$P_t^{(m)} = \frac{1}{\omega_t(z)} \begin{vmatrix} D_1V_t^{(0)} - C_1V_t^{(1)} & D_1V_{t+1}^{(0)} - C_1V_{t+1}^{(1)} \\ C_1U_t^{(0)} - D_1U_t^{(1)} & C_1U_{t+1}^{(0)} - D_1U_{t+1}^{(1)} \end{vmatrix},$$

(ii) $Q_t^{(m)}$ is derived from $P_t^{(m)}$ by replacing $D_1$ by $D_2$, $C_1$ by $C_2$, and finally changing the sign,

(iii) $\omega_t(z) = |\, U_t^{(1)},\ U_{t+1}^{(0)} \,|$.

For $\mathrm{Im}\{\Phi_m(z)\}$, $C_1$, $D_1$, and $C_2$, $D_2$ are replaced by

$$C_1^* = \sqrt{\left\{\frac{1+\sin\{(m+1)\theta\}}{2r\cos(\theta)}\right\}}, \qquad D_1^* = \frac{\cos\{(m+2)\theta\}-\sin(\theta)}{2C_1^*\cos(\theta)};$$

$$C_2^* = \sqrt{\left\{\frac{1-\sin\{(m+1)\theta\}}{2r\cos(\theta)}\right\}}, \qquad D_2^* = -\frac{\cos\{(m+2)\theta\}+\sin(\theta)}{2C_2^*\cos(\theta)}.$$

If one of the parameters $C_1, C_2, C_1^*, C_2^*$ is zero, then proceed as follows:

(a) For the case $\sin\{(m+1)\theta\} = 0$, use

$$\operatorname{Re}\{\Phi_m(z)\} = \frac{\cos\{(m+2)\theta\}}{r^m}\, l.\ i.\ s\ \{P_t^{(m)}(r;\theta)\},$$

where

$$P_t^{(m)}(r;\theta) = \frac{1}{\omega_t(z)}\begin{vmatrix} U_t^{(1)} & U_{t+1}^{(1)} \\ V_t^{(0)} & V_{t+1}^{(0)} \end{vmatrix},$$

and for the complementary decreasing sequence

$$\frac{\cos\{(m+2)\theta\}}{r^{m+2}\cos^2(\theta)}\, l.\ d.\ s\ \{Q_t^{(m)}(r;\theta)\},$$

where

$$Q_t^{(m)}(r;\theta) = \frac{1}{\omega_t(z)}\begin{vmatrix} r\sin(\theta)V_t^{(0)}+V_t^{(1)} & r\sin(\theta)V_{t+1}^{(0)}+V_{t+1}^{(1)} \\ U_t^{(0)}+r\sin(\theta)U_t^{(1)} & U_{t+1}^{(0)}+r\sin(\theta)U_{t+1}^{(1)} \end{vmatrix}.$$

(b) For the case $\cos\{(m+1)\theta\} = 0$, use the increasing sequence with $\cos\{(m+2)\theta\}/r^m$ replaced by $-\sin\{(m+2)\theta\}/r^m$, and the decreasing sequence with the factor outside the braces replaced by $-\sin\{(m+2)\theta\}/\{r^{m+2}\cos^2(\theta)\}$.

*The Fundamental Components*

The sequences $\{U_t^{(0)}\}$, $\{U_t^{(1)}\}$, $\{V_t^{(0)}\}$ and $\{V_t^{(1)}\}$ are set-up from the recurrences

$$U_t^{(0)} = -C_{t-1}^{(m)}U_{t-2}^{(0)} + r^2U_{t-1}^{(1)}, \quad U_t^{(1)} = -2yU_{t-1}^{(1)} - C_{t-1}^{(m)}U_{t-2}^{(1)} - U_{t-1}^{(0)},$$
$$V_t^{(0)} = -C_{t-1}^{(m)}V_{t-2}^{(0)} + V_{t-1}^{(1)}, \quad V_t^{(1)} = -2yV_{t-1}^{(1)} - C_{t-1}^{(m)}V_{t-2}^{(1)} - r^2V_{t-1}^{(0)}$$
$$(t \geqslant 2)$$

with initiators

| $s$ | $U_s^{(0)}$ | $U_s^{(1)}$ | $V_s^{(0)}$ | $V_s^{(1)}$ |
|---|---|---|---|---|
| 0 | -1 | 0 | 0 | 0 |
| 1 | 0 | 1 | 0 | $C_0^{(m)}$ |

Note that only the first few partial numerators $C_0^{(m)}$, $C_1^{(m)}$, are known in the general case (Shenton and Bowman, 1971, (24b)); however, general expressions are known for $\psi_2(x)$ (Stieltjes, 1918) and also for $\psi_1(x)$ (Shenton, unpublished). For the latter

$$\psi_1(x) = \frac{1}{x} + \frac{1}{2x^2} + \frac{1}{x^2}\left\{\frac{1/6}{x+}\ \frac{C_1^{(1)}}{x+}\ \cdots\right\}$$

where $C_{s-1}^{(1)} = s^2(s^2-1)/\{4(4s^2-1)\}$, $(s = 2, 3, \cdots)$.

**Numerical Example.** Let $m = 1$, $z = 1 + i\sqrt{3}$, $r = 2$, and $\theta = \pi/3$. Then

$$C_1 = \tfrac{1}{2}\sqrt{3}, \quad D_1 = -1, \quad C_2 = \tfrac{1}{2}, \quad D_2 = -\sqrt{3},$$

For the *real part* we have for the *coefficients in the determinants* $P_t^{(m)}(r;\theta)$:

| | | | | | |
|---|---|---|---|---|---|
| $\sqrt{(3)}/12$ | $-0.\dot{3}$ | 0.503119521 | $-0.1\dot{7}$ | | -1.609082689 |
| 1 | 0.173205081 | -4.714285710 | 13.69145024 | -18.0 | |

(change sign of final determinant value), and for the *coefficients in the determinants* $Q_t^{(m)}(r;\theta)$:

| | | | | |
|---|---|---|---|---|
| 1/12 | 0 | -0.376190476 | 1.154700536 | -1.607792206 |
| $\sqrt{3}$ | -3.9 | 5.691024110 | -1.428571490 | -19.83985464 |

Table 4.3

| $s$ | $U_s^{(0)}$ | $U_s^{(1)}$ | $V_s^{(0)}$ | $V_s^{(1)}$ |
|---|---|---|---|---|
| 0 | -1 | 0 | 0 | 0 |
| 1 | 0 | 1 | 0 | $0.1\dot{6}$ |
| 2 | 4.2 | $-3.46410\dot{1}\dot{6}$ | $0.1\dot{6}$ | -0.577350269 |
| 3 | -13.85640646 | 7.28571430 | -0.577350269 | 1.247619048 |
| 4 | 25.14285720 | -8.08290382 | $1.0\dot{8}$ | -1.462620685 |
| 5 | -11.33706004 | -8.181818080 | -0.587847550 | -1.179220764 |

Table 4.4a

| $t$ | $P_t^{(1)}$ | $Q_t^{(1)}$ |
|---|---|---|
| 1 | -0.0213294 | -0.0193452 |
| 2 | -0.0213259 | -0.0210796 |
| 3 | -0.0212490 | -0.0211917 |
| 4 | -0.0212129 | -0.0211921 |
| True value | -0.02120402 | |

Table 4.4b

| $t$ | $P_t^{(1)*}$ | $Q_t^{(1)*}$ |
|---|---|---|
| 1 | -0.001851216 | 0.000132911 |
| 2 | -0.001059332 | -0.000813027 |
| 3 | -0.001057682 | -0.001000342 |
| 4 | -0.001043183 | -0.001022385 |
| True value | -0.00102873 | |

The *real part* of

$$\Phi_1(1 + i\sqrt{3}) = \operatorname{Re}\left\{-\left[\frac{e^{-i\theta}}{r} + \frac{e^{-2i\theta}}{2r^2}\right] + \psi_1(z)\right\}$$

follows from Table 4.4a.

For the *imaginary part* we have for the *coefficients in the determinant* $P_t^{(m)*}(r;\theta)$,

$$-(2+\sqrt{3})\begin{bmatrix} V_t^{(0)} + \tfrac{1}{2}V_t^{(1)} & V_{t+1}^{(0)} + \tfrac{1}{2}V_{t+1}^{(1)} \\ \tfrac{1}{2}U_t^{(0)} + U_t^{(1)} & \tfrac{1}{2}U_{t+1}^{(0)} + U_{t+1}^{(1)} \end{bmatrix}:$$

| | | | | |
|---|---|---|---|---|
| 1/12 | -0.122008468 | 0.046459255 | 0.357578546 | -1.177457932 |
| 1 | -1.364101616 | 0.357511070 | 4.488524780 | -13.85034810 |

and for the *coefficients in the determinants* $Q_t^{(m)}(r;\theta)$,

$$-(2-\sqrt{3})\begin{bmatrix} V_t^{(0)} - \tfrac{1}{2}V_t^{(1)} & V_{t+1}^{(0)} - \tfrac{1}{2}V_{t+1}^{(1)} \\ \tfrac{1}{2}U_t^{(0)} - U_t^{(1)} & \tfrac{1}{2}U_{t+1}^{(0)} - U_{t+1}^{(1)} \end{bmatrix}:$$

| | | | | |
|---|---|---|---|---|
| -1/12 | 0.455341801 | -1.201159793 | 1.820199231 | 0.001762832 |
| 1 | 5.566101616 | -14.21391753 | 20.65433242 | 2.51328806 |

The *imaginary part* of

$$\Phi_1(z) = \operatorname{Im}\left\{\psi_1(z) - \frac{e^{-i\theta}}{r} - \frac{e^{-2i\theta}}{2r^2}\right\}$$

follows from Table 4.4b.

**Remarks on Validity**

The distribution functions (weight functions given in (1.67) and (1.68)) for $\ln\Gamma(z)$ and the polygamma functions relate to determined Stieltjes moment problems. For example, omitting the first two terms in $\psi_m(z)$, the series coefficients in the remaining series are the moments of a bounded non-decreasing function with infinitely many points of increase on $(0, \infty)$.

**Further Example 4.5**

Sequences for $\ln\Gamma(z)$ and $\psi(z)$

We have

$$\ln\Gamma(z) = (z - \tfrac{1}{2})\ln z - z + \tfrac{1}{2}\ln(2\pi) + J(z),$$

where

$$J(z) = z\int_0^\infty \frac{d\,\sigma(t)}{t + z^2},$$

with

$$\sigma(t) = \frac{1}{2\pi}\int_0^t \frac{1}{\sqrt{x}}\{\ln(1 - e^{-2\pi\sqrt{x}})^{-1}\}dx\,. \qquad (\mathrm{Re}(z) > 0)$$

Moreover

$$J(z) = \frac{zB_0}{z^2 + A_1 -}\;\frac{B_1}{z^2 + A_2 -}\;\cdots$$

where

$$B_0 = a_0, \quad B_s = a_{2s-1}a_{2s}, \quad s \geqslant 1;$$
$$A_1 = a_1, \quad A_s = a_{2s-2} + a_{2s-1}, \quad s \geqslant 2;$$

($a_0 = 1/12$, $a_1 = 1/30$, $a_2 = 53/210$, $a_3 = 195/371$, $a_4 = 22999/22737$, etc). A slight change of notation shows that

$$\int_0^\infty \frac{d\,\sigma(t)}{t + Z} = \frac{B_0}{Z + A_1 -}\;\frac{B_1}{Z + A_2 -}\;\cdots.$$

But

$$\mathrm{Re}\{J(z)\} = x\int_0^\infty \frac{(t + x^2 + y^2)}{(t + z^2)(t + \bar{z}^2)}d\,\sigma(t),$$

$$\mathrm{Im}\{J(z)\} = y\int_0^\infty \frac{(t - x^2 - y^2)}{(t + z^2)(t + \bar{z}^2)}d\,\sigma(t),$$

where $z = x + iy$, $x > 0$, and $y \neq 0$. We can set up now monotonic sequences of approximants to these components using Section 4.5 with $z_1 = z^2$, $z_2 = \bar{z}^2$, and equations (4.41), (4.19) and (4.20). We find

$$\mathrm{Re}\{J(z)\} = x\{l.\ i.\ s.\ P_m(r;\theta)\} = x\{l.\ d.\ s.\ Q_m(r;\theta)\}$$

where

$$P_m(r;\theta) = \frac{1}{\omega_m(z)}\begin{vmatrix} d_1V_m^{(0)} - c_1V_m^{(1)} & d_1V_{m+1}^{(0)} - c_1V_{m+1}^{(1)} \\ c_1U_m^{(0)} - d_1U_m^{(1)} & c_1U_{m+1}^{(0)} - d_1U_{m+1}^{(1)} \end{vmatrix},$$

$$Q_m(r;\theta) = \frac{1}{\omega_m(z)}\begin{vmatrix} d_2V_m^{(0)} - c_2V_m^{(1)} & d_2V_{m+1}^{(0)} - c_2V_{m+1}^{(1)} \\ -c_2U_m^{(0)} + d_2U_m^{(1)} & -c_2U_{m+1}^{(0)} + d_2U_{m+1}^{(1)} \end{vmatrix},$$

with

$$c_1 = \frac{\surd\{|\sin(\theta)| - \sin^2(\theta)\}}{r\,|\sin(2\theta)|}, \qquad d_1 = \frac{2c_1^2r^2\cos(2\theta) + 1}{2c_1};$$

$$c_2 = \frac{\surd\{|\sin(\theta)| + \sin^2(\theta)\}}{r\,|\sin(2\theta)|}, \qquad d_2 = \frac{2c_2^2r^2\cos(2\theta) - 1}{2c_2}.$$

For the *imaginary part* we have similarly

$$y^{-1}\mathrm{Im}\{J(z)\} = l.\ i.\ s.\ P_m^*(r;\theta) = l.\ d.\ s.\ Q_m^*(r;\theta),$$

for which $c_1$, $d_1$ etc. are replaced by

$$c_1^* = \frac{\surd\{\cos(\theta) + \cos^2(\theta)\}}{r\,|\sin(2\theta)|}, \qquad d_1^* = \frac{2c_1^{*2}r^2\cos(2\theta) + 1}{2c_1^*};$$

$$c_2^* = \frac{\surd\{\cos(\theta) - \cos^2(\theta)\}}{r\,|\sin(2\theta)|}, \qquad d_2^* = \frac{2c_2^{*2}r^2\cos(2\theta) - 1}{2c_2^*}.$$

The fundamental entities are defined in

$$U_s^{(0)} = A_sU_{s-1}^{(0)} - B_{s-1}U_{s-2}^{(0)} + r^4U_{s-1}^{(1)},$$
$$U_s^{(1)} = \{A_s + 2r^2\cos(2\theta)\}U_{s-1}^{(1)} - B_{s-1}U_{s-2}^{(1)} - U_{s-1}^{(0)};$$
$$V_s^{(0)} = A_{s-1}^{(0)} - B_{s-1}V_{s-2}^{(0)} + V_{s-1}^{(1)},$$
$$V_s^{(1)} = \{A_s + 2r^2\cos(2\theta)\}V_{s-1}^{(1)} - B_{s-1}V_{s-2}^{(1)} - r^4V_{s-1}^{(0)}.$$

Table 4.5

| $s$ | $U_s^{(0)}$ | $U_s^{(1)}$ | $V_s^{(0)}$ | $V_s^{(1)}$ |
|---|---|---|---|---|
| 0 | -1 | 0 | 0 | 0 |
| 1 | -1/30 | 1 | 0 | 1/12 |
| 2 | 24.982247978 | -5.188679246 | 1/12 | -0.435167715 |
| 3 | -66.51864975 | -7.504220150 | -0.224417989 | -0.617170022 |
| 4 | -624.8213913 | 89.79538218 | -2.08895459 | 7.553848138 |

Table 4.6a *Real parts*

| $m$ | $P_m$ | ln Γ |
|---|---|---|
| 1 | 0.016905756 | -1.876093669 |
| 2 | 0.016920588 | -1.876078837 |
| 3 | 0.016920591 | -1.876078834 |
| $m$ | $Q_m$ | ln Γ |
| 1 | 0.016921553 | -1.876077872 |
| 2 | 0.016920979 | -1.876078446 |
| 3 | 0.016920658 | -1.876078767 |

Table 4.6b *Imaginary parts*

| $m$ | $2P^*_m$ | ln Γ |
|---|---|---|
| 1 | -0.033376754 | 0.129635518 |
| 2 | -0.033366135 | 0.129646136 |
| 3 | -0.033366003 | 0.129646268 |
| $m$ | $2Q^*_m$ | ln Γ |
| 1 | -0.033360957 | 0.129651314 |
| 2 | -0.033365744 | 0.129646527 |
| 3 | -0.033365936 | 0.129646334 |

with initiators

| $s$ | $U_s^{(0)}$ | $U_s^{(1)}$ | $V_s^{(0)}$ | $V_s^{(1)}$ |
|---|---|---|---|---|
| 0 | -1 | 0 | 0 | 0 |
| 1 | $-A_1$ | 1 | 0 | $B_0$ |

Numerical example. Let $z = 1 + 2i$, $r = \surd 5$, $\sin(\theta) = 2/\surd 5$, $\sin(2\theta) = 4/5$, and $\cos(2\theta) = -3/5$. Then

$$c_1 = \surd(2\surd 5 - 4)/4, \qquad d_1 = (10 - 3\surd 5)/(8c_1);$$
$$c_2 = \surd(2\surd 5 + 4)/4, \qquad d_2 = (-10 - 3\surd 5)/(8c_2);$$
$$c_1^* = \surd(\surd 5 + 1)/4, \qquad d_1^* = (5 - 3\surd 5)/(16c_1^*);$$
$$c_2^* = \surd(\surd 5 - 1)/4, \qquad d_2^* = (-5 - 3\surd 5)/(16c_2^*),$$

and we find the values of Table 4.5.

The approximants $\{P_m\}$ etc. to $J(z)$, and $\ln\Gamma(z)$ for $z = 1 + 2i$ follow from Tables 4.6a and b.

Hence, at this stage

$$-1.87607883 < \mathrm{Re}\{\ln\Gamma(1 + 2i)\} < -1.87607877,$$
$$0.12964627 < \mathrm{Im}\{\ln\Gamma(1 + 2i)\} < 0.12964634.$$

**Further Example 4.6**

Use the c.f. form (Example 2.5a)

$$\int_0^\infty \frac{(t+1)e^{-t}\,dt}{t+z} = \frac{2}{z+1.5-}\;\frac{1.75}{z+B_2-}\;\frac{A_2}{z+B_3-}\cdots\frac{A_{22}}{z+B_{23}-}\cdots.$$

to set up second order sequences for the real part of the modified exponential integral $e^z E_1(z)$. Now

$$\mathrm{Re}\{e^z E_1(z)\} = \int_0^\infty \frac{(t+x)e^{-t}\,dt}{(t+x)^2+y^2}.$$

If $z = 1 + i$

$$\mathrm{Re}\{e^z E_1(z)\} = \int_0^\infty \frac{\{(t+1)e^{-t}\}dt}{(t+1)^2+1} = \hat{R},$$

then from (4.5) we have the sequences

$$\hat{R}_r = \underset{r\to\infty}{l.\ i.\ s.} \begin{vmatrix} V_r^{(0)} & V_{r+1}^{(0)} \\ U_r^{(1)} & U_{r+1}^{(1)} \end{vmatrix} \div \begin{vmatrix} U_r^{(0)} & U_{r+1}^{(0)} \\ U_r^{(1)} & U_{r+1}^{(1)} \end{vmatrix}$$

$$\hat{R}_r^* = \underset{r\to\infty}{l.\ d.\ s.} \begin{vmatrix} xV_r^{(0)} - V_r^{(1)} & -xV_{r+1}^{(0)} + V_{r+1}^{(1)} \\ xU_r^{(1)} - U_r^{(0)} & -xU_{r+1}^{(1)} + U_{r+1}^{(0)} \end{vmatrix} \div \begin{vmatrix} U_r^{(0)} & U_{r+1}^{(0)} \\ U_r^{(1)} & U_{r+1}^{(1)} \end{vmatrix}$$

and with $x = y = 1$, $\quad (C_2(t) = t^2 + 2t + 2)$

$$\begin{cases} U_r^{(0)} = B_r U_{r-1}^{(0)} - A_{r-1} U_{r-1}^{(0)} + 2U_{r-1}^{(1)}, \\ U_r^{(1)} = (2 + B_r) U_{r-1}^{(1)} - A_{r-1} U_{r-2}^{(1)} - U_{r-1}^{(0)}, \\ V_r^{(0)} = B_r V_{r-1}^{(0)} - A_{r-1} V_{r-2}^{(0)} + V_{r-1}^{(1)}, \\ V_r^{(1)} = (2 + B_r) V_{r-1}^{(1)} - A_{r-1} V_{r-2}^{(1)} - 2V_{r-1}^{(0)}. \end{cases}$$

(See (3.19), (3.20) with $n = 2$.)

Initiators:

| $r$ | $U_r^{(0)}$ | $U_r^{(1)}$ | $V_r^{(0)}$ | $V_r^{(1)}$ |
|---|---|---|---|---|
| 0 | -1 | 0 | 0 | 0 |
| 1 | -1.5 | 1 | 0 | 2 |
| 2 | -1.7142855 | 7.142857 | 2 | 11.285714 |
| 3 | 12.82351480 | 51.23529297 | 22.705882 | 71.91176294 |
| 4 | 221.3778798 | 405.3588640 | 225.1004790 | 527.1770283 |
| 5 | 2728.188184 | 3565.987544 | 2290.769838 | 4373.694941 |
| 6 | 32865.40321 | 34660.21033 | 24838.66783 | 40401.33911 |

Approximants to $\mathrm{Re}\{e^z E_1(z)\}$, $z = 1+i$

| $r$ | $\hat{R}_r$ | $\hat{R}_r^*$ | $r$ | $\hat{R}_r$ | $\hat{R}_r^*$ |
|---|---|---|---|---|---|
| 1 | 0.2222 | 0.6111 | 10 | 0.41053 | 0.41067 |
| 2 | 0.3328 | 0.4411 | 11 | 0.41057 | 0.41064 |
| 3 | 0.3791 | 0.4151 | 12 | 0.41058 | 0.41062 |
| 4 | 0.3978 | 0.4113 | 13 | 0.410589 | 0.410611 |
| 5 | 0.4053 | 0.4108 | 14 | 0.410591 | 0.410604 |
| 6 | 0.4084 | 0.4108 | 15 | 0.410592 | 0.410599 |
| 7 | 0.40969 | 0.41079 | 16 | 0.4105924 | 0.4105966 |
| 8 | 0.41022 | 0.41075 | 17 | 0.4105924 | 0.4105949 |
| 9 | 0.41044 | 0.41070 | 18 | 0.4105924 | 0.4105939 |
| | | | 19 | 0.4105924 | 0.4105934 |

($\mathrm{Re}\{e^z E_1(z)\} = 0.4105925$ from Handbook of Mathematical Functions N.B.S. 55, 1964. For $r = 23$, we have the bounds 0.4105925, 0.4105926).

**Comment:** The example is an illustration of the use of the "generalized odd part" (Chapter 2) and 2nd order c.f.s. (The function could easily be computed from the c.f. with complex $n$, or directly from the integral formulation; but these approaches are not relevant in the present context.) Another possibility would be to sum the series for the real part

$$\sum_{s=0}^{\infty}\{(-1)^s / |z|^{2s+2}\}\int_0^{\infty}(t+x)(t^2+2tx)^s\,dt \;,$$

which, for $z = 1+i$, becomes

$$\frac{2}{|z|^2}-\frac{14}{|z|^4}+\frac{296}{|z|^6}-\frac{12768}{|z|^8}+\frac{933504}{|z|^{10}}-\frac{103706880}{|z|^{12}}$$
$$+\frac{16290293760}{|z|^{14}}-\frac{3.438384445\times10^{12}}{|z|^{16}}+\cdots$$

$$=\frac{2}{|z|^2+7-}\;\frac{99}{|z|^2+47.02020202-}\;\frac{15752.31915}{|z|^2+126.2234224-}$$
$$\frac{7942.501832}{|z|^2+244.8195055-}\cdots$$

$$=\frac{2}{|z|^2+}\;\frac{7}{1+}\;\frac{14.14285714}{|z|^2+}\;\frac{32.87734488}{1+}\;\frac{47.91238224}{|z|^2+}$$
$$\frac{78.31104016}{1+}\;\frac{101.4225046}{|z|^2+}\;\frac{143.3970009}{1+}\cdots .$$

Assessment of continued fraction with $z = 1 + i$

| | |
|---|---|
| 1 | $<1$ |
| 2 | $>0.\dot{2}$ |
| 3 | $<0.6975$ |
| 4 | $>0.2865$ |
| 5 | $<0.6006$ |
| 6 | $>0.3172$ |
| 7 | $<0.5530$ (Mean $=0.4440$) |
| 8 | $>0.3350$ |

Again consider $\mathrm{Re}\{e^z E_1(z)\}$ for a more favorable value of the argument, namely $x = y = 5$. In this case the basic c.f. is

$$\int_0^\infty \frac{e^{-t}(t+5)dt}{t+n} = \frac{6}{n+\frac{7}{6}-}\;\frac{\frac{47}{36}}{n+\frac{943}{282}-}\;\frac{\frac{10704}{2209}}{n+\frac{113987}{20962}-}\;\cdots$$

$$= \frac{6}{n+b_1^*-}\;\frac{a_1^*}{N+b_2^*-}\;\cdots$$

and for the 2nd order c.f. components we have

$$\begin{cases} U_r^{(0)} = b_r^* U_{r-1}^{(0)} - a_{r-1}^* U_{r-2}^{(0)} + 50 U_{r-1}^{(1)}, \\ U_r^{(1)} = (b_r^* + 10) U_{r-1}^{(1)} - a_{r-1}^* U_{r-2}^{(1)} - U_{r-1}^{(0)}, \\ V_r^{(0)} = b_r^* V_{r-1}^{(0)} - a_{r-1}^* V_{r-2}^{(0)} + V_{r-1}^{(1)}, \\ V_r^{(1)} = (b_r^* + 10) V_{r-1}^{(1)} - a_{r-1}^* V_{r-2}^{(1)} - 50 V_{r-1}^{(0)}. \end{cases}$$

Initiators:

| $r$ | $U_r^{(0)}$ | $U_r^{(1)}$ | $V_r^{(0)}$ | $V_r^{(1)}$ |
|---|---|---|---|---|
| 0 | -1 | 0 | 0 | 0 |
| 1 | -7/6 | 1 | 0 | 6 |
| 2 | 47.40425532 | 14.51063830 | 6 | 80.06382978 |
| 3 | 988.9696414 | 171.7623320 | 112.690583 | 906.9349780 |

Approximants to $\mathrm{Re}\{e^z E_1(z)\}$, $z = 5 + i5$

| $r$ | $R_r$ (Increasing) | $R_r^*$ (Decreasing) |
|---|---|---|
| 1 | 0.0932642 | 0.0981347 |
| 2 | 0.0973940 | 0.0976385 |
| 3 | 0.0976141 | 0.0976335 |
| 4 | 0.0976261 | 0.0976282 |
| 5 | 0.0976267 | 0.0976269 |
| 6 | 0.0976267 | 0.0976267 |

Comment: Convergence is now more rapid due to larger $|z|$.

**Further Example 4.7**

For the c.f. $\frac{b_1}{z+}\ \frac{b_2}{1+}\ \cdots$ $(\equiv \chi_s(z)/\omega_s(z))$ there are series developments

(a) $\sum_{s=1}^{\infty}(-1)^{s+1}B_s/(\omega_{s-1}\omega_s)$

(b) $\sum_{s=0}^{\infty}B_{2s+1}/(\omega_{2s}\,\omega_{2s+2})$

(c) $b_1/z - z\sum_{s=1}^{\infty}B_{2s}/(\omega_{2s-1}\omega_{2s+1})$. $\qquad (B_s = b_1 b_2 \cdots b_s)$

For 2nd order c.f.s there are the following:

$$f(z_1, z_2) = \sum_{s=0}^{\infty} B_{2s+1}W_{2s+2}^2/(k_{2s}k_{2s+2}),$$

$$= b_1/(z_1z_2) - \sum_{s=0}^{\infty} B_{2s+2}W_{2s+3}W_{2s+3}^*/(k_{2s+1}k_{2s+3}),$$

$$= \sum_{s=1}^{\infty}(-1)^{s-1}B_sW_sW_{s+1}/(k_{s-1}k_s),$$

$$= \lim_{s\to\infty}\left\{\frac{|\chi_{2s}(z_1), \omega_{2s+2}(z_2)|^+ + |\chi_{2s}(z_2), \omega_{2s+2}(z_1)|^+ + 2B_{2s+1}}{(z_2 - z_1)^2 k_{2s}}\right\}$$

$$= \lim_{s\to\infty}\frac{|\chi_{2s+1}(z_1),\omega_{2s+3}(z_2)|^+ + |\chi_{2s+1}(z_2),\omega_{2s+3}(z_1)|^+ - (z_1+z_2)B_{2s+2}}{(z_2 - z_1)^2 k_{2s+1}}$$

where

$$F(z_1 z_2) = \int_0^\infty \frac{d\psi(t)}{(t+z_1)(t+z_2)},$$

$$W_s = \{\omega_s(z_2) - \omega_s(z_1)\}/(z_2 - z_1),$$

$$W_s^* = \{z_1\omega_s(z_2) - z_2\omega_s(z_1)\}/(z_2 - z_1).$$

$$k_s = |\,\omega_s(z_1),\, \omega_{s+2}(z_2)\,|/(z_2 - z_1).$$

Furthermore,

$$\int_0^\infty \frac{(t+z_1+z_2)d\psi(t)}{(t+z_1)(t+z_2)} = \sum_{s=0}^\infty \frac{B_{2s+1}W_{2s+2}\overline{W}_{2s+2}}{k_{2s}k_{2s+2}},$$

$$= \frac{b_1(b_2+z_1+z_2)}{z_1z_2} - z_1z_2\sum_{s=0}^\infty \frac{B_{2s+2}W_{2s+3}^2}{k_{2s+1}k_{2s+3}},$$

$$= \frac{B_1W_1\overline{W}_2}{k_0k_1} - \frac{B_2\overline{W}_2W_3}{k_1k_2} + \frac{B_3W_3\overline{W}_4}{k_2k_3} - \frac{B_4\overline{W}_4W_5}{k_3k_4} + \cdots,$$

$$= \lim_{s\to\infty}\left\{\frac{z_2\,|\,\chi_{2s+1}(z_1),\omega_{2s+3}(z_2)\,|^+ + z_1\,|\,\chi_{2s+1}(z_2),\omega_{2s+3}(z_1)\,|^+ - 2z_1z_2B_{2s+2}}{(z_2-z_1)^2k_{2s+1}}\right\},$$

$$= \lim_{s\to\infty}\left\{\frac{z_2\,|\,\chi_{2s}(z_1),\omega_{2s+2}(z_2)\,|^+ + z_1\,|\,\chi_{2s}(z_2),\omega_{2s+2}(z_1)\,|^+ + (z_1+z_2)B_{2s+1}}{(z_2-z_1)^2k_{2s}}\right\}.$$

$$(\overline{W}_s = \{z_2\omega_s(z_2) - z_1\omega_s(z_1)\}/(z_2-z_1))$$

$$\int_0^\infty \frac{(t+z_1+z_2)\,t\,d\psi(t)}{(t+z_1)(t+z_2)} = \sum_{s=0}^\infty \frac{B_{2s+2}W_{2s+1}^*W_{2s+1}}{k_{2s+1}k_{2s+3}}.$$

**Further Example 4.8**

Higher order c.f.s may be derived formally from the Gauss c.f. for the hypergeometric function

$$F(a,1,c;x) = 1 + \frac{ax}{c} + \frac{a(a+1)x^2}{c(c+1)} + \cdots.$$

We have, as the basic c.f.

$$\frac{1}{z}F(a,1,c;-\frac{1}{z}) = \frac{1}{z+}\;\frac{a/c}{1+}\;\frac{b_2}{z+}\;\frac{b_3}{1+}\;\cdots,$$

where

$$b_{2s+2} = (s+1)(c-a+s)/\{(c+2s)(c+2s+1)\}, \quad (s=0,1,\cdots)$$
$$b_{2s+1} = (a+s)(c-1+s)/\{(c+2s-1)(c+2s)\}. \quad (s=1,2,\cdots)$$

Also there is the integral representation

$$F(a,b,c;z) = \frac{\Gamma(c)}{\Gamma(b)\Gamma(c-b)}\int_0^1 t^{b-1}(1-t)^{c-b-1}(1-zt)^{-a}\,dt .$$
$$(c > b > 0; \text{ or } \operatorname{Re}(c) > \operatorname{Re}(b) > 0)$$

In the present context we choose $a$ to be unity, so that $a = 2, 3, \cdots$, correspond to c.f. developments of order 2, 3, $\cdots$, and derivatives of the function. Generalized c.f.s would arise from linear forms of this kind. Also there are many modifications associated with truncations of the basic c.f.; for example truncating one term relates to

$$\frac{1}{F(a,1,c;-\frac{1}{z})} - 1 = \frac{a/c}{z+}\ \frac{b_2}{1+}\ \frac{b_3}{z+}\ \cdots .$$

Many functions are related to the hypergiometric series. Wall (1948, Chapter 18) gives

$$F(1,1,2;-z) = \frac{1}{z}\ln(1+z),$$
$$F(-k,1,1;-z) = (1+z)^k,$$
$$zF(\tfrac{1}{2},\tfrac{1}{2},\tfrac{3}{2};z^2) = \arcsin(z),$$
$$zF(\tfrac{1}{2},1,\tfrac{3}{2};-z^2) = \arctan(z).$$

The Gauss c.f. relates to the ratio

$$F(a,b+1,c+1;z)/F(a,b,c;z).$$

Wall gives c.f.s for several related functions. For example,

$$\frac{\Omega(a,b;-z)}{\Omega(a,b-1;-z)}$$
$$= \frac{1}{1+}\ \frac{az}{1+}\ \frac{bz}{1+}\ \frac{(a+1)z}{1+}\ \frac{(b+1)z}{1+}\ \frac{(a+2)z}{1+}\ \frac{(b+2)z}{1+}\ \cdots$$
$$= \frac{\displaystyle\int_0^\infty \frac{e^{-u}u^{a-1}du}{(1+zu)^b}}{\displaystyle\int_0^\infty \frac{e^{-u}u^{a-1}du}{(1+zu)^{b-1}}} \qquad (\operatorname{Re}(a) > 0,\ z \text{ not on the neg. real axis})$$

where

$$\Omega(a,b;z) \sim 1 + ab\frac{z}{1!} + a(a+1)b(b+1)\frac{z^2}{2!} + \cdots .$$

A similar result is given by Perron (1957, p291).

The flavor of other similar results may be gathered from (Perron, Chapter 6)

(i) $$\gamma - (1-n+\beta)x - \frac{F_1(-n,\beta,\gamma;x)}{F_1(-n+1,\beta+1,\gamma+1;x)}$$
$$= \frac{(n-1)(\beta+1)\omega}{\gamma+1-(3-n+\beta)x-}\ \frac{(n-2)(\beta+2)\omega}{\gamma+2-(5-n+\beta)x-}\ \cdots\ \frac{1\cdot(\beta+n-1)\omega}{\gamma+n-1-(n-1+\beta)x}$$
$$(\omega = x - x^2;\ \beta \neq 0,-1,\cdots,-(n-1);\ \gamma+n \neq 0,-1,-2,\cdots)$$

where

$$F_1(a,b,c;x) = F(a,b,c;x)/\Gamma(c);$$

(ii)

$$a\left\{\frac{\int_1^\infty u^{-b-c}(u-1)^{b+c-a-1}e^{b/u}\,du}{\int_1^\infty u^{-b-c-1}(u-1)^{b+c-a-1}e^{b/u}\,du}\right\} = c + \frac{(a+1)b}{c+1+}\ \frac{(a+2)b}{c+2+}\ \frac{(a+3)b}{c+3+}\ \cdots$$

$$= a\left\{\frac{\int_0^1 t^{a-1}(1-t)^{b+c-a-1}e^{bt}\,dt}{\int_0^1 t^{a}(1-t)^{b+c-a-1}e^{bt}\,dt}\right\}.$$

$$(\mathrm{Re}(b+c) > \mathrm{Re}(a) > 0,\ b \neq 0)$$

**Further Example 4.9**

Example 3.4 gives approximants to $z^{-1}\tan^{-1}z^{-1}$ $(z = 1.5)$ based on the c.f. for $\ln(1+1/z)$, and using a forth order recurrence for the numerator and denominator of a 2nd order c.f. The basic c.f. being

$$\frac{1}{z+\frac{1}{2}-}\ \frac{\frac{1^4}{1\cdot 2^2\cdot 3}}{z+\frac{1}{2}-}\ \frac{\frac{2^4}{3\cdot 4^2\cdot 5}}{z+\frac{1}{2}-}\ \frac{\frac{3^4}{5\cdot 6^2\cdot 7}}{z+\frac{1}{2}-}\ \cdots,$$

we can use (4.32) and (4.33) for the $U$'s and $V$'s.

The recursions with $z_1 = 1.5i$ and $z_2 = -1.5i$ are

$$U_s^{(1)} = \tfrac{1}{2}U_{s-1}^{(1)} - b_{s-1}U_{s-2}^{(1)} - U_{s-1}^{(0)}, \qquad (b_s = s^2/\{4(4s^2-)\})$$
$$U_s^{(0)} = \tfrac{1}{2}U_{s-1}^{(0)} - b_{s-1}U_{s-2}^{(0)} + y^2U_{s-1}^{(1)}, \qquad (y = 1.5)$$
$$V_s^{(0)} = \tfrac{1}{2}V_{s-1}^{(0)} - b_{s-1}V_{s-2}^{(0)} + V_{s-1}^{(1)},$$
$$V_s^{(1)} = \tfrac{1}{2}V_{s-1}^{(1)} - b_{s-1}V_{s-2}^{(1)} - y^2V_{s-1}^{(0)}.$$

First few component values:

| $s$ | $U_s^{(0)}$ | $U_s^{(1)}$ | $V_s^{(0)}$ | $V_s^{(1)}$ |
|---|---|---|---|---|
| 0 | -1 | 0 | 0 | 0 |
| 1 | -1/2 | 1 | 0 | 1 |
| 2 | 25/12 | 1 | 1 | 1/2 |
| 3 | 3.325 | -1.65 | 1 | $-2.0\dot{6}$ |
| 4 | -2.183928571 | -4.214285714 | -1.630952380 | -3.315476190 |
| 5 | -10.78521825 | 0.181547619 | $-4.19\dot{4}$ | 2.143121691 |

and the bounds are

| $s$ | Lower | Upper |
|---|---|---|
| 1 | 0.387097 | 0.401434 |
| 2 | 0.391867 | 0.392232 |
| 3 | 0.391994 | 0.392003 |
| 4 | 0.3920017 | 0.3920020 |

**Further Example 4.10**

From Further Example 2.28, consider

$$G(z_1, z_2; k; \sigma) = \frac{H_{z_2}(k;\sigma) - H_{z_1}(k;\sigma)}{z_1 - z_2} = \int_0^\infty \frac{(t+k)d\,\sigma(t)}{(t+z_1)(t+z_2)},$$

with $C_2(t) = t^2 + p_1 t + p_2$. An increasing sequence of approximants to $G$ is given by

$$J_r(z_1, z_2) = |U_r^{(1)}, V_{r+1}^{(0)}| \div |U_r^{(1)}, U_{r+1}^{(0)}|,$$

where $z_1, z_2$ are both positive (recursions for the components are given in (4.19), (4.20). If we take $z_1$ (or $z_2$) $= k$, then $J_r(\cdot,\cdot)$ will yield an increasing sequence to the integral $G(k, z_2; k; \sigma)$.

# 5.

# ALTERNATIVE SUMMATION ALGORITHMS

## 5.1 Introduction

The need for back-up assessments for sums of divergent series has been stressed and illustrated in previous chapters. A very brief description of alternative algorithms including accelerating processes is given in Bowman and Shenton (1988). The Levin (1973) algorithms (examples are given in Chapter 2) has been found to be quite effective with moments of the sample standard deviation ($\sqrt{m_2}$ being preferred) when sampling is from one of the standard populations (exponential, logistic, half-normal). Our experience provides evidence that the algorithm is asymptotic in the sense that there is a "best" sequence member to use, this being signaled by results for small samples ($n = 2, 3, 4$ in general, and if valid).

Borel and modified Borel algorithms are also useful. We give a brief description here.

## 5.2 Levin and Shanks' Nonlinear Transformation

For the series

$$S(n) \sim e_0 + e_1/n + \cdots \tag{5.1}$$

let

$$A_j = \sum_{s=0}^{j-1} a_s, \qquad (a_s = e_s/n^s \neq 0)$$

be a partial sum, $n$ given. The Levin t-algorithm (basically of Shanks (1955) type) uses the approximation

$$A_{j+1}^{(r)} = \alpha_r + a_j\{b_0^{(r)} + b_1^{(r)}/(j+1) + \cdots + b_{r-1}^{(r)}/(j+1)^{r-1}\},$$
$$(j = 0, 1, \cdots, r;\ r = 2, 3, \cdots) \tag{5.2}$$

involving $r+1$ parameters. The basic point is the assumption that $a_j$ is approximately $A_{j+1}^{(r)}$ for $j \to \infty$, so that the "integral part" $\alpha_r$ is isolated. The solution is

$$\alpha_r = \frac{\sum_{j=0}^{r} (-1)^j \binom{r}{j} \left( \frac{j+1}{r+1} \right)^{r-1} \frac{A_{j+1}}{a_j}}{\sum_{j=0}^{r} (-1)^j \binom{r}{j} \left( \frac{j+1}{r+1} \right)^{r-1} \frac{1}{a_j}} . \tag{5.3}$$

The algorithm can be applied to the series, truncated versions of the series, the inverse series, and truncated version of the inverse series. It has a simple numerical form but is mathematically rather intractable; very few theorems are known for their convergency properties when applied to divergent series. (Theorems would be of interest but are not essential since in applications only a finite number of coefficients in general will be available.) However, results are known for convergent alternating series and oscillatory and monotone sequences (Sidi, 1979).

We have studied the Levin $t$-algorithm applied to the factorial series

$$I_m(x) \sim 1 - m!x + (2m)!x^2 - (3m)!x^3 + \cdots . \tag{5.4}$$
$$(m = 1, 2, \cdots)$$

Results for $m$, $x = 1, 2, 3$ are shown in Table 5.1 (Levin also quotes this case); the work was done using a multiple precision package due to J. Sullivan (Oak Ridge National Laboratory) on IBM System 3033, effectively carrying 50 decimal digits. It will be seen that after about 28 terms the Levin sequence for $I_1(1)$ starts to deteriorate, and similarly, for $I_1(2)$ and $I_1(3)$. Allowance in the tabulation is made for loss of accuracy, the underscored digits being unreliable. For the triple factorial series, we found that the best approximant was $\alpha_{24}$ which only yielded three correct decimal digits (0.628) to $I_3(1)$. Loss of accuracy is due to the ultimate appearance of changing sign patterns in the numerator of the algorithm. Another point to notice from this example is that the sequences may exhibit oscillatory tendencies (especially for the double and triple factorial cases) making it more difficult to come to terms with error analysis.

## 5.3 The Borel-Padé Algorithm

Series which diverge as fast or faster than the triple factorial series ((5.4) with $m = 3$) can first of all be transformed by using an integral transform, and then summed under the integral sign using a suitable algorithm (Padé, for example). Thus

Table 5.1a Levin Approximants to $I_1(n)$

| | $n=1$ | $n=2$ | $n=3$ |
|---|---|---|---|
| 1 | .5000000000000000000000000 | .6666666666666666666666667 | .7500000000000000000000000 |
| 2 | .6153846153846153846153846 2 | .7333333333333333333333333 | .7924528301886792452830188 7 |
| 3 | .5942857142857142857142857 1 | .7209302325581395648837209 3 | .7852494577006507592190889 4 |
| 4 | .5961389961389961389961390 0 | .7228594838181073330602212 2 | .7863951247460805250273451 7 |
| 5 | .5964837354971884119866182 6 | .7226517066174605812031587 5 | .7862348666904034761886848 9 |
| 6 | .5963305940229372176572988 9 | .7226526496697874581603731 1 | .7862521356005330641066373 7 |
| 7 | .5963423759099341987766738 3 | .7226584545809723552767106 2 | .7862513852317099034871317 0 |
| 8 | .5963494074721149753334549 5 | .7226571387528311678701660 4 | .7862511597575803315559279 5 |
| 9 | .5963473706781095943666008 5 | .7226572022620312020738310 9 | .7862512296560993475245512 6 |
| 10 | .5963471997564564943442109 1 | .7226572449769613602107697 9 | .7862512207683257341668703 3 |
| 11 | .5963473867802734070571483 9 | .7226572330336188166785993 5 | .7862512204473643719421672 5 |
| 12 | .5963473732321486337895320 2 | .7226572333178172129080685 0 | .7862512208360542944493481 5 |
| 13 | .5963473585482896088248890 5 | .7226572339109540230801404 4 | .7862512207639372422361659 3 |
| 14 | .5963473617350670774121394 8 | .7226572337784926951542821 4 | .7862512207632796387763575 6 |
| 15 | .5963473627786017369375662 9 | .7226572337673598599905649 7 | .7862512207666186089683580 9 |
| 16 | .5963473623352718426193686 3 | .7226572337779777008203503 8 | .7862512207659494737069452 4 |
| 17 | .5963473622719213038072841 3 | .7226572337768142341251255 6 | .7862512207659221586869521 9 |
| 18 | .5963473623266789482159258 1 | .7226572337762680777857477 7 | .7862512207659625827597155 8 |
| 19 | .5963473623288863349165521 3 | .7226572337764468367313809 0 | .7862512207659559688657297 8 |
| 20 | .5963473623223202726448229 4 | .7226572337764589221209147 5 | .7862512207659549741059720 1 |
| 21 | .5963473623225548801611808 2 | .7226572337764428459166730 8 | .7862512207659555583974595 6 |
| 22 | .5963473623233467938223101 3 | .7226572337764443674975280 0 | .7862512207659555080464541 1 |
| 23 | .5963473623232678124221967 9 | .7226572337764454910185402 2 | .7862512207659554803441458 7 |
| 24 | .5963473623231699623460971 2 | .7226572337764451943579668 4 | .7862512207659554887504561 9 |
| 25 | .5963473623231852200454717 3 | .7226572337764451364467820 1 | .7862512207659554891070593 7 |

Table 5.1a Continued

| | $n = 1$ | $n = 2$ | $n = 3$ |
|---|---|---|---|
| 26 | .596347362323197743440041648 | .722657233776445171527613337 | .786251220765955488457460060 |
| 27 | .596347362323195130300024437 | .722657233776445172244016616 | .786251220765955488545071123 |
| 28 | *.596347362323193275458700047* | *.722657233776445168824741127* | .786251220765955488577522722 |
| 29 | .5963473623231913 3810913324 | .722657223776445168788250030 | .786251220765955488564288155 |
| 30 | .596347362323182852140339092 | .722657233776445167595318849 | *.786251220765955488565317715* |
| 31 | .596347363323097842284660016 | .722657233776445162637431131 | .786251220765955488553043321 |
| 32 | .596347362322987591994121101 | .722657233776445152211668810 | .786251220765955488544139994 |
| 33 | .596347362321443902116319946 | .722657233776445111495890040 | .786251220765955488573101140 |
| 34 | .596347362310799563919986699 | .722657233776444906240373302 | .786251220765955488372989555 |
| 35 | .596347362344297742875933372 | .722657233776445143628449906 | .786251220765955486365487255 |
| 36 | .596347362195927532305451169 | .722657233776441844763210047 | .786251220765955486826134112 |
| 37 | .596347363663371135120775591 | .722657233776455709368218829 | .786251220765955492077817665 |
| 38 | .596347366328404609761204487 | .722657233776496219758800006 | .786251220765955484728477741 |
| 39 | .596347384204042251772998844 | .722657233776610980192342214 | .786251220765955491306467721 |
| 40 | .596347254574577299032433376 | .722657233775616552615755506 | .786251220765955103399724558 |
| 41 | .596347373713639611636311136 | .722657233778184239707014436 | .786251220765955970756653373 |
| 42 | .596345589463370495550636629 | .722657233771112962446259227 | .786251220765950348407805559 |
| 43 | .596311321524896846894127727 | .722657233594856488570283345 | .786251220765939869240087737 |
| 44 | .596024068050961556545992214 | .722657232865913032412529962 | .786251220765877456714899915 |
| 45 | .595885046972247057461202220 | .722657232236667790623575443 | .786251220765891800108360046 |
| Exact | .5963473623231941 | .7226572337764451 6939 | .7862512207659554885 6616 |

**Table 5.1b** Levin Approximants to $I_2(n)$

| | $n = 0$ | $n = 2$ | $n = 3$ |
|---|---|---|---|
| 1 | .3333333333333333333333333 | .5000000000000000000000000 | .6000000000000000000000000 |
| 2 | .6000000000000000000000000 | .7272727272727272727272727 3 | .7846153846153846153846153 8 |
| 3 | .6610978520286396181384248 2 | .7391577624135763670647391 6 | .7796467619848612279226240 5 |
| 4 | .6332555261449641534841949 2 | .7138990365505101171660578 1 | .7617564336811536860807781 8 |
| 5 | .6143887738443828056491489 7 | .7095761989993439432980949 2 | .7622991201511530638197214 0 |
| 6 | .6159392640159957656216842 7 | .7142710705557531494397654 1 | .7656605722502416595805380 2 |
| 7 | .6217979286196432464615102 9 | .7161674388010205339638612 2 | .7658898561076583497472779 0 |
| 8 | .6234415744869859162583118 0 | .7153527213575121097815869 0 | .7651206757552215200534214 7 |
| 9 | .6221650483247072490029606 7 | .7145946315730156114294509 9 | .7649175869545959562236339 7 |
| 10 | .6210365745014010293080178 7 | .7146127052177344697867928 5 | .7650855594405535520180398 7 |
| 11 | .6209727741739631263150614 2 | .7148612126077602104496674 6 | .7651883753119506492128616 0 |
| 12 | .6213680305026174259782066 7 | .7149463944785400500775023 8 | .7651664639318844442367642 0 |
| 13 | .6216003965901611014732368 2 | .7148939552127460357410773 2 | .7651265762167810403556148 3 |
| 14 | .6215691923411611315202491 1 | .7148412966415914377212897 5 | .7651200719712308348298002 1 |
| 15 | .6214554694021806263737530 2 | .7148370962002461260268402 2 | .7651313214050737137823230 8 |
| 16 | .6214009615598353284628398 7 | .7146550543597425207691422 3 | .7651380478022067732011729 8 |
| 17 | .6214163851288586759956990 4 | .7148655120233476465081736 6 | .7651367830537909310806489 5 |
| 18 | .6214500811784104404387838 3 | .7148637569721326279395422 9 | .7651337418556514387541477 5 |
| 19 | .6214651814334438689534001 3 | .7148586034768377663096736 0 | .7651328300843293494081030 4 |
| 20 | .6214600175283495260031313 8 | .7148564949540322465263011 4 | .7651335682695330434641658 0 |
| 21 | .6214495376456268389282673 9 | .7148574545184379604619362 5 | .7651342899810391570767836 7 |
| 22 | .6214446047732559286329842 2 | .7148589079379912485207469 8 | .7651343634611719478325618 2 |
| 23 | .6214460234068882755979530 5 | .7148593635855789204687382 5 | .7651341096722708297869898 8 |
| 24 | .6214494105258811275608705 8 | .7148590085737673564365817 9 | .7651339422017515682293514 5 |
| 25 | .6214512354934165820274655 5 | .7148585849616795630762392 0 | .7651339549728203370055962 8 |

Table 5.1b Continued

| | $n = 0$ | $n = 2$ | $n = 3$ |
|---|---|---|---|
| 26 | .62145093489547290550737169 | .71485847236687961010315116 | .76513403291959280053638224 |
| 27 | *.62144959375563764370236479* | .71485858897375473262924417 | .76513407290577570221249799 |
| 28 | .62144794267161587011688545 | .71485871002585399284758522 | .76513406053595029373194579 |
| 29 | .62145316269835388131292434 | *.71485863441994810163447412* | *.76513402483046897245082017* |
| 30 | .62126099525696852871857971 | .71485721477514630134897116 | .76513396205818645081362744 |
| 31 | .61781800153525503446768433 | .71481932237472537136990416 | .76513340307243006094764800 |
| 32 | .59992569293486765184215496 | .71467545084386786350779609 | .76513567718168603510541625 |
| 33 | .23840925157461948078240345 | .71188181241512727390850468 | .76512598848209749564993961 |
| Exact | .621449624236 | .714858685475 | .765134039965 |

Table 5.1c Levin Approximants to $I_3(n)$

| | $n = 1$ | $n = 2$ | $n = 3$ |
|---|---|---|---|
| 1 | .14285714285714285714285714 | .25000000000000000000000000 | .33333333333333333333333333 |
| 2 | .38653366583541147132169576 | .55319148936170212765957447 | .64609053497942386831275720 |
| 3 | .60707779624481975279969227 | .73058813457751904802671004 | .78441073890826506610430692 |
| 4 | .70410193891351825638529546 | .77015265688443818519033407 | .79800913709013856053869416 |
| 5 | .70247362733213886938759351 | .74307533741525996994038394 | .76551266009853024439356284 |
| 6 | .66225396740582071011553051 | .70465241966195600749686936 | .73408859635205028352135556 |
| 7 | .62325992226458565763773274 | .68183720796398015150159561 | .72199043315241813088528623 |
| 8 | .60333265785114413629273971 | .67998177796863103524310814 | .72714079343049014535401841 |
| 9 | .60424372553141167191113117 | .69079051402075717491436817 | .73821912551863800777363881 |
| 10 | .61735303530656541697058525 | .70252673864162598235057964 | .74574142422160962725749705 |

Table 5.1c

| | $n = 1$ | $n = 2$ | $n = 3$ |
|---|---|---|---|
| 11 | .63121969759923754988100208 | .70816733332055407739517497 | .74685900671712472755839474 |
| 12 | .63851780962195447709960120 | .70719673554537476096145622 | .74373009577060396516738047 |
| 13 | .63817889401042802317985676 | .70283423351533632601184899 | .73989256874224212065435605 |
| 14 | .63323813867272702753777818 | .69867482160026624840057597 | .73773977228225972463464182 |
| 15 | .62765221102435991318190287 | .69675846341011089177931348 | .73777291689106087254387124 |
| 16 | .62417853031585411633977804 | .69722145340972400883074510 | .73913527708752162062475059 |
| 17 | .62366046943363786230788041 | .69897638505778713520049722 | .74061203994350840086367471 |
| 18 | .62535664267538967048185603 | .70070684934509607681593117 | .74140018551655635152281534 |
| 19 | .62780519304518974071000241 | .70158884691425246245297691 | .74135158213846117112853082 |
| 20 | .62970172459287914764444897 | .70149726484861746554637613 | .74078932160086094181920210 |
| 21 | .63040918487872099092687009 | .70080394823447111866276503 | .74016868533730721153305885 |
| 22 | .63000117187284926773205575 | .70002614924790567348910995 | .73980921331744820488182116 |
| 23 | .62898940641607522834677187 | .69954260405946233005196726 | .73979357810914651776781808 |
| 24 | *.62796588832427103942713862* | .69947386241305668974528306 | .74001441026395410222809774 |
| 25 | .62723368942138997356030268 | .69970761016220912219205821 | *.74029176154918484177876900* |
| 26 | .62927626763522464153523517 | *.70015943729602475875410560* | .74046687789002253961311836 |
| 27 | .59570542356756911122675379 | .69915663617458148800850653 | .74083638351078544110194137 |
| 28 | .47254567558254453022104781 | .69334338579041740988994582 | .73564032620885123277490223 |
| Exact | .628255438980 | .700140134043 | .740286060738 |

$$S(n) \sim e_0 + e_1/n + \cdots \tag{5.5}$$

$$\sim \frac{1}{m}\int_0^\infty e^{-t^{1/m}} t^{1/m-1}\left\{e_0 + \left(\frac{t}{n}\right)\frac{e_1}{\Gamma(m+1)} + \left(\frac{t}{n}\right)^2 \frac{e_2}{\Gamma(2m+1)} + \cdots\right\} dt$$

and this relates to a Borel sum of order $m$ ($m$ usually a positive integer or positive half-integer). Using the Padé approach, approximants to $S(n)$ are now, for example,

$$B_{m,n}(r-1 \mid r) = m^{-1}\int_0^\infty e^{-t^{1/m}} t^{1/m-1}\left\{\sum_{s=1}^{r} \frac{A_s^{(r)}}{1 + x_s^{(r)} t/n}\right\} dt \tag{5.6}$$

which uses $2r$ coefficients of the series for $S(n)$. Under certain conditions the zeros $x_1^{(r)}, x_2^{(r)}, \cdots x_r^{(r)}$ are real and positive, and the sequence $\{B_{m,n}\}$ may converge to yield a sum assessment. However, in our applications only a finite set of coefficients is available, so that the approach can only be used tentatively; we have not so far found it particularly useful. The examples (Table 5.2) show that negative zeros can arise and also negative residues. There is then the possibility of the sum (when $m = 1$)

$$B_{1,n}(r-1 \mid r) = \sum_{x_s > 0} E_1\left(\frac{n}{x_s}\right)\frac{A_s}{x_s} e^{n/x_s} + \sum_{x_s > 0} Ei\left(\frac{n}{x_s}\right)\frac{A_s^*}{x_s} e^{-n/x_s}$$

where the summations refer to positive and negative zeros, and $E_1(\cdot)$, $Ei(\cdot)$ are the usual exponential integrals. For example, the mean value of the standard deviation ($\sqrt{m_2}$) in exponential sampling has the approximants

$$B_{1,2}(9 \mid 10) = 0.4883\,,$$

$$B_{1,10}(9 \mid 10) = 0.8763\,,$$

in reasonable agreement with entries given in Section 2.8.2. Similarly, for a modified Student's $t$ ($t^* = (m_1' - \mu_1')/\sqrt{m_2}$), we find using seven components (Table 5.3)

$$B_{1,10}(6 \mid 7) = -0.1357$$

in comparison with the simulation result (200,000 samples of ten) -0.1353. Since the approximate standard deviation of this is 0.0011, the closeness should not be accepted with too much enthusiasm.

**Table 5.2** $E(\sqrt{m_2})$ **from the Exponential.** $E(\sqrt{m_2}) \sim \sum A_s \int_0^\infty \frac{e^{-t}\,dt}{1+tx_s/n}$

| Degree | | s = 1 | s = 2 | s = 3 | s = 4 | s = 5 | s = 6 |
|---|---|---|---|---|---|---|---|
| 2 | $x_s$ | 1.4638 | 23.9209 | | | | |
| | $A_s$ | 0.9984 | $1.6136^{-3}$ | | | | |
| 3 | $x_s$ | 1.4314 | 18.1468 | -12.8096 | | | |
| | $A_s$ | 0.9972 | $3.4676^{-3}$ | $-7.4785^{-4}$ | | | |
| 4 | $x_s$ | 1.3039 | 12.9236 | 33.8838 | -3.2865 | | |
| | $A_s$ | 1.0112 | $8.7308^{-3}$ | $8.7964^{-5}$ | $-1.9990^{-2}$ | | |
| 5 | $x_s$ | 1.1518 | 9.9170 | 25.1494 | 47.8691 | -1.6724 | |
| | $A_s$ | 1.0539 | $1.5808^{-2}$ | $4.7167^{-4}$ | $3.2744^{-6}$ | $-7.0158^{-2}$ | |
| 6 | $x_s$ | 1.0165 | 7.9987 | 20.1080 | 37.0082 | 61.7075 | -1.0839 |
| | $A_s$ | 1.1135 | $2.3391^{-2}$ | $1.2293^{-3}$ | $3.1033^{-5}$ | $1.1790^{-7}$ | $-1.3872^{-1}$ |
| 10 | $x_s$ | 0.6859 | 4.4406 | 11.0505 | 20.1977 | 31.7028 | 45.8913 |
| | $A_s$ | 1.3600 | $6.0884^{-2}$ | $7.2370^{-3}$ | $8.6274^{-4}$ | $7.0346^{-5}$ | $2.881^{-6}$ |

| Degree | | s = 7 | s = 8 | s = 9 | s = 10 |
|---|---|---|---|---|---|
| 10 | $x_s$ | 63.6604 | 86.5604 | 118.2437 | -0.4594 |
| | $A_s$ | $4.6239^{-8}$ | $2.0790^{-9}$ | $7.2196^{-10}$ | $-4.2903^{-10}$ |

($1.6136^{-3}$ represents $1.6136\times10^{-3}$).

**Table 5.3** $E(t^*)$ **from the Exponential Sampling**

| Degree | | s = 1 | s = 2 | s = 3 | s = 4 | s = 5 | s = 6 | s = 7 |
|---|---|---|---|---|---|---|---|---|
| 2 | $x_s$ | 9.8853 | -2.1226 | | | | | |
| | $A_s$ | 0.1147 | -1.1147 | | | | | |
| 3 | $x_s$ | 4.6471 | 26.6306 | -1.5904 | | | | |
| | $A_s$ | 0.2851 | $0.4642^{-2}$ | -1.2595 | | | | |
| 4 | $x_s$ | 2.5845 | 18.0868 | 40.6791 | -1.2960 | | | |
| | $A_s$ | 0.4891 | $0.1513^{-1}$ | $0.3003^{-3}$ | -1.5046 | | | |
| 5 | $x_s$ | 1.5463 | 13.4738 | 30.4508 | 54.3601 | -1.1046 | | |
| | $A_s$ | 0.7316 | $0.2769^{-1}$ | $0.1629^{-2}$ | $0.1655^{-4}$ | -1.7610 | | |
| 6 | $x_s$ | 0.9452 | 10.6021 | 24.4316 | 42.4088 | 68.1466 | -0.9677 | |
| | $A_s$ | 1.0234 | $0.4025^{-1}$ | $0.4026^{-2}$ | $0.1549^{-3}$ | $.07624^{-6}$ | -2.0678 | |
| 7 | $x_s$ | 0.5634 | 8.6690 | 20.3360 | 35.2345 | 54.5192 | 82.1368 | -0.8634 |
| | $A_s$ | 1.3805 | $0.5197^{-1}$ | $0.7132^{-2}$ | $0.5460^{-3}$ | $0.1180^{-4}$ | $0.3095^{-7}$ | -2.4402 |

(Shenton and Bowman, 1977b)

## 5.4 Modified Borel: One Component (1cB)

There are simpler versions of the Borel approach which we have found useful. For one thing, they are less demanding from a computational point of view. Consider

$$S(n) \sim \int_0^\infty e^{-t} t^{a-1} \sum_{s=0}^\infty (ht/n)^s k_s(a,h)\, dt , \tag{5.7}$$

$$(a,h>0; \quad n=1,2,\cdots)$$

where

$$k_s(a,h) = e_s / \{h^s \Gamma(a+s)\}. \qquad (s=0,1,\cdots)$$

Define $n/h=N$ and $t/(N+t)=x$, so that $t/N = x/(1-x)$, and $1-x=N/(N+t)$. Then formally

$$S(n) \sim \int_0^\infty \frac{e^{-t} t^{a-1}}{1+t/N} \sum_{s=0}^\infty k_s(a,h) \frac{x^s}{(1-x)^{s+1}}\, dt ,$$

or

$$S(n) \sim \int_0^\infty \frac{e^{-t} t^{a-1}}{1+t/N} \sum_{s=0}^\infty \left\{\frac{t}{(n+t)}\right\}^s K_s(a,h)\, dt \tag{5.8}$$

where

$$K_s(a,h) = \sum_{r=0}^{s} \binom{s}{r} \frac{e_r}{h^r \Gamma(a+r)} . \tag{5.9}$$

Approximants to the series sum $S(n)$ are

$$D_r(n;a,h) = N \sum_{s=0}^{r-1} K_s(a,h)\, \Psi_s(N;a), \qquad (r=1,2,\cdots) \tag{5.10a}$$

consisting of $r$ terms and using the coefficients $e_0, e_1, \cdots, e_{r-1}$, where $K_s$ is given in (5.9) and

$$\Psi_s(N;a) = \int_0^\infty \frac{e^{-t} t^{a+s-1}}{(N+t)^{s+1}}\, dt . \tag{5.10b}$$

Clearly $\Psi_s$ is positive, and the sequence $\{\Psi_s\}$ monotonically decreases to zero since

$$\Psi_s(N;a) < \epsilon^s \int_0^\infty \frac{e^{-t}t^{a-1}}{N+t}\,dt \,. \qquad (0<\epsilon<1)$$

If the modified series coefficients $K_s$ tend to zero, or tend to stablize then the technique may provide useful assessments for comparison purposes.

**Computation of $\Psi_s$.** There are three possibilities useful for checking purposes:

(a) Use direct quadrature. The integrand is a smooth function unless $0<a<1$, in which case care must be taken at the origin; however, this case rarely occurs. Care is also needed in choosing the upper limit of integration, especially so if $s$ is large. Actually at present we are not likely to use $s$ greater than 30 - often much less.

(b) Use the recursive scheme given in the next section. Here, since sums involving alternating signs are involved, loss of accuracy needs careful attention. A crude warning signal of trouble is either the disruption of monotonicity or a switch over to negative values.

(c) Use the matrix approach described in Section 5.8. This is readily programmed and only requires two starting values involving quadratures.

By integration by parts there is the recurrence for $\Psi_s$, namely

$$s\Psi_s(N;a) = (N+2s+a-2)\Psi_{s-1}(N;a) - (a+s-2)\Psi_{s-2}(N;a), \qquad (s=2,3,\cdots) \qquad (5.11)$$

with

$$\Psi_0(N;a) = \int_0^\infty \frac{e^{-t}t^{a-1}}{N+t}\,dt \,, \qquad (a>0)$$

$$\Psi_1(N;a) = (N+a)\Psi_0(N;a) - \Gamma(a).$$

In terms of the exponential integral,

$$\Psi_0(N;1) = e^N E_1(N),$$

$$\Psi_0(N;2) = 1 - Ne^N E_1(N)$$

and similarly for $a=3,4,\cdots$ .

Interestingly enough, there is a polynomial form for $\Psi_s$. For example

$$2\Psi_2(N;a) = \{(N+a)(N+a+2)-a\}\Psi_0(N;a) - (N+a+2)\Gamma(a),$$

and in general

$$\Psi_s(N;a) = P_s(N)\Psi_0(N;a) + P^*_{s-1}(N) \tag{5.12}$$

where $P_s(\cdot)$, $P^*_{s-1}(\cdot)$ are polynomials of degrees $s$ and $s-1$ (see Section 5.9.1).

There is now a link with the c.f. for $\Psi_0(N;a)$ given in Chapter 1 (see (1.32)). If

$$\Psi_s(N;a) = \Phi_s(N;a)/s!$$

then from (5.11)

$$\Phi_s(N;a) = (N+2s+a-2)\,\Phi_{s-1}(N;a) \\ -(s-1)(a+s-2)\Phi_{s-2}(N;a), \qquad (s = 2, 3, \cdots)$$

with

$$\Phi_i(N;a) = \Psi_i(N;a). \qquad (i = 0, 1)$$

Clearly, in terms of $\psi_s/\omega_s$ (the $s$ th convergent of (1.32))

$$\Psi_s(N;a) = \{\omega_s\Psi_0(N;a) - \chi_s\Gamma(a)\}/s!$$

with $\chi_0 = 0$, $\omega_0 = 1$. Since the c.f. converges to $\Psi_0(N;a)/\Gamma(a)$, this expression shows the distinct possibility of loss of accuracy problems.

Other choices of the basic integral (5.10b) have been considered (Shenton and Bowman, 1976; Bowman and Shenton, 1976) but we lack clear guidelines in general application. The choice just discussed should be considered for series with alternating signs and magnitudes not in excess of the single factorial series. The choice of $k$ and $a$ should be directed towards stabilizing the sequence $\{K_s\}$, or at least avoiding increasing values of $|K_s|$, $s = 0, 1, \cdots$.

An approximation to the basic $\{\Psi_s\}$ sequence arises from the saddle-point approach. We have

$$\Psi_s(N;a) \sim \surd(2\pi)\exp\{h_s(t)\}/\{-H_s(t)\}, \tag{5.13}$$

where

$$\begin{aligned} t &= [a-N-2+\surd\{N^2+N(4s+2a)+(2-a)^2\}]/2, \\ h_s(t) &= -t+(a+s-1)\ln(t)-(s+1)\ln(N+t), \\ H_s(t) &= (1-a-s)/t^2+(s+1)/(N+t)^2. \end{aligned}$$

A less accurate assessment is

$$\Psi_s(N;a) \sim \sqrt{\pi}\exp\{-2\sqrt{(sN)}\}/(sN)^{(3-2a)/4}, \tag{5.14}$$

showing the dependence on $sN$ and the damping effect induced by increasing $a$.

**Example 5.1**

Find the sum $S(n)$ of the series

$$S(n) \sim \sum_{s=0}^{\infty} \frac{(-1)^s s!}{n^s(s+1)}$$

for $n = 1$ and $n = 10$.

Take $a = h = 1$. Then

$$K_s(1,1) = 1 - \frac{1}{2}\binom{s}{1} + \frac{1}{3}\binom{s}{2} - \cdots + \frac{(-1)^s}{s+1}\binom{s}{s} = \frac{1}{s+1}.$$

From NBS (1964) tables for the exponential integral $E_1(x)$

$$\Psi_0(1;1) = 0.596347361.$$

Similarly,

$$\Psi_0(10;1) = 0.0915633339.$$

1cB approximants are:

| $r$ | $n = 1$ $D_r(1;1,1)$ | $r$ | $n = 10$ $D_r(10;1,1)$ |
|---|---|---|---|
| 1 | 0.596347361 | 1 | 0.915633339 |
| 3 | 0.721766643 | 3 | 0.954939063 |
| 5 | 0.738563725 | 4 | 0.955403452 |
| 7 | 0.742768571 | 5 | 0.955487315 |
| 10 | 0.744495225 | 6 | 0.955505205 |
| 15 | 0.745064223 | 7 | 0.955508958 |
| 20 | 0.745162694 | | |
| 25 | 0.745185482 | | |
| 30 | 0.745191363 | | |

The calculations were carried out on a programmable calculator (restricted to 10 digit output) and machine error in the recurrence (5.11) is unavoidable. Actually, $\Psi_s(N;a)$ becomes negative at $s = 35$ for $n = 1$, and at $s = 8$ for $n = 10$, and of course, by definition $\Psi_s > 0$, and indeed, $\Psi_s$ is monotonic decreasing. Alternative approaches are clearly needed; since this is merely an illustration we content ourselves with using the check

$$S(n) = \int_0^\infty \frac{e^{-t}\ln(1+t/n)dt}{(t/n)}$$

which by quadrature (using about 200 steps on the interval 0.1E-06 to 20 with a Simpson formula) gives the approximations

$$S(1) = 0.745199, \qquad S(10) = 0.955512.$$

**Example 5.2.**

Bender and Wu (1969) give the first 75 terms in the series for an eigenvalue ($E(\lambda)$) of the differential equation

$$(-d^2/dx^2 + x^2/4 + \lambda x^4)\Phi(x) = E(\lambda)\Phi(x),$$

with the boundary condition

$$\lim_{x^2\to\infty} \Phi(x) = 0,$$

the series being

$$E_0(\lambda) = m/2 + \sum_{n=1}^{\infty} mA_n(\lambda/m^3)^n$$

with (see their Table 1) for example, $A_1 = 0.75$, $A_2 = -2.6$, $A_3 = 20.8$, $A_4 = -241.3$, $A_5 = 3581.0$, $A_{20} = -7.69E26$, $A_{50} = -2.13E100$, $A_{75} = 7.51E143$, the entries being reduced here from their 12 digit output. The terms alternate in sign and from their equation (E2).

$$A_s \sim (\surd 6/\pi^{3/2})3^s\Gamma(s+½).$$

To apply 1cB consider

$$\{E_0(\lambda) - m/2\}/m - 0.5 = \sum_{s=1}^{\infty} A_s/n^s. \qquad (n = m^3/\lambda)$$

Thus with $a = 1.5$, $h = 3$, the series becomes

$$\int_0^\infty e^{-t}\surd t(k_0 + k_1 t/N + k_2 t^2/N^2 + \cdots)dt, \qquad (N = n/3)$$

where for example,

$$\begin{array}{ll} k_0 = \phantom{-}0.282095 & K_0 = 0.282095 \\ k_1 = -0.219407 & K_1 = 0.062688 \\ k_2 = \phantom{-}0.231945 & K_2 = 0.075226 \end{array}$$

| | |
|---|---|
| $k_3 = -0.256099$ | $K_3 = 0.063610$ |
| $k_4 = 0.281539$ | $K_4 = 0.053280$ |
| $k_5 = -0.304871$ | $K_5 = 0.046344$ |
| $k_6 = 0.324925$ | $K_6 = 0.041632.$ |

$\Phi_s(N;1.5)$ would go to zero about as fast as $\sqrt{\pi}\exp\{-2\sqrt{(sN)}\}$, and one might be able to assess an asymptote to $K_s$ using a numerical approach.

## 5.5 Modified Borel: Two Components (2cB)

### 5.5.1 Definition of 2cB

In this case we use the form

$$S(n) \sim e_0 + e_1/n + \cdots \tag{5.15}$$

$$\sim \int_0^\infty e^{-t}t^{a-1}\{k_o + k_1(h/n)t^2 + k_2(h/n)^2t^4 + \cdots\}\,dt$$

where

$$k_s = e_s/\{h_s\Gamma(a+2s)\}. \qquad (a,h>0)$$

Defining $x = t^2/(N+t^2)$, we find as for 1cB, the approximants

$$F_r(n;a,h) = N\sum_{s=0}^{r-1} K_s(a,h)\Phi_s(N;a) \tag{5.16}$$

where

(i) $K_s(a,h) = \sum_{r=0}^{s}\binom{s}{r}\dfrac{e_r}{h^r\Gamma(a+2r)}$,

(ii) $\Phi_s(N;a) = \displaystyle\int_0^\infty \frac{e^{-t}t^{a+2s-1}}{(N+t^2)^{s+1}}\,dt$,

with recurrence

$$\begin{aligned} 4s(s+1)\Phi_{s+1}(N;a) &= 2s(6s+2a-3)\Phi_s(N;a) \\ &\quad - \{12s^2+8s(a-3)+a^2-7a+12+N\}\Phi_{s-1}(N;a) \\ &\quad + (2s+a-3)(2s+a-4)\Phi_{s-2}(N;a), \quad (s=2,3,\cdots;N=n/k) \end{aligned} \tag{5.17a}$$

and

(a) $\Phi_0(N;a) = \displaystyle\int_0^\infty e^{-t}t^{a-1}(N+t^2)^{-1}dt$,

(b) $\Phi_1(N;a) = \{a\,\Phi_0(N;a) - \Phi_0(N;a+1)\}/2$,

(c) $\Phi_2(N;a) = \{(4a+6)\Phi_1(N;a) - (a^2+a+N)\Phi_0(N;a) + \Gamma(a)\}/8$.

Since $\Phi_2(N;a)$ can be expressed as a linear function of $\Phi_0$ and $\Phi_1$, what is the structure of $\Phi_s(N;a)$? Using (5.17a) one can show that

$$\Phi_s(N;a) = N\Pi_{s_2}^{(0)}(N) + \Pi_{s_0}^{(1)}(N)\,\Phi_0(N;a) + \Pi_{s_1}^{(2)}(N)\Phi_1(N;a), \qquad (s \geqslant 3) \qquad (5.17b)$$

where $s_i = [(s-i-1)/2]$, $(i = 1, 2, 3)$, and $[x]$ is the integer part of $x$.

**Proof of the recurrence for $\Phi_s$.**

$$\begin{aligned} 2s\,\Phi_s &= -\int_0^\infty e^{-t}t^{a+2s-2}d\,(N+t^2)^{-s} \\ &= \int_0^\infty e^{-t}\{-t^{a+2s-2} + (a+2s-2)t^{a+2s-3}\}\,(N+t^2)^{-s}\,dt \qquad (5.18) \\ &= (a+2s-s)\,\Phi_{s-1} - \Psi_s\,, \qquad (s \geqslant 1) \end{aligned}$$

where

$$\Psi_s = \int_0^\infty e^{-t}t^{a+2s-2}(N+t^2-N)(N+t^2)^{-s}\,dt\,.$$

But

$$\begin{aligned} \Psi_s &= \int_0^\infty e^{-t}t^{a+2s-4}(N+t^2-N)(N+t^2)^{-s}\,dt\,, \\ &= \Psi_{s-1} - N\int_0^\infty e^{-t}t^{a+2s-4}(N+t^2)^{-s}\,dt\,, \qquad (s \geqslant 2) \\ &= \Psi_{s-1} + \frac{N}{2(s-1)}\int_0^\infty e^{-t}t^{a+2s-5}d\,(N+t^2)^{1-s}\,, \\ &= \Psi_{s-1} - \frac{N}{2(s-1)}\int_0^\infty e^{-t}\{-t^{a+2s-5} + (a+2s-5)t^{a-2s-6}\} \\ &\qquad \times (N+t^2)^{1-s}\,dt\,, \qquad (s \geqslant 3) \\ &= \Psi_{s-1} + \frac{N}{2(s-1)}\Phi_{s-2} - \frac{N(a+2s-5)}{2(s-1)}\int_0^\infty e^{-t}t^{a+2s-6}(N+t^2)^{1-s}\,dt\,, \\ &= \Psi_{s-1} + \frac{N}{2(s-1)}\Phi_{s-2} - \frac{(a+2s-5)}{2(s-1)} \\ &\qquad \times \int_0^\infty e^{-t}t^{a+2s-6}(N+t^2-t^2)(N+t^2)^{1-s}\,dt\,, \end{aligned}$$

$$= \Psi_{s-1} + \frac{N}{2(s-1)}\Phi_{s-2} - \frac{(a+2s-5)}{2(s-1)}\Psi_{s-2} + \frac{(a+2s-5)}{2(s-1)}\Psi_{s-1};$$

i.e.,

$$\Psi_s = \frac{(a+4s-7)}{2(s-1)}\Psi_{s-1} - \frac{(a+2s-5)}{2(s-1)}\Psi_{s-2} + \frac{N\Phi_{s-2}}{2(s-1)}.$$

$$(s > 3) \qquad (5.19)$$

Substituting in (5.18) with simplification leads to (5.17a).

**Example 5.3**

Evaluate

$$S(n) \sim \sum_{s=0}^{\infty} \frac{(-1)^s(2s+1)!}{n^s(s+1)}.$$

Take $a = h = 1$. Then

$$k_s = \frac{(-1)^s(2s+1)}{s+1}, \quad \text{and} \quad K_s = \frac{-1}{s+1}. \qquad (s > 0, \quad K_0 = 1)$$

Recurrence:

$$4s(s+1)\Phi_{s+1} - 2s(6s-1)\Phi_s + (12s^2 - 16s + 6 + n)\Phi_{s-1}$$
$$- (2s-2)(2s-3)\Phi_{s-2} = 0. \qquad (s = 2, 3, \cdots)$$

With $n = 10$, we have for example,

$$\Phi_0(10;1) = 0.08842513, \qquad \Phi_1(10;1) = 0.0077600935,$$
$$\Phi_2(10;1) = 0.0020624210.$$

Approximants:

| $r$ | $F_r(10;1,1)$ | $r$ | $F_r(10;1,1)$ | $r$ | $F_r(10;1,1)$ |
|---|---|---|---|---|---|
| 5 | 0.835806 | 22 | 0.835054 | 35 | 0.8250519 |
| 6 | 0.835466 | 23 | 0.835054 | 40 | 0.8350518 |
| 7 | 0.835296 | 24 | 0.835053 | 50 | 0.83505171 |
| 10 | 0.835118 | 25 | 0.835053 | 55 | 0.83505169 |
| 11 | 0.835097 | 26 | 0.835053 | 65 | 0.83505168 |
| 12 | 0.835084 | 27 | 0.835052 | 70 | 0.83505168 |
| 20 | 0.835055 | 28 | 0.8350524 | | |
| 21 | 0.835055 | 30 | 0.8350522 | | |

(Quadrature on $S(n) = n\int_0^\infty (e^{-t}/t)\ln(1+t^2/n)\,dt$ gives $S(10) = 0.835052$.)

### 5.5.2 An Application of 2cB

Consider the series for the mean value of the standard deviation in exponential sampling, i.e., $E(\surd m_2)$. Take $a = h = 1$. Then with $n = 10$ the fundamental entities for (5.17a) are:

**Table 5.4 Components for $E(\surd m_2)$ in Exponential Sampling**

| $r$ | $\Phi_r(10;1)$ | $K_r(1,1)$ | $r$ | $\Phi_r(10;1)$ | $K_r(1,1)$ |
|---|---|---|---|---|---|
| 1 | 0.088425131 | 1.000 | 12 | 0.000016007 | -3.2996958 |
| 2 | 0.007760093 | 0.250 | 13 | 0.000011657 | -3.5717578 |
| 3 | 0.002062421 | -0.24479167 | 14 | 0.000008648 | -3.8386213 |
| 4 | 0.00803075 | -0.69453125 | 15 | 0.000006519 | -4.1014305 |
| 5 | 0.000351057 | -1.0950345 | 16 | 0.000004984 | -4.3611912 |
| 6 | 0.000204305 | -1.4621632 | 17 | 0.000003859 | -4.6188017 |
| 7 | 0.000119093 | -1.8049979 | 18 | 0.000003021 | -4.8750741 |
| 8 | 0.000073002 | -2.1289686 | 19 | 0.000002390 | -5.1307506 |
| 9 | 0.000047933 | -2.4378299 | 20 | 0.000001907 | -5.3865164 |
| 10 | 0.000032314 | -2.7344478 | 21 | 0.000001535 | -5.6430091 |
| 11 | 0.000022457 | -3.0211123 | | | |

Approximants from these and additional values are given in Table 5.5; corresponding results are given for $n = 2, 3, 4, 5, 20$, and 50.

**Table 5.5 2cB Approximants to $E(\surd m_2)$ in Exponential Sampling ($a = h = 1$)**

| $r$ | $n$=2 | $n$=3 | $n$=4 | $n$=5 | $n$=10 | $n$=20 | $n$=50 |
|---|---|---|---|---|---|---|---|
| 4 | 7.13761 | 7.67962 | 8.03022 | 8.28150 | 8.93025 | 9.37962 | 9.7230519 |
| 5 | 6.94084 | 7.53167 | 7.91517 | 8.18850 | 8.88853 | 9.36483 | 9.7205505 |
| 6 | 6.76470 | 7.40588 | 7.82015 | 8.11400 | 8.85865 | 9.35566 | 9.7193519 |
| 21 | 5.66596 | 6.74172 | 7.37835 | 7.80170 | 8.77015 | 9.33713 | 9.7179474 |
| 22 | 5.63770 | 6.72788 | 7.37062 | 7.79701 | 8.76941 | 9.33706 | 9.7179463 |
| 23 | 5.61170 | 6.71534 | 7.36371 | 7.79286 | 8.76879 | 9.33700 | 9.7179455 |
| 24 | 5.58772 | 6.70395 | 7.35750 | 7.78917 | 8.76825 | 9.33696 | 9.7179450 |
| 25 | 5.56554 | 6.69357 | 7.35190 | 7.78587 | 8.76779 | 9.33692 | 9.7179445 |

(Entries are $10F_r(n;1,1)$. The 5-point Shanks' extrapolates (see B5, p.222, Bowman and Shenton, 1988) are: $n = 2$, 0.5292; $n = 3$, 0.6606; $n = 4$, 0.7304; $n = 5$; 0.7774; $n = 10$, 0.8765. Recall the true values 0.5, 0.6506, 0.7271, for $n = 2, 3, 4$ respectively; also the c.f. approximant 0.971794, for $n = 50$.)

Comments:

A simulation study for $n = 10$ gave the 95% confidence interval $0.873 < E(\surd m_2) < 0.879$. A Stieltjes c.f. (see "Further Examples" to this chapter) gave the plausible bounds (an assumption is made that the partial numerators in the c.f. will continue to increase monotonically)

$$n = 5: \quad 0.7720 < E(\sqrt{m_2}) < 0.7790$$

$$n = 10: \quad 0.87620 < E(\sqrt{m_2}) < 0.87669.$$

Altogether, taking into account the Levin sequences in Table 2.3, there is remarkable consistency. Notice also that the Shanks' extrapolates are acceptable; i.e., they do not project values already used in the extrapolate. If we have to settle for preferred values, then the Levin $\alpha_{16}$ would be our choice; here we are possibly getting an error of at most one digit in the fourth decimal place for $n \geqslant 4$, and slightly more accuracy the larger the value of $n$, possibly not exceeding 6 to 8 decimal digits in any case.

### 5.5.3 Asymptote for $\Phi_s(N;a)$

It is important to have an approximation to the quantities which form a fundamental part of the sum $F_r(n;a,h)$. Now

$$\Phi_s(N;a) = \int_0^\infty e^{\alpha(t)}dt$$

where

$$\alpha(t) = -t + (a+2s-1)\ln(t) - (s+1)\ln(N+t^2).$$

A saddle point approximation for $s \to \infty$ is

$$\Phi_s(N;a) \sim \sqrt{(2\pi)}\exp\{\alpha(t_0)\}/\sqrt{\{-\alpha^{(2)}(t_0)\}} \tag{5.20}$$

where

$$\alpha^{(2)}(t) = -\frac{(a+2s-1)}{t^2} + \frac{2(s+1)}{(N+t^2)} - \frac{4(s+1)N}{(N+t^2)^2};$$

here

$$t_0 = x + (a-3)/3$$

where

$$\begin{aligned} x &= Q_s(N;a) - H(N;a)/Q_s(N;a), \\ H(N;a) &= N/3 - (a-3)^2/9, \\ Q_s(N;a) &= \sqrt[3]{([-G_s(N;a) + \sqrt{\{G_s^2(N;a) + 4H^3(N;a)\}}]/2)}, \\ G_s(N;a) &= 2(3-a)^3/27 - 2N(a+3s)/3. \end{aligned}$$

A simple approximation is

$$\Phi_s(N;a) \sim \sqrt{(2\pi/3)}\exp\{-3(sN/4)^{1/3}\}/(2sN)^{(5-2a)/6}. \tag{5.21}$$

A guide to the value of $\Phi_s$ (Table 5.6) can be used, along with similar ones, to size up the "convergence" for the sequence $F_r(N;a,h)$; a typical term then follows after multiplication by $NK_s$ (see Section 5.5.1).

**Table 5.6 Asymptotics for $\Phi_s(N;a,h)$; Formula (5.20) ($a = h = 1$)**

| $s$ | $n=2$ | $n=5$ | $n=10$ | $n=15$ | $n=20$ | $n=25$ | $n=50$ |
|---|---|---|---|---|---|---|---|
| 10 | 1.47-03 | 1.68-04 | 2.15-05 | 5.61-06 | 2.05-06 | 8.40-07 | 4.18-08 |
| 15 | 5.72-04 | 4.91-05 | 4.79-06 | 1.03-06 | 3.19-07 | 1.14-07 | 3.46-09 |
| 20 | 2.73-04 | 1.88-05 | 1.48-06 | 2.71-07 | 7.43-08 | 2.40-08 | 4.92-10 |
| 25 | 1.48-04 | 8.48-06 | 5.56-07 | 8.94-08 | 2.21-08 | 6.51-09 | 9.65-11 |
| 30 | 8.67-05 | 4.25-06 | 2.38-07 | 3.42-08 | 7.66-09 | 2.10-09 | 2.35-11 |

### 5.5.4 Loss of Accuracy in Using 2cB

The sequence $\Phi_s$ decreases monotonically as $s$ increases, so that loss of accuracy is possible, and particularly so if the method of calculation involves sums with alternating signs. Four methods of computing the first hundred terms for the case $n = 20$, $a = h = 1$ are shown in Table 5.7. The rapid loss of accuracy when the recurrence (5.17a) is used in single precision arithmetic is noteworthy. Direct quadrature holds accuracy surprisingly well. As another alternative for checking purposes, the polynomial structure (5.17b) may be considered.

## 5.6 Modified Borel: Three Components (3cB)

The approximants are now of the form

$$G_r(n;a,h) = N\sum_{s=0}^{r-1} K_s(a,h)\Omega_s(N;a) \tag{5.22}$$

where

$$S(n) \sim \int_0^\infty e^{-t}t^{a-1}\left\{k_0 + k_1\left(\frac{h}{n}\right)t^3 + k_2\left(\frac{h}{n}\right)^2 t^6 + \cdots\right\}dt$$

and

$$k_r = e_r/\{h^r\Gamma(a+3r)\}, \qquad K_s(a,h) = \sum_{r=0}^{s}\binom{s}{r}k_r. \qquad (a,h>0)$$

Moreover,

$$\Omega_s(N;a) = \int_0^\infty \frac{e^{-t}t^{a+3s-1}}{(N+t^3)^{s+1}}\,dt, \tag{5.23}$$

Table 5.7 2cB; $\Phi_s(20;1)$ by Four Methods

| | Quadrature | Rec. 1 | Rec. 2 | Rec. 3 |
|---|---|---|---|---|
| 0 | 0.464770447836D-01 | 0.46477042E-01 | 0.464770447836D-01 | 0.464770447836Q-01 |
| 1 | 0.267861333457D-02 | 0.26786118E-02 | 0.267861333457D-02 | 0.267861333457Q-02 |
| 2 | 0.536393513428D-03 | 0.53647161E-03 | 0.536393513428D-03 | 0.536393513428Q-03 |
| 3 | 0.168901837748D-03 | 0.16904622E-03 | 0.168901837748D-03 | 0.168901837748Q-03 |
| 4 | 0.675313606327D-04 | 0.67697838E-04 | 0.675313606328D-04 | 0.675313606327Q-04 |
| 5 | 0.313336593255D-04 | 0.31467891E-04 | 0.313336593255D-04 | 0.313336593255Q-04 |
| 6 | 0.161045783553D-04 | 0.16155056E-04 | 0.161045783553D-04 | 0.161045783553Q-04 |
| 7 | 0.892238506155D-05 | 0.88485176E-05 | 0.892238506155D-05 | 0.892238506156Q-05 |
| 8 | 0.523657911582D-05 | 0.50132421E-05 | 0.523657911580D-05 | 0.523657911584Q-05 |
| 9 | 0.321747328044D-05 | 0.28370368E-05 | 0.321747328039D-05 | 0.321747328045Q-05 |
| 10 | 0.205223155389D-05 | 0.15246351E-05 | 0.205223155381D-05 | 0.205233155390Q-05 |
| 11 | 0.135045437746D-05 | 0.70190924E-06 | 0.135045437735D-05 | 0.135045437746Q-05 |
| 12 | 0.912469930245D-06 | 0.18319935E-06 | 0.912469930106D-06 | 0.912469930229Q-06 |
| 13 | 0.630721156157D-06 | -0.12793106E-06 | 0.630721156002D-06 | 0.630721156130Q-06 |
| 14 | 0.444691043616D-06 | -0.28413461E-06 | 0.444691043454D-06 | 0.444691043578Q-06 |
| 15 | 0.319040559119D-06 | -0.31629236E-06 | 0.319040558962D-06 | 0.319040559071Q-06 |
| 16 | 0.232459950494D-06 | -0.24465350E-06 | 0.232459950353D-06 | 0.232459950437Q-06 |
| 17 | 0.171733629072D-06 | -0.84568228E-07 | 0.171733628961D-06 | 0.171733629008Q-06 |
| 18 | 0.128460421886D-06 | 0.15038574E-06 | 0.128460421817D-06 | 0.128460421817Q-06 |
| 19 | 0.971808995199D-07 | 0.44695906E-06 | 0.971808995044D-07 | 0.917808994503Q-07 |
| 20 | 0.742767137166D-07 | 0.79146196E-06 | 0.742767137617D-08 | 0.742767136459Q-07 |
| 25 | 0.219986903596D-07 | 0.27114238E-05 | 0.219986908103D-07 | 0.219986903600Q-07 |
| 30 | 0.766294971185D-08 | 0.35812236E-05 | 0.766295046085D-08 | 0.766294985347Q-08 |
| 35 | 0.300319418990D-08 | 0.19788931E-05 | 0.300319482238D-08 | 0.300319447339Q-08 |
| 40 | 0.128796128283D-08 | -0.23976554E-05 | 0.128796123411D-08 | 0.128796161448Q-08 |
| 45 | 0.593280234444D-09 | -0.86340453E-05 | 0.593279031257D-09 | 0.593280446542Q-09 |
| 50 | 0.289688765051D-09 | -0.14380873E-04 | 0.289686252917D-09 | 0.289688666955Q-09 |
| 55 | 0.148493572599D-09 | -0.17439816E-04 | 0.148490059289D-09 | 0.148493015024Q-09 |
| 60 | 0.793215162026D-10 | -0.15456390E-04 | 0.793177940457D-10 | 0.793204506068Q-10 |
| 65 | 0.439027197129D-10 | -0.70483638E-05 | 0.438999621884D-10 | 0.439012382881Q-10 |
| 70 | 0.250626744171D-10 | 0.78521462E-05 | 0.250622286071D-10 | 0.250610217245Q-10 |
| 75 | 0.147025894146D-10 | 0.27809423E-04 | 0.147057034318D-10 | 0.147011426608Q-10 |
| 80 | 0.883629836210D-10 | 0.49963099E-04 | 0.884383133532D-11 | 0.883551888001Q-11 |
| 85 | 0.542695941793D-11 | 0.70349939E-04 | 0.543912195485D-11 | 0.542732227522Q-11 |
| 90 | 0.339879176765D-11 | 0.84410640E-04 | 0.341495862563D-11 | 0.340070332320Q-11 |
| 95 | 0.216662610060D-11 | 0.87614811E-04 | 0.218625203260D-11 | 0.217034069978Q-11 |
| 100 | 0.140363924899D-11 | 0.76116106E-04 | 0.142229557219D-11 | 0.140918617163Q-11 |

(Rec. 1, 2, and 3 refer to the use of the recurrence (5.17a) in single, double, and double-double precision on IBM 3033. A desk calculator gave $\Phi_{20}$ = 0.7419523305-07, $\Phi_{30}$ = 0.6816060090-08, $\Phi_{40}$ = 0.1462156755-08, $\Phi_{45}$ = 0.2107326364-08, the last one being clearly wrong. The asymptote (5.20) gives $\Phi_{100}$ = 0.104-11. Initializers were $\Phi_0(20;1)$ = 0.464770447236-01 (tolerance 0.704-15), and $\Phi_0(20;2)$ = 0.41119818114-01 (tolerance 0.629-15).)

with recurrence

$$\begin{aligned} 27(s-1)s(s+1)\Omega_{s+1}(N;a) &= 9(s-1)s(12s+3a-12)\Omega_s(N:a) \\ &+ Q_s(a)\Omega_{s-1}(N;a) + \{N+R_s(a)\}\Omega_{s-2}(N;a) \qquad (5.24) \\ &- (3s+a-7)^{(3)}\Omega_{s-3}(N;a), \qquad (s=3,4,\cdots) \end{aligned}$$

where

$$\begin{aligned} \Omega_0(N;a) &= \int_0^\infty e^{-t}t^{a-1}dt\,/\,(N+t^3), \qquad (a>0) \\ \Omega_1(N;a) &= \{a\,\Omega_0(N;a) - \Omega_0(N;a+1)\}/\,3, \\ \Omega_2(N;a) &= \{a(a+3)\Omega_0(N;a) - 2(a+2)\Omega_0(N;a+1) \\ &\qquad + \Omega_0(N;a+2)\}/\,18, \\ \Omega_3(N;a) &= [\{N+a(a+3)(a+6)\}\Omega_0(N;a) \\ &\quad - (3a^2+21a+28)\Omega_0(N;a+1) + 3(a+4)\Omega_0(N;a+2) \\ &\qquad - \Gamma(a)]/\,162. \end{aligned}$$

There are now three basic functions of $N$ and $a$ upon which $\Omega_s$ depends, namely $\Omega_0(N;a)$, $\Omega_0(N;a+1)$, and $\Omega_0(N;a+2)$; the function $\Omega_s$ is a linear function of these with polynomial coefficients, the precise structure being given in (5.54). The expression (5.24) may be derived by a three-stage integration by parts on (5.23), or heuristically by assuming a structure

$$p_s^*\Omega_{s+1} + Q_s^*\Omega_s + R_s^*\Omega_{s-1} + (N+S_s^*)\Omega_{s-2} + T_s^*\Omega_{s-3} = 0, \qquad (s>3)$$

and setting up equations for the coefficients by filtering out coefficients of powers $N^{-1}$ in its asymptotic expansion ($N\to\infty$); the assumption is that the variable coefficients are functions of $s$ only, the only term involving $N$ being the coefficient of $\Omega_{s-2}$.

### Example 5.4

Graffi, Grecchi, and Turchetti (1971) study the generalized anharmonic oscillator $p^2+x^2+\beta x^8$. The coefficients of the perturbation expansion for an energy eigenvalue (c/f example 5.2) diverge about as fast as $C\sigma^s(3s)!/\sqrt{s}$, where $C$ is a constant (Graffi, et al., quote the value $\sigma=\{8\Gamma(4/3+\tfrac{1}{2})/3\Gamma(1+4/3)\}^3$ ). They give the first 37 terms (to 25 significant digits), the coefficients alternating in sign without exception (see their Table 1).

We estimated $\sigma$ from their 25th and 37th terms and found the value 1.67954458 quite different from their formula. We are indebted to

Dr. George Baker (personal communication, 1978) for the reference to Brezinski, and his quotation of the value

$$\sigma = \frac{1}{4}\left\{\frac{\Gamma(8/3)}{\Gamma^2(4/3)}\right\} = 1.67931907 \text{ approx.}$$

The actual series is $E_0(\beta) = 1 + \beta\{A_1 - A_2\beta + A_3\beta^2 - \cdots\}$ where for example

$$A_1 = 105/16,\ A_2/A_1 = 1.6075E02,\ A_3/A_1 = 1.195086E05,$$
$$8^9 A_{10}/A_1 = 2.5749\cdots E41,\ 8^{36}A_{37}/A_2 = 2.607\cdots E219.$$

We consider

$$y(\beta) \sim A_1 - A_2\beta + \cdots$$
$$= A_1\sigma\int_0^\infty e^{-t}t^3\{k_0 + k_1(t\sigma\beta) + k_2(t\sigma\beta)^2 + \cdots\}dt\ ,$$

where

$$k_s = A_{s+1}/\{\sigma^{s+1}A_1(3s+3)!\}, \quad (\sigma = 1.67931817318)$$

so that

| $s$ | $(-1)^{s-1}k_{s-1}$ | $s$ | $(-1)^{s-1}k_{s-1}$ |
|---|---|---|---|
| 1 | 0.992466283808Q-01 | 19 | 0.296353242953Q-01 |
| 2 | 0.791685169563Q-01 | 20 | 0.288965578621Q-01 |
| 3 | 0.695403928395Q-01 | 21 | 0.282104007124Q-01 |
| 4 | 0.620708029421Q-01 | 22 | 0.275708946102Q-01 |
| 5 | 0.562693156419Q-01 | 23 | 0.269729845285Q-01 |
| 6 | 0.517358166944Q-01 | 24 | 0.264123500498Q-01 |
| 7 | 0.481133162782Q-01 | 25 | 0.258852736670Q-01 |
| 8 | 0.451467403524Q-01 | 26 | 0.253885368145Q-01 |
| 9 | 0.426639586409Q-01 | 27 | 0.249193369903Q-01 |
| 10 | 0.405480096461Q-01 | 28 | 0.244752210957Q-01 |
| 11 | 0.387173481522Q-01 | 29 | 0.240540313730Q-01 |
| 12 | 0.371134775966Q-01 | 30 | 0.236538612230Q-01 |
| 13 | 0.356933103840Q-01 | 31 | 0.232730188385Q-01 |
| 14 | 0.344243576706Q-01 | 32 | 0.229099970701Q-01 |
| 15 | 0.332816164078Q-01 | 33 | 0.225634483022Q-01 |
| 16 | 0.322454977449Q-01 | 34 | 0.222321633824Q-01 |
| 17 | 0.313004121136Q-01 | 35 | 0.219150538555Q-01 |
| 18 | 0.304337798044Q-01 | 36 | 0.216113669045Q-01 |
| | | 37 | 0.213195225271Q-01 |

From these we derive the sequence $\{K_s\}$:

| | $K_s$ | | $K_s$ |
|---|---|---|---|
| 0 | 0.992466283808Q-01 | 9 | 0.712107105468Q-02 |
| 1 | 0.200781114245Q-01 | 10 | 0.681622803340Q-02 |
| 2 | 0.104499873077Q-01 | 11 | 0.647671316348Q-02 |
| 3 | 0.829145308835Q-02 | 12 | 0.611797023634Q-02 |
| 4 | 0.780102146626Q-02 | 13 | 0.575266529866Q-02 |
| 5 | 0.771070408869Q-02 | 14 | 0.539069446808Q-02 |
| 6 | 0.766351113430Q-02 | 15 | 0.503948170010Q-02 |
| 7 | 0.755752682275Q-02 | 16 | 0.470436141499Q-02 |
| 8 | 0.737377913890Q-02 | 17 | 0.438895862491Q-02 |

From the quadrature with $\beta = 0.01$, $N = 59.5474519 = 1/(\beta\sigma)$, we have for the initial $\Omega$'s

| $a$ | $\Omega_0(N;a)$ | Tolerance |
|---|---|---|
| 6 | 0.620133618366D+00 | 0.939897891746D-14 |
| 5 | 0.166694183102D+00 | 0.262704161432D-14 |
| 4 | 0.547803724889D-01 | 0.951403830647D-15 |

A comparison of the $\Omega$'s computed by quadrature and the recurrence (5.24) is as follows:

| $s$ | Recurrence $\Omega_s$ | Quadrature $\Omega_s$ |
|---|---|---|
| 0 | 0.547803724889Q-01 | 0.547803724889Q-01 |
| 1 | 0.174757689512Q-01 | 0.174757689512Q-01 |
| 2 | 0.853632504624Q-02 | 0.853632504621Q-02 |
| 3 | 0.501668027124Q-02 | 0.501668027124Q-02 |
| 4 | 0.326996353393Q-02 | 0.326996353393Q-02 |
| 5 | 0.227837903824Q-02 | 0.227837903824Q-02 |
| 6 | 0.166353510911Q-02 | 0.166353510911Q-02 |
| 7 | 0.125765190006Q-02 | 0.125765190006Q-02 |
| 8 | 0.976844586927Q-03 | 0.976844586927Q-03 |
| 9 | 0.775340637170Q-03 | 0.775340637170Q-03 |
| 10 | 0.626440636023Q-03 | 0.626440636024Q-03 |
| 11 | 0.513728132218Q-03 | 0.513728132219Q-03 |
| 12 | 0.426668690153Q-03 | 0.426668690154Q-03 |
| 13 | 0.358258877385Q-03 | 0.358258877386Q-03 |
| 14 | 0.303701798852Q-03 | 0.303701798853Q-03 |
| 15 | 0.259627073070Q-03 | 0.259627073072Q-03 |
| 16 | 0.223613806505Q-03 | 0.223613806507Q-03 |
| 17 | 0.193889256228Q-03 | 0.193889256230Q-03 |
| 18 | 0.169133031662Q-03 | 0.169133031665Q-03 |
| 19 | 0.148346682422Q-03 | 0.148346682425Q-03 |
| 20 | 0.130764911060Q-03 | 0.130764911063Q-03 |

Sums of the series $y(\beta) = \{E_0(\beta) - 1\}/\beta$ from (5.22) are with $n = N$:

| $r$ | $A_1\sigma G_r(n;4,1)$ | $r$ | $A_1\sigma G_r(n;4,1)$ |
|---|---|---|---|
| 1 | 3.56787852157 | 14 | 3.94325583654 |
| 2 | 3.79814380785 | 15 | 3.94433022516 |
| 3 | 3.85668425335 | 16 | 3.94518885340 |
| 4 | 3.88398134559 | 17 | 3.94587920225 |
| 5 | 3.90072166341 | 18 | 3.94643765258 |
| 6 | 3.91225060209 | 19 | 3.94689223021 |
| 7 | 3.92061681823 | 20 | 3.94726462968 |
| 8 | 3.92685430252 | 32 | 3.948993 |
| 9 | 3.93158129505 | 33 | 3.949045 |
| 10 | 3.93520461915 | 34 | 3.949092 |
| 11 | 3.93800678186 | 35 | 3.9491339 |
| 12 | 3.94019030264 | 36 | 3.9491724 |
| 13 | 3.94190334243 | | |

From Graffi, et al., the corresponding third order Borel approximants are 3.82402, 3.92632, 3.94385, 3.94889, and 3.94917, the last one using twelve terms of the series (see their Table IV). The approximants are still increasing slightly, but more terms would require a careful consideration of loss of accuracy.

## 5.7 A More General 2cB Model

Consider

$$S(n) \sim e_0 + e_1/n + \cdots \tag{5.25}$$

$$\sim \int_0^\infty \frac{e^{-t}t^{a-1}}{\Gamma(a)} \left\{k_0 + \frac{k_1(bt+B^2t^2)}{n} + \frac{k_2(bt+B^2t^2)^2}{n^2} + \cdots\right\} dt$$

with

$$e_s = \frac{k_s}{\Gamma(a)} \int_0^\infty e^{-t}t^{a-1}(bt+B^2t^2)^s\, dt\,. \qquad (a>0,\ B>0) \tag{5.26}$$

Then under the transformation

$$(bt+B^2t^2)/n = x/(1-x),$$

with

$$x = \frac{bt+B^2t^2}{n+bt+B^2t^2}, \qquad 1-x = \frac{n}{n+bt+B^2t^2},$$

$$S(n) \sim \frac{n^p}{\Gamma(a)}\int_0^\infty \frac{e^{-t}t^{a-1}}{(n+bt+B^2t^2)^{p+1}} \times \left\{\frac{k_0}{(1-x)^{p+1}} + \frac{k_1 x}{(1-x)^{p+2}} + \cdots\right\} dt\,, \qquad (p > -1)$$

$$\sim \frac{n^p}{\Gamma(a)}\int_0^\infty \frac{e^{-t}t^{a-1}}{(n+bt+B^2t^2)^{p+1}} \{K_0^{(p)} + K_1^{(p)}x + \cdots\} dt\,;$$

i.e., approximants to $S(n)$ are

$$F_r^{(p)}\{n;a,b,B\} = n^p \sum_{s=0}^{r-1} K_s^{(p)}\Theta_s^{(p)}(n;a,b,B) \tag{5.27}$$

where

$$\begin{cases} K_s^{(p)} = \displaystyle\sum_{r=0}^{s} \frac{k_r\Gamma(p+s+1)}{\Gamma(p+r+1)(s-r)!}\,, \\ \Theta_s^{(p)}(n;a,b,B) = \displaystyle\frac{1}{\Gamma(a)}\int_0^\infty \frac{e^{-t}t^{a-1}(bt+B^2t^2)^s}{(n+bt+B^2t^2)^{s+p+1}}\,dt\,, \end{cases}$$

and $n+bt+Bt^2>0$ for $0 \leqslant t < \infty$. Special cases are

(i) $b=0$, reducing to 2cB,

(ii) $B=0$, reducting to 1cB.

Note that in (5.25) with $b=0$, $B$ plays the role of $h$ in the treatment of 2cB in 5.6.1 (see (5.15)). Thus the introduction of $b$ allows for a perturbation. Again the value of $p$ will usually be small. Altogether there are now four parameter $a$, $b$, $B$, and $p$ at our disposal, instead of $a$ and $h$ in the more specialized 2cB model.

We now show that there is a fourth order recurrence for $\Theta_s^{(p)}(\cdot)$. This suggests the basic structure of this function and also could be used for checking purposes. Throughout the next Sections 5.7.1 - 5.7. we assume restrictions on the parameters for the validity of the integrals.

### 5.7.1 First Phase for the Recurrence

$$\Theta_s^{(p)} - k\,\Theta_{s-1}^{(p)} = \int_0^\infty \frac{e^{-t}t^{a-1}}{\Gamma(a)} \left[\frac{\tau^s}{T^{s+p+1}} - k\frac{\tau^{s-1}}{T^{s+p}}\right] dt\ ,$$

$$(T = n + bt + B^2t^2,\ \tau = T - n)$$

$$= (1-k)\int_0^\infty \frac{e^{-t}t^{a-1}}{\Gamma(a)} \frac{\tau^{s-1}}{T^{s+p+1}} \left[Bt + \frac{b}{2B}\right]^2 dt\ , \qquad (5.28)$$

if $k = b^2/(b^2 - 4nB^2)$, $b^2 \neq 4nB^2$. Hence, preparing for integration by parts

$$(s+p)(\Theta_s^{(p)} - k\,\Theta_{s-1}^{(p)}) = \lambda\int_0^\infty \frac{e^{-t}t^{a-1}}{\Gamma(a)}\tau^{s-1}\left[Bt + \frac{b}{2B}\right] d\left[\frac{1}{T^{s+p}}\right]$$

where

$$\lambda = -(1-k)/2B = 2nB/(b^2 - 4nB^2),$$

and so

$$(s+p)(\Theta_s^{(p)} - k\,\Theta_{s-1}^{(p)}) = (s-1)\{\Theta_{s-1}^{(p)} - k\,\Theta_{s-2}^{(p)}\} - \lambda B\,\Theta_{s-1}^{(p)} + \lambda\Phi_s^{(p)} \qquad (5.29)$$

where

$$\Phi_s^{(p)} = \frac{1}{\Gamma(a)}\int_0^\infty e^{-t}\left[Bt + \frac{b}{2B}\right] t^{a-2}\tau^{s-1}\frac{\{t-(a-1)\}}{T^{s+p}}\,dt\ . \qquad (5.30a)$$

There are two basic functions here. Let $(s = 2, 3, \cdots)$

$$\Phi_s^{(p)} = \alpha_s - (a-1)\beta_s \qquad (5.30b)$$

where

$$\alpha_s = \frac{1}{\Gamma(a)}\int_0^\infty \frac{e^{-t}t^{a-1}\tau^{s-1}}{T^{s+p}}\left[Bt + \frac{b}{2B}\right] dt\ ,$$

$$B_s = \frac{1}{\Gamma(a)}\int_0^\infty \frac{e^{-t}t^{a-2}}{T^{s+p}}\tau^{s-1}\left[Bt + \frac{b}{2B}\right] dt\ , \qquad (5.30c)$$

### 5.7.2 Second Phase for the Function $\alpha_s$

We have

$$(s+p-1)\alpha_s = -\frac{1}{\Gamma(a)}\int_0^\infty \frac{e^{-t}t^{a-1}}{2B}\,T^{s-1}d\left(\frac{1}{T^{s+p-1}}\right),$$

$$= \frac{(s-1)}{\Gamma(a)}\int_0^\infty \frac{e^{-t}t^{a-1}T^{s-2}}{T^{s+p-1}}\left(Bt+\frac{b}{2B}\right)dt$$

$$-\frac{1}{2B\,\Gamma(a)}\int_0^\infty \frac{e^{-t}t^{a-1}T^{s-1}}{T^{s+p-1}}\,dt + \frac{(a-1)}{2B\,\Gamma(a)}\int_0^\infty \frac{e^{-t}t^{a-2}T^{s-1}}{T^{s+p-1}}\,dt\ ;$$

i.e.,

$$(s+p-1)\alpha_s = (s-1)\,\alpha_{s-1} - \gamma_s + (a-1)\delta_s \tag{5.31}$$

where

$$2B\,\gamma_s = \frac{1}{\Gamma(a)}\int_0^\infty \frac{e^{-t}t^{a-1}T^{s-1}}{T^{s+p-1}}\,dt\ , \tag{5.32a}$$

$$2B\,\delta_s = \frac{1}{\Gamma(a)}\int_0^\infty \frac{e^{-t}t^{a-2}T^{s-1}}{T^{s+p-1}}\,dt\ . \tag{5.32b}$$

Now

$$2B\,\delta_s = \frac{1}{\Gamma(a)}\int_0^\infty \frac{e^{-t}t^{a-1}T^{s-2}(B^2t+b)}{T^{s+p-1}}\,dt\ ,$$

or

$$2\delta_s = \frac{1}{\Gamma(a)}\int_0^\infty \frac{e^{-t}t^{a-1}T^{s-2}\{Bt+b/(2B)+b/(2B)\}}{T^{s+p-1}}\,dt\ ,$$

$$= \alpha_{s-1} + \frac{b}{2B}\,\Theta_{s-2}^{(p)}. \tag{5.33}$$

Similarly,

$$2B\,\gamma_s = \frac{1}{\Gamma(a)}\int_0^\infty \frac{e^{-t}t^{a-1}T^{s-2}(T-n)}{T^{s+p-1}}\,dt\ ,$$

$$= 2B\,\gamma_{s-1} - n\,\Theta_{s-2}^{(p)}. \tag{5.34}$$

Hence from (5.31)

$$(s+p-1)\alpha_s - (2s+p-3)\alpha_{s+1} + (s-2)\alpha_{s-2} - (a-1)(\alpha_{s-1} - \alpha_{s-2})/2 = \frac{(a-1)b}{4B}\{\Theta_s^{(p)} - \Theta_{s-3}^{(p)}\} + \frac{n}{2B}\,\Theta_{s-2}^{(p)},$$

and on using the difference operator notation,

$$(s+p-1)\,\nabla\alpha_s - (s + \frac{a}{2} - \frac{5}{2})\,\nabla\alpha_{s-1} = \frac{(a-1)b}{4B}\{\Theta_{s-2}^{(p)} - \Theta_{s-3}^{(p)}\} + \frac{n}{2B}\Theta_{s-2}^{(p)}. \tag{5.35}$$

But from (5.30c),

$$\beta_s = B\,\Theta_{s-1}^{(p)} + \frac{b}{n}(\delta_s - \delta_{s-1}) \tag{5.36}$$

and so from (5.33)

$$\beta_s = B\,\Theta_{s-1}^{(p)} + \frac{b}{2n}\{\alpha_{s-1} - \alpha_s + \frac{b}{2B}\Theta_{s-2}^{(p)} - \frac{b}{2B}\Theta_{s-1}^{(p)}\};$$

i.e., from (5.30b)

$$\frac{b}{2n}\,\nabla\alpha_s = \frac{b^2}{4nB}\{\Theta_{s-2}^{(p)} - \Theta_{s-1}^{(p)}\} + B\,\Theta_{s-1}^{(p)} - \frac{(\alpha_s - \Phi_s^{(p)})}{(a-1)}. \qquad (a \neq 1)$$

Rearranging leads to

$$\Phi_s^{(p)} = \alpha_s + (a-1)\left\{\frac{b}{2n}\,\nabla\alpha_s - \frac{b^2}{4nB}\Theta_{s-2}^{(p)} - \left(B - \frac{b^2}{4nB}\right)\Theta_{s-1}^{(p)}\right\}.$$

From this

$$\Phi_s^{(p)} - \Phi_{s-1}^{(p)} = \nabla\alpha_s\left(1 + \frac{b(a-1)}{2n}\right) - \frac{b(a-1)}{2n}\,\nabla\alpha_{s-1} + H_s^{(p)}$$

where

$$H_s^{(p)} = -(a-1)\left\{\left(B - \frac{b^2}{4nB}\right)\Theta_{s-1}^{(p)} - \left(B - \frac{b^2}{2nB}\right)\Theta_{s-2}^{(p)} - \frac{b^2}{4nB}\Theta_{s-3}^{(p)}\right\}. \tag{5.37}$$

Now this is a relation between $\nabla\alpha_s$ and $\nabla\alpha_{s-1}$, parallel to that in (5.35). Thus we have

$$\text{from (5.35):}\quad A_s^{(1)}\,\nabla\alpha_s + A_s^{(2)}\,\nabla\alpha_{s-1} = B_s^{(1)},$$

$$\text{from (5.37):}\quad C_1\,\nabla\alpha_s + C_2\,\nabla\alpha_{s-1} = B_s^{(2)}.$$

Note that $C_1$ and $C_2$ are independent of $s$ so that the two equations cannot be dependent in general. Clearly,

$$\nabla\alpha_s = (C_2 B_s^{(1)} - B_s^{(2)} A_s^{(2)}) / (A_s^{(1)} C_2 - A_s^{(2)} C_1)$$

where the denominator is independent of $\Phi_s^{(p)}$ and $\Theta_s^{(p)}$. Hence we derive the fourth order recurrence

$$\frac{A_s^{(1)}(C_2 B_s^{(1)} - B_s^{(2)} A_s^{(2)})}{A_s^{(1)} C_2 - A_s^{(2)} C_1} + \frac{A_s^{(2)}(C_2 B_{s-1}^{(1)} - B_{s-1}^{(2)} A_{s-1}^{(2)})}{A_{s-1}^{(1)} C_2 - A_{s-1}^{(2)} C_1} = B_s^{(1)} \tag{5.38}$$

where

$$A_s^{(1)} = s + p - 1, \qquad A_s^{(2)} = -(s + a/2 - s/2);$$

$$B_s^{(1)} = \frac{(a-1)b}{4B}\{\Theta_{s-2}^{(p)} - \Theta_{s-3}^{(p)}\} + \frac{n}{2B}\Theta_{s-2}^{(p)};$$

$$B_s^{(2)} = \Phi_s^{(p)} - \Phi_{s-1}^{(p)} - H_s^{(p)},$$

$$C_1 = 1 + b(a-1)/(2n), \qquad C_2 = -b(a-1)/(2n).$$

For a check (i); with $b = 0,\ B \neq 0$,

$$A_s^{(1)} = s + p - 1, \qquad A_s^{(2)} = -\{s + (a-5)/2\},$$

$$B_s^{(1)} = n\,\Theta_{s-2}^{(p)}/(2B),$$

$$B_s^{(2)} = 2B\{-(s+p)\Theta_s^{(p)} + (2s + p + a/2 - 2)\Theta_{s-1}^{(p)} - (s + a/2 - 2)\Theta_{s-2}^{(p)}\},$$

$$C_1 = 1, \qquad C_2 = 0, \qquad k = 0, \qquad \lambda = -1/(2B),$$

$$H_s^{(p)} = -(a-1)B\{\Theta_{s-1}^{(p)} - \Theta_{s-2}^{(p)}\},$$

$$\nabla_s = A_s^{(1)} C_2 - A_s^{(2)} C_1 = s + (a-5)/2,$$

$$\Phi_s^{(p)} = 2B\{(s - ½)\Theta_{s-1}^{(p)} - (s+p)\Theta_s^{(p)}\}.$$

Substituting in (5.38), we find

$$(s+p)(s+p-1)\Theta_s^{(p)} - ½(s+p-1)(6s+2p+2a-9)\Theta_{s-1}^{(p)} + \{3s^2 + s(2a+2p-12) + p(a-9/2) + a^2/4 + n/(4B^2) - 15a/4 + 12\}\Theta_{s-2}^{(p)} - \{(2s+a-5)(2s+a-6)/4\}\Theta_{s-3}^{(p)} = 0. \quad (s = 3, 4, \cdots) \tag{5.39}$$

This can be checked directly by first principles or against (5.17) with $p = 0$.

For a check (ii); with $a = 1$, $B = 0$ (using (5.38)),

$$A_s^{(1)} = s + p - 1, \qquad A_s^{(2)} = -(s-2),$$

$$2BB_s^{(1)} = n\,\Theta_{s-2}^{(p)}, \qquad B_s^{(2)} = \Phi_s^{(p)} - \Phi_{s-1}^{(p)},$$

$$C_1 = 1, \qquad C_2 = 0, \qquad k = 1,$$

$$2nB\,\Phi_s^{(p)} \sim (s+p)\Theta_s^{(p)} - (2s+p-1)\Theta_{s-1}^{(p)} + (s-1)\Theta_{s-2}^{(p)},$$

$$V_s = s - 2, \qquad H_s^{(p)} = 0.$$

Hence from (5.38)

$$(s+p-1)\,B_s^{(2)} - (s-2)\,B_{s-1}^{(2)} = B_s^{(1)}. \tag{5.40}$$

But if we assume the correct recurrence is

$$(s+p)\Theta_s^{(p)} = (n/b + 2s + p - 1)\,\Theta_{s-1}^{(p)} - (s-1)\Theta_{s-2}^{(p)}$$

then

$$(2nB/b^2)\,\Phi_s^{(p)} \sim (n/b)\Theta_{s-1}^{(p)},$$

and (5.40) leads to

$$(s+p-1)\Theta_{s-1}^{(p)} - (2s+p-3+n/b)\Theta_{s-2}^{(p)} + (s-2)\Theta_{s-3}^{(p)} = 0$$

as it should.

Note that (5.38) could turn out to be a degenerate recurrence, but the algebraic complication is a deterrent to further probes.

## 5.8 Alternative Forms for the Modified Borel Models

### 5.8.1 One-Component Borel

From (5.12) we may define

$$D_r(n;a,h) = N\sum_{s=0}^{r-2} a_s^{(r)} N^s + \Psi_0(N;a)\sum_{s=0}^{r-1} \frac{(-1)^s B_s^{(r)} N^{s+1}}{s!} \tag{5.41}$$

with

$$D_1(n;a,h) = \Psi_0(N;a)\,B_0^{(1)},$$

where $\Psi_s(N;a)$ is defined in (5.10b).

In the interests of simplicity, take $r=4$; generalization to $r>4$ is straightforward. The model is

$$D_4(n;a,h) = N(a_0^{(4)} + a_1^{(4)}N + a_2^{(4)}N^2) \qquad (5.42)$$
$$+ N\Psi_0(N;a)\left\{B_0^{(4)} - \frac{NB_1}{1!} + \frac{N^2B_2^{(4)}}{2!} - \frac{N^3B_3^{(4)}}{3!}\right\}.$$

Equating coefficients of powers of $N^{-s}$ ($s=0,1,2,3$) in $D_4(n;a,h)$ and the series $S(n) \sim e_0 + e_1/n + \cdots$, we find the equations

$$\begin{bmatrix} 1 & a & \binom{a+1}{2} & \binom{a+2}{3} \\ 1 & a+1 & \binom{a+2}{2} & \binom{a+3}{3} \\ 1 & a+2 & \binom{a+3}{2} & \binom{a+4}{3} \\ 1 & a+3 & \binom{a+4}{2} & \binom{a+5}{3} \end{bmatrix} \begin{bmatrix} B_0^{(4)} \\ B_1^{(4)} \\ B_2^{(4)} \\ B_3^{(4)} \end{bmatrix} = \begin{bmatrix} k_0 \\ -k_1 \\ k_2 \\ -k_3 \end{bmatrix} \qquad (5.43)$$

where $k_s = e_s/\{h^s\Gamma(a+s)\}$, and $N = n/h$. Premultiplying by the binomial matrix

$$\begin{bmatrix} 1 & -3 & 3 & -1 \\ 1 & -2 & 1 & 0 \\ 1 & -1 & 0 & 0 \\ 1 & 0 & 1 & 0 \end{bmatrix}$$

leads to

$$\begin{bmatrix} 0 & 0 & 0 & -1 \\ 0 & 0 & 1 & a+2 \\ 0 & -1 & -(a+1) & -\binom{a+2}{2} \\ 1 & a & \binom{a+1}{2} & \binom{a+2}{3} \end{bmatrix} \begin{bmatrix} B_0^{(4)} \\ B_1^{(4)} \\ B_2^{(4)} \\ B_3^{(4)} \end{bmatrix} = \begin{bmatrix} k_3 \\ k_2 \\ k_1 \\ k_0 \end{bmatrix} \qquad (5.44)$$

where $K_0 = k_0$, $K_1 = k_0 + k_1$, $K_2 = k_0 + 2k_1 + k_2$, and $K_3 = k_0 + 3k_1 + 3k_2 + k_3$. For the generalization, the coefficients in the successive rows (starting at the last) are those in the expansion of $(1-t)^{-a}$, $(-t)(1-t)^{-a-1}$, $(-t)^2(1-t)^{-a-2}$, etc. The coefficients $a_0^{(4)}$,

$a_1^{(4)}$ are found from the coefficients in $N$ and $N^2$. Thus

$$\begin{aligned} a_0^{(4)} &= \frac{B_1^{(4)}\,\Gamma(a)}{1!} + \frac{B_2^{(4)}\,\Gamma(a+1)}{2!} + \frac{B_3^4\,\Gamma(a+2)}{3!}, \\ a_1^{(4)} &= -\left\{\frac{B_2^{(4)}\,\Gamma(a)}{2!} + \frac{B_2^{(4)}\,\Gamma(a+1)}{3!}\right\}, \\ a_2^{(4)} &= \frac{B_3^{(4)}\,\Gamma(a)}{3!}. \end{aligned} \tag{5.45}$$

### Example 5.5

Consider Example 5.1. With $a = h = 1$, we have for $D_3(n\,;1,1)$ the matrix equation

$$\begin{bmatrix} 0 & 0 & 1 \\ 0 & -1 & -2 \\ 1 & 1 & 1 \end{bmatrix} \begin{bmatrix} B_0^{(3)} \\ B_1^{(3)} \\ B_2^{(3)} \end{bmatrix} = \begin{bmatrix} 1/3 \\ 1/2 \\ 1 \end{bmatrix}$$

leading to

$$\begin{aligned} B_2^{(3)} &= 1/3, \quad B_1^{(3)} = -7/6, \quad B_0^{(3)} = 11/6, \\ a_0^{(3)} &= -1, \quad a_1^{(3)} = -1/6. \end{aligned}$$

Thus

$$D_3(n\,;1,1) = n\left(-1-\frac{n}{6}\right) + \left(\frac{11}{6} + \frac{7n}{6} + \frac{n^2}{6}\right)\int_0^\infty \frac{e^{-t}\,dt}{(1+t/n)} \tag{5.46}$$

from which, using $\Psi_0(1,1) = 0.596347361$ and $\Psi_0(10,1) = 0.0915633339$, we find $D_3(1;1) = 0.721766643$, and $D_3(10;1) = 0.954939060$ in agreement with the recursive approach. Note that the different segments of the 4×4 matrix in (5.44) and its extension are invariant with respect to $a$; i.e., once the value of $a$ is fixed and the matrix set up for a given $r$, extensions do not require any recalculations.

### 5.8.2 Two-Component Borel

For this, the approximants are

$$\begin{aligned} F_r(n\,;a,h) = {} & N\sum_{s=0}^{R_3} a_s^{(r)} N^s + N\Phi_0(N;a)\sum_{s=0}^{R_1}(-1)^s\frac{N^s B_{2s}^{(r)}}{(2s)!} \\ & + N\Phi_0(N;a+1)\sum_{s=0}^{R_2}(-1)^s\frac{N^s B_{2s+1}^{(r)}}{(2s+1)!}, \end{aligned} \tag{5.47}$$

where $R_i = [(r-i)/2]$, $(i = 1, 2, 3)$; if $R_i < 0$, the summatory term is taken to be zero. The expression (5.47) corresponds to the polynomial form given in section 5.6.1. The parameters $B_0^{(r)}, B_1^{(r)}, \cdots$, are determined from the equivalence of powers of $N^{-1}$ in (5.47) and the series for $S(n)$. Thus

$$\begin{bmatrix} 1 & \binom{a}{1} & \cdots & \binom{a+r-2}{r-1} \\ 1 & \binom{a+2}{1} & \cdots & \binom{a+r}{r-1} \\ \vdots & \vdots & \vdots & \vdots \\ 1 & \binom{a+2r-2}{1} & \cdots & \binom{a+3r-4}{r-1} \end{bmatrix} \begin{bmatrix} B_0^{(r)} \\ B_1^{(r)} \\ \vdots \\ B_{r-1}^{(r)} \end{bmatrix} = \begin{bmatrix} k_0 \\ -k_1 \\ \vdots \\ (-1)^{r-1}k_{r-1} \end{bmatrix} \tag{5.48}$$

where $k_s = e_s / \{h^s \Gamma(a+2s)\}$, and $s = 0, 1, \cdots$. Premultiplying (5.48) by a matrix of binomial coefficients with alternating signs leads to

$$\underset{\sim}{A}_r \underset{\sim}{B}_r = \underset{\sim}{K}_r , \tag{5.49}$$

where

(i) $\underset{\sim}{K}_r = [K_{r-1}, K_{r-2}, \cdots, K_0]'$ (row transpose)

with $K_s = \sum_{r=0}^{s} \binom{s}{r} k_r$ ;

(ii) $\underset{\sim}{B}_r = [B_0^{(r)}, B_1^{(r)}, \cdots, B_{r-1}^{(r)}]'$;

(iii)

$$\begin{bmatrix} 0 & 0 & 0 & \cdots & 0 & 0 & A_{r-1,r-1} \\ 0 & 0 & 0 & \cdots & 0 & A_{r-2,r-2} & A_{r-2,r-1} \\ 0 & 0 & 0 & \cdots & A_{r-3,r-3} & A_{r-3,r-2} & A_{r-3,r-1} \\ \vdots & \vdots & \vdots & \vdots & \vdots & \vdots & \vdots \\ 0 & A_{1,1} & A_{1,2} & \cdots & A_{1,r-3} & A_{1,r-2} & A_{1,r-1} \\ A_{0,0} & A_{0,1} & A_{0,2} & \cdots & A_{0,r-3} & A_{0,r-2} & A_{0,r-1} \end{bmatrix} = \underset{\sim}{A}_r$$

whose elements are determined from the recursive scheme

$$2(t-s)A_{s,t} = (3t-2+2s)A_{s,t-1}-(t+a-2)A_{s,t-2}. \qquad (5.50)$$
$$(t>s,\ s,t=0,1,\cdots;\ A_{s,t}=0,\ t<s;\ A_{s,s}=(-2)^s).$$

Moreover, for the first polynomial in (5.47)

$$a_s^{(r)} = (-1)^s \sum_{m=2}^{R(r,s)} B_{m+2s}^{(r)} \frac{\Gamma(a+m-2)}{(m+2s)!}, \qquad (5.51)$$

where $s=0,\ 1,\ \cdots,\ [(r-3)/2]$; $R_{r,s}=r-1-2s$, $(r=3,4,\cdots)$; $a_s^{(r)}=0$, $(r=1,2)$. Note that the column totals are generated by $(1-t)^{2-a}$.

**Illustration:** For $r=8$, $a=1$ we have for the coefficients

$$\begin{bmatrix} 0 & 0 & 0 & 0 & 0 & 0 & 0 & -128 \\ 0 & 0 & 0 & 0 & 0 & 0 & 64 & 640 \\ 0 & 0 & 0 & 0 & 0 & -32 & -272 & -1312 \\ 0 & 0 & 0 & 0 & 16 & 112 & 456 & 1408 \\ 0 & 0 & 0 & -8 & -44 & -146 & -377 & -833 \\ 0 & 0 & 4 & 16 & 41 & 85 & 155 & 259 \\ 0 & -2 & -5 & -9 & -14 & -20 & -27 & -35 \\ 1 & 1 & 1 & 1 & 1 & 1 & 1 & 1 \end{bmatrix} \begin{bmatrix} B_0^{(8)} \\ B_1^{(8)} \\ B_2^{(8)} \\ B_3^{(8)} \\ B_4^{(8)} \\ B_5^{(8)} \\ B_6^{(8)} \\ B_7^{(8)} \end{bmatrix} = \begin{bmatrix} K_7 \\ K_6 \\ K_5 \\ K_4 \\ K_3 \\ K_2 \\ K_1 \\ K_0 \end{bmatrix} \qquad (5.52)$$

where for example

$$K_0 = e_0/\Gamma(a),$$
$$K_1 = e_0/\Gamma(a)+e_1/\{h\,\Gamma(a+2)\},$$
$$K_2 = e_0/\Gamma(a)+2e_1/\{h\,\Gamma(a+2)\}+e_2/\{h^2\Gamma(a+4)\};$$

and

$$\Phi_0(N;i) = \int_0^\infty \frac{e^{-t}t^{i-1}dt}{N+t^2}. \qquad (i=a,a+1;\ a>0)$$

(Again, note that segments of the triangular matrix (5.52), structured from the bottom left-hand corner are functions of $a$ only. For example, without changing the matrix, we have

$$\begin{bmatrix} 0 & 0 & 0 & -8 \\ 0 & 0 & 4 & 16 \\ 0 & -2 & -5 & -9 \\ 1 & 1 & 1 & 1 \end{bmatrix} \begin{bmatrix} B_0^{(4)} \\ B_1^{(4)} \\ B_2^{(4)} \\ B_3^{(4)} \end{bmatrix} = \begin{bmatrix} K_3 \\ K_2 \\ K_1 \\ K_0 \end{bmatrix} \qquad (r=4,\ a=1)$$

Moreover

$$a_s^{(8)} = (-1)^s \sum_{m-2}^{7-2s} B_{m+2s}^{(8)} \frac{\Gamma(m-1)}{(m+2s)!}, \qquad (s = 0, 1, 2)$$

and the model is

$$\begin{aligned} F_8(n;1,h) &= N(a_0^{(8)} + Na_1^{(8)} + N^2 a_2^{(8)}) \\ &+ N\Phi_0(N;1)\left\{B_0^{(8)} - \frac{NB_2^{(8)}}{2!} + \frac{N^2 B_4^{(8)}}{4!} - \frac{N^3 B_6^{(8)}}{6!}\right\} \\ &+ N\Phi_0(N;2)\left\{\frac{B_1^{(8)}}{1!} - \frac{NB_3^{(8)}}{3!} + \frac{N^2 B_5^{(8)}}{5!} - \frac{N^3 B_7^{(8)}}{7!}\right\}. \end{aligned}$$

**Example 5.6**

Let

$$y_m(n) \sim \sum_{s=0}^{\infty} (-1)^s s^m (2s)!/\, n^s . \qquad (n>0,\ m = 1, 2, \cdots)$$

Then the 2cB sums are

$$\begin{aligned} 2y_1(n) &= n\{-\Phi_0(n;1) + \Phi_0(n;2)\}, &(5.53)\\ y_2(n) &= n\{(1-n)\Phi_0(n;1) - 3\Phi_0(n;2) + 1\}, \\ 8y_3(n) &= n\{(6n-1)\Phi_0(n;1) - (n-7)\Phi_0(n;2) - 5\}, \\ 16y_4(n) &= n\{(n^2-25n+1)\Phi_0(n;1) + (10n-15)\Phi_0(n;2) + 17 - n\}. \end{aligned}$$

These examples indicate that the double factorial series with a typical term modified by a polynomial multiplier can be summed by the 2cB algorithm.

**Example 5.7**

Consider the mean values of the standard deviation $E(\sqrt{m_2})$ in exponential sampling (the series is given in Table 2.2, and Levin approximants in Table 2.3). With $a = h = 1$, $k_s = e_s/(2s)!$; thus

$$\begin{aligned} k_0 &= 1, & K_0 &= 1; \\ k_1 &= -0.75, & K_1 &= 0.25; \\ k_2 &= 0.255208333, & K_2 &= -0.244791667; \\ k_3 &= -0.210156250, & K_3 &= -0.694531251; \\ k_4 &= 0.214340549, & K_4 &= -1.095034453. \end{aligned}$$

Using the 5×5 lower segment of (5.52) we find:

$$B_4^{(5)} = -0.06843965331\,,$$
$$B_3^{(5)} = 0.4632344996\,,$$
$$B_2^{(5)} = -1.212629469\,,$$
$$B_1^{(5)} = 1.301095998\,,$$
$$B_0^{(5)} = 0.5167386240\,.$$

$$a_0^{(5)} = -0.5348122890\,,$$
$$a_1^{(5)} = 0.002851652221\,,$$

Now

$$\Phi_0(2;1) = 0.357429\,, \qquad \Phi_0(2;2) = 0.228573\,,$$
$$\Phi_0(10;1) = 0.0884251\,, \qquad \Phi_0(10;2) = 0.0729049\,,$$
$$\Phi_0(50;1) = 0.01933046\,, \qquad \Phi_0(50;2) = 0.0181798\,.$$

and from (5.50)

$$F_5(2;1,1) = 0.694083\,,$$
$$F_5(10;1,1) = 0.888850\,,$$
$$F_5(50;1,1) = 0.972065\,.$$

Compared to the values in Table 5.5 there are slight discrepancies, the explanation being that Table 5.5 was computed using double precision arithmetic on IBM 3033, whereas the three calculations here were obtained on a desk calculator.

### 5.8.3 Three-Component Borel

For series diverging about as fast as the triple factorial ($\sum(-1)^s(3s)!$), we can use a 3cB model

$$\begin{aligned} G_r(n;a,h) = {} & N \sum_{s=0}^{R_4} a_s^{(r)} N^s + N\,\Omega_0(N;a) \sum_{s=0}^{R_1} (-1)^s \frac{N^s B_{3s}^{(r)}}{(3s)!} \\ & + N\,\Omega_0(N;a+1) \sum_{s=0}^{R_2} (-1)^s \frac{N^s B_{3s+1}^{(r)}}{(3s+1)!} \\ & + N\,\Omega_0(N;a+2) \sum_{s=0}^{R_3} (-1)^s \frac{N^s B_{3s+2}^{(r)}}{(3s+2)!}\,, \end{aligned} \tag{5.54}$$

where

$$R_i(r) = [(r-i)/3], \qquad (i = 1,2,3,4)$$

$$\Omega_0(N;a) = \int_0^\infty \frac{e^{-t}t^{a-1}}{N+t^3}\,dt\;;$$

and if $r - i < 0$ the corresponding summatory expression is taken to be zero.

The parameters $B_0^{(r)}, B_1^{(r)}, \cdots$, are determined from

$$\underset{\sim}{A}_r \underset{\sim}{B}_r = \underset{\sim}{K}_r ,$$

where

(i) $\underset{\sim}{K}_r = [K_{r-1}, K_{r-2}, \cdots, K_0]'$,

$$\text{with} \quad K_s = \sum_{r=0}^{s} \binom{s}{r} k_r, \quad k_r = e_r / \{h^r \Gamma(a+3r)\} ;$$

(ii) $\underset{\sim}{B}_r = [B_0^{(r)}, B_1^{(r)}, \cdots, B_{r-1}^{(r)}]'$;

(iii) $\underset{\sim}{A}_r$ has exactly the same form defined in (5.50) but the elements now following the recurrence

$$3(t-s)A_{s,t} = (6t+3a-6)A_{s,t-1} - (4t+3a-8)A_{s,t-2} + (t+a-3)A_{s,t-3} .$$

$$(t > s, \; s,t = 0,1, \cdots; \; A_{s,s} = (-3)^s, \; A_{s,t} = 0, \; t < s) \qquad (5.55)$$

Moreover

$$a_s^{(r)} = (-1)^s \sum_{m=3}^{r-3s-1} B_{m+3s}^{(r)} \frac{\Gamma(a+m-3)}{(m+3s)!} .$$

$$(s = 0, 1, \cdots, [(r-4)/3], \; r > 3; \; a_s^{(r)} = 0, \; r = 1,2,3)$$

The elements in the matrix $A_r$ may be checked by noting that the column totals are generated by $(1-t)^{3-a}$.

**Illustration:** For $a = 1$, a segment of the triangular matrix is

$$\begin{bmatrix} 0 & 0 & 0 & 0 & 0 & 0 & 729 \\ 0 & 0 & 0 & 0 & 0 & -243 & -2673 \\ 0 & 0 & 0 & 0 & 81 & 729 & 3753 \\ 0 & 0 & 0 & -27 & -189 & -783 & -2484 \\ 0 & 0 & 9 & 45 & 141 & 351 & 757 \\ 0 & -3 & -9 & -19 & -34 & -55 & -83 \\ 1 & 1 & 1 & 1 & 1 & 1 & 1 \end{bmatrix} .$$

In particular for $r = 4$, $a = h = 1$,

$$\begin{bmatrix} 0 & 0 & 0 & -27 \\ 0 & 0 & 9 & 45 \\ 0 & -3 & -9 & -19 \\ 1 & 1 & 1 & 1 \end{bmatrix} \begin{bmatrix} B_0^{(4)} \\ B_1^{(4)} \\ B_2^{(4)} \\ B_3^{(4)} \end{bmatrix} = \begin{bmatrix} K_3 \\ K_2 \\ K_1 \\ K_0 \end{bmatrix}$$

where

$$K_0 = e_0, \qquad K_1 = e_0 + e_1/3!,$$
$$K_2 = e_0 + 2e_1/3! + e_2/6!, \qquad K_3 = e_0 + 3e_1/3! + 3e_2/6! + e_3/9!$$

for the series $S(n) \sim e_0 + e_1/n + \cdots$.

Moreover,

$$G_4(n;1,1) = na_0^{(4)} + n\,\Omega_0(n;1)\left[\frac{B_0^{(4)}}{0!} - \frac{nB_3^{(4)}}{3!}\right] + n\,\Omega_0(n;2)\frac{B_1^{(4)}}{1!} + n\,\Omega_0(n;3)\frac{B_2^{(4)}}{2!}$$

with

$$a_0^{(4)} = \frac{B_3^{(4)}}{3!}, \qquad \Omega_0(n;i) = \int_0^\infty \frac{e^{-t}t^{i-1}dt}{n+t^3}. \qquad (i = 1, 2, 3)$$

## Further Examples

### Further Example 5.1

Sum the series

$$S(n) \sim \sum_{s=0}^{\infty} (-1)^s (2s)!/\{s!\,n^s\}$$

for $n = 1, 5, 10$. Use $F_r(n, 1)$ to derive:

| | $n=1$ | $n=5$ | $n=10$ |
|---|---|---|---|
| $F_6$ | 0.537559 | 0.786575 | 0.865250 |
| $F_{22}$ | 0.547724 | 0.787557 | 0.865409 |
| $F_{24}$ | 0.546998 | 0.787516 | 0.865403 |
| $F_{25}$ | 0.546630 | 0.787496 | 0.865401 |

Note that

$$S(n) = \sqrt{(n\pi)}\exp(n/4)P_r\{x > \sqrt{(n/2)}\},$$

where $P_r(x > y) = (2\pi)^{-1/2} \int_y^\infty \exp(-t^2/2)dt$ so that from normal probability integral tables $S(1) = 0.545641$, $S(5) = 0.787439$, and $S(10) = 0.865393$.

$$K_s = 1 - \frac{s}{1!\,1!} + \frac{s(s-1)}{2!\,2!} - \cdots \sim J_0(2\sqrt{s})$$

in terms of a Bessel function. Hence

$$K_s \sim \{1/(\pi\sqrt{s})\}^{1/2}\cos(2\sqrt{s} - \pi/4)$$

so an increment will be small whenever $2\sqrt{s} = (4m+3)\pi/4$ $m = 1, 2, \cdots$; moreover sequences of terms will be one-signed.

**Further Example 5.2**

For the series

$$S(n) \sim 1 + \sum_{s=1}^{\infty} \frac{(-1)^s (2s)!\sqrt{(2s)}}{n^s}$$

apply the 2cB algorithm with $a = h = 1$. Show that

$$S(n) = 1 + \frac{1}{\sqrt{\pi}} \int_0^\infty \int_0^\infty \frac{e^{-2x-y}(y^2 - y^3)dx\,dy}{\sqrt{x}(n + y^2 e^{-2x})},$$

and

$$K_s(1,1) = 1 + \sum_{r=1}^{s} (-1)^r \binom{s}{r} \sqrt{(2r)}, \qquad (s = 1, 2, \cdots)$$

$$= 1 - s(2/\pi)^{1/2} g(s),$$

where $sg(s) = \frac{1}{2}\int_0^\infty \frac{(1 - e^{-x})^s dx}{x^{3/2}}$.

Show $K_s$ increases to unity as $s$ ($\geqslant 1$) increases and that

$$K_s \sim 1 - \left(1 - \frac{\gamma}{\ln(s)}\right)\sqrt{\left(\frac{2}{\pi \ln(s)}\right)}, \qquad (s \to \infty;\ \gamma \text{ is Euler's const}).$$

**Further Example 5.3**

If $S(n) \sim 1 - 1/n + 2/n^2 - \cdots$, then with $a = h = 1$, show that for 2cB

$$K_s = \sum_{r=0}^{s} (-1)^r \binom{s}{r} \frac{r}{(2r)!},$$

with generating function

$$\sum_{s=0}^{\infty} K_s t^s = (1-t)^{-1}\cos\{t/(1-t)\}^{1/2}. \qquad (|t|<1)$$

Deduce that

$$K_{s+1} = \frac{3s+1/2}{s+1}K_s - \frac{s(3s-1/2)}{(s+1/2)(s+1)}K_{s-1} + \frac{s(s-1)}{(s+1/2)(s+1)}K_{s-2}.$$
$$(s=0,1,\cdots;\ K_s=0,\ s<0,\ K_0=1)$$

Is this recurrence less susceptible to error? By using Hankel's contour integral for a reciprocal of the gamma function, show that

$$K_s = \frac{i}{2\pi}\int_{(0+)} \frac{e^{-t}}{(-t)}\left(1-\frac{1}{t^2}\right)^s dt,$$

and deduce as asymptote,

$$K_s = \frac{e^{3\lambda_s^2/2}}{\sqrt{(3\pi)}}\left\{\frac{\alpha_s(12\lambda_s+2\lambda_s^{-1})+\beta_s(6\lambda_s-\lambda_s^{-1})}{12\lambda_s^2+3/4}\right\}, \qquad (s\to\infty)$$

where

$$\lambda_s = (s/4)^{1/6}, \qquad \alpha_s = \sqrt{3}\cos(\theta_s)/2,$$
$$\beta_s = \sin(\theta_s), \qquad \theta_s = 3\sqrt{3}\lambda_s^2/2.$$

The dominant term in $nK_s\theta_s$ is now

$$(2/3)(n)^{1/6}\sin(\theta_s+\pi/3)\exp\{3(s/4)^{1/3}/2-3(ns/4)^{1/3}\}$$

which tends to zero in a delayed oscillatory manner; for example,

| | $n=1$ | $n=2$ | $n=5$ | $n=10$ |
|---|---|---|---|---|
| $F_{44}$ | 0.4768 | 0.6622 | 0.8332 | 0.9091 |
| $F_{45}$ | 0.4843 | 0.6652 | 0.8338 | 0.9092 |
| $F_{46}$ | 0.4874 | 0.6669 | 0.8343 | 0.9094 |
| $F_{47}$ | 0.4632 | 0.6569 | 0.8321 | 0.9089 |
| $F_{\infty}$ | 0.5 | $0.666\dot{6}$ | $0.833\dot{3}$ | $0.\dot{9}\dot{0}$ |

($F_s$ is the $s$ th approximant to $n/(n+1)$).

Approximate the sums $S(1)$, $S(2)$, $S(5)$, and $S(10)$ using the 2cB and Levin algorithms.

### Further Example 5.4

Discuss the 2cB algorithm applied to the series

$$S(n) \sim \sum(-1)^s s!/n^s .$$

Show that for $a = h = 1$

$$K_{s+1} = \frac{(4s + \frac{1}{2})}{2s+1} K_s - \frac{2s}{2s+1} K_{s-1},$$

$$(s = 0, 1, \cdots ; \quad K_s = 0, \ s < 0, \ K_0 = 1)$$

and that

$$K_s \sim 8\sqrt{e} \left\{ \cos(\sqrt{s}) - \frac{11}{96} \frac{\sin(\sqrt{s})}{\sqrt{s}} \right\}. \qquad (s \to \infty)$$

Numerically assess for the sequence $K_s$ run lengths of terms with the same sign.

Similarly, for the case $a = 2$, $h = 1$, and the series $S(n)$, show that

$$\sum_{s=0}^{\infty} K_s^* \theta^s / s! = e^{\theta} \sin(\sqrt{\theta}) / \sqrt{\theta}$$

and

(i) $K_s^* = 2(K_s - K_{s+1})$,

(ii) $K_s^* \sim 8\sqrt{e} \left\{ \frac{\sin(\sqrt{s})}{\sqrt{s}} + \frac{35}{96} \frac{\cos(\sqrt{s})}{s} \right\}. \qquad (s \to \infty)$

### Further Example 5.5

In sampling from the exponential density, the series for $E(\sqrt{m_2'})$ (the sample second noncentral moment) is as follows

| $s$ | $e_s$ | $s$ | $e_s$ |
|---|---|---|---|
| 0 | 0.141421356237 01 | 11 | -0.337811045188 18 |
| 1 | -0.883883476483 00 | 12 | 0.995750166879 20 |
| 2 | 0.239753392996 00 | 13 | -0.343210069231 23 |
| 3 | -0.283049872664 02 | 14 | 0.136768644105 26 |
| 4 | 0.803728840054 03 | 15 | -0.624046568991 28 |
| 5 | -0.394975907547 05 | 16 | 0.323285203789 31 |
| 6 | 0.290537740985 07 | 17 | -0.188740410477 34 |
| 7 | -0.294869749253 09 | 18 | 0.123362666427 37 |
| 8 | 0.391923251749 11 | 19 | -0.897366996713 39 |
| 9 | -0.657928910787 13 | 20 | 0.722610280744 42 |
| 10 | 0.135824192646 16 | 21 | -0.641036314529 45 |

(i) Derive the 2cB ($a = h = 1$) approximants:

| $r$ | $F_r(n;1,1)$ $n = 1$ | $n = 2$ | $n = 3$ | $n = 4$ | $n = 5$ |
|---|---|---|---|---|---|
| 21 | 1.014366145 | 1.153368418 | 1.218021831 | 1.256187492 | 1.281584661 |
| 22 | 1.012495873 | 1.152658349 | 1.217674166 | 1.255993367 | 1.281466898 |
| 23 | 1.010793586 | 1.152025994 | 1.217369224 | 1.255825157 | 1.281365927 |
| 24 | 1.009244446 | 1.151462570 | 1.217401548 | 1.255679292 | 1.281279209 |
| 25 | 1.007834905 | 1.150960337 | 1.216866364 | 1.255552572 | 1.281204621 |
| True Value* | 1.000000 | 1.14794 | 1.215395 | 1.254784 | ----- |

(*Derived by Lam (1978))

(ii) Consider the validity of the Shank's 3 point and 5 point extrapolates (see expressions B4 and B5 of Bowman and Shenton, 1988).

(iii) Derive Levin's $t$-algorithm approximants $\{\alpha_r\}$:

| $r$ | $n = 1$ | $n = 2$ | $n = 3$ | $n = 4$ | $n = 5$ |
|---|---|---|---|---|---|
| 1 | 0.8702852690 | 1.077496047 | 1.170388687 | 1.223103621 | 1.244507935 |
| 2 | 1.089409261 | 1.136647047 | 1.245180459 | 1.223103621 | 1.294653833 |
| 3 | 1.027459865 | 1.153381120 | 1.216224916 | 1.254328842 | 1.279979204 |
| 4 | 0.9638253490 | 1.133348392 | 1.208280833 | 1.250221961 | 1.278375485 |
| 5 | 0.9530368555 | 1.135998938 | 1.211124699 | 1.252923498 | 1.279859642 |
| 6 | 0.9706537995 | 1.143009324 | 1.214176311 | 1.254401658 | 1.280633884 |
| 7 | 0.9907612009 | 1.147194097 | 1.215392232 | 1.254819922 | 1.280791751 |
| 8 | 1.002757413 | 1.148507289 | 1.215594469 | 1.254847375 | 1.280789976 |
| 9 | 1.004081276 | 1.148614429 | 1.215567916 | 1.254835458 | 1.280787787 |
| 10 | 1.007019335 | 1.142423639 | 1.215523014 | 1.254824320 | 1.280784440 |
| 11 | 1.005645701 | 1.148267540 | 1.215433641 | 1.254814720 | 1.28077979 |
| 12 | 1.004216153 | 1.142146587 | 1.215465732 | 1.254804368 | 1.280775189 |
| 13 | 1.002130322 | 1.147931933 | 1.215414046 | 1.254787267 | 1.280768507 |
| 14 | 1.001775470 | 1.147932627 | 1.215419664 | 1.254790766 | 1.280770465 |
| 15 | *1.001367044* | *1.147912293* | *1.215418101* | *1.254790816* | *1.280770555* |
| 16 | 0.9947280197 | 1.147392705 | 1.215321238 | 1.254763866 | 1.280761063 |
| 17 | 1.004822209 | 1.149301126 | 1.215594556 | 1.254826944 | 1.280778865 |
| 18 | 1.014657413 | 1.148680490 | 1.215533854 | 1.254817590 | 1.280778401 |
| 19 | 0.9364885888 | 1.144569222 | 1.214939163 | 1.254681704 | 1.280737595 |
| 20 | 1.033220249 | 1.149301126 | 1.215594556 | 1.254826944 | 1.280778865 |

(This output is reduced to 9 decimal digits from 26 digit output on CDC machine; series coefficients were inputed to 12 significant digits. Notice that $\alpha_{15}$ is the best Levin approximant.)

(iv) A simulation run of 30 cycles, each of 10,000 samples of five, resulted in a mean value of $\sqrt{m_2'}$ of 1.2814 and a standard deviation of 0.0046. The 90% confidence region for the mean of the simulation run is $1.2800 < E(\sqrt{m_2'}) < 1.2828$. Compare the 2cB and Levin assessments.

**Further Example 5.6**

For $E(\sqrt{m_2})$ from the exponential density (Table 2.2) and the 2cB model

$$F_4(n;1,1) = n\left\{a_0^{(4)} + \left(B_0^{(4)} - \frac{nB_2^{(4)}}{2!}\right)\Phi_0(n;1) + \left(B_1^{(4)} - \frac{nB_3^{(4)}}{3!}\right)\Phi_0(n;2)\right\},$$

show that

$$B_0^{(4)} = 0.816162, \quad B_1^{(4)} = 0.505485, \quad B_2^{(4)} = -0.408464,$$
$$B_3^{(4)} = 0.086816, \quad a_0^{(4)} = -0.189762,$$

and (for comparisons see Table 5.5)

$$F_4(2;1,1) = 0.713761,$$
$$F_4(10;1,1) = 0.893024,$$
$$F_4(50;1,1) = 0.972310.$$

## APPENDIX I

## A SERIAL CORRELATION COEFFICIENT WHOSE MOMENTS SUFFER SURGES

### A1. Introduction

Shenton and Johnson(1965) considered the moments of

$$\hat{\alpha} = \sum_1^n x_t x_{t-1} / \sum_1^{n-1} x_t^2, \qquad (t = 1, 2, \cdots, n) \tag{A1.1}$$

where

$$x_t = \alpha x_{t-1} + \epsilon_t, \qquad (x_0 = 0)$$

is a discrete first-order auto-regressive scheme with the $\epsilon$'s independent normal variates from $N(0, 1)$, and $\hat{\alpha}$ the maximum likelihood estimator of $\alpha$. For example, they show that

$$E(\hat{\alpha} - \alpha) = \int_0^\infty (D_\alpha U_0) dq, \tag{A1.2a}$$

$$E(\hat{\alpha} - \alpha)^2 = \int_0^\infty (D_\alpha^2 U_1 + U_0) dq \tag{A1.2b}$$

where $D_\alpha \equiv d / d\alpha$, and

$$U_r = \frac{1}{r!} \frac{q^r}{\sqrt{\{D_n(q)\}}}. \qquad (r = 1, 2, \cdots) \tag{A1.2c}$$

There are similar expressions for higher moments. The function $D_n(q)$ is defined by the second-order recursion

$$D_n = (1 + \alpha^2 + 2q) D_{n-1} - \alpha^2 D_{n-2}, \qquad (n = 2, 3, \cdots) \tag{A1.3}$$

with $D_0 = D_1 = 1$.

The authors state that for $n \geqslant 6,\ \alpha^2 \leqslant 1$.

$$E(\hat{\alpha} - \alpha) = \frac{-2(n-2)\alpha}{(n+1)^{[2]}} + \frac{12\alpha^3}{(n+5)^{[3]}} + \frac{18(n+8)\alpha^5}{(n+9)^{[4]}} \qquad \text{(A1.4)}$$

$$+ \frac{24(n+10)(n+12)\alpha^7}{(n+13)^{[5]}} + \frac{30(n+12)(n+14)(n+16)\alpha^9}{(n+17)^{[6]}} + \cdots ,$$

where $n^{[s]} \equiv n(n-2)\cdots(n-2s+2)$. They did not seem to be intrigued by the obvious (but incorrect) generalization displayed by the pattern of terms which is the subject of our present study.

Let us consider examples of numerical evidence from output done in 1963-5.

Table A1.1 Terms in $\alpha$-series for $E(\hat{\alpha} - \alpha)$

| Terms | $\alpha = 0.6\ n = 6$ | $\alpha = 1.0\ n = 5$ | $\alpha = 1.0\ n = 20$ |
|---|---|---|---|
| 1 | -0.137143 | -0.25 | -0.09022556 |
| 2 | 0.003740 | 0.025 | 0.00099378 |
| 3 | 0.001015 | 0.01741071 | 0.00111944 |
| 4 | 0.000279 | 0.01264880 | 0.00115054 |
| 5 | 0.000079 | -0.01316879⊙ | 0.00113277 |
| 6 | -0.000042 | 0.00221604 | 0.00109027 |
| 7 | 0.000002 | 0.00093038 | 0.00103614 |
| 8 | 0.000000 | 0.00009511 | 0.00097763 |
| 9 | | -0.00045106 | 0.00091873 |
| 10 | | 0.00013234• | 0.00086158 |
| 11 | | 0.00003039 | 0.00080731 |
| 12 | | -0.00000990 | 0.00075641 |
| 13 | | -0.00001277 | 0.00070904 |
| 14 | | 0.00000638 | 0.00066517 |
| 15 | | 0.00000052 | 0.00062465 |
| 16 | | -0.00000072 | 0.00058726 |
| 17 | | -0.00000026 | 0.00055279 |

(Computed on IBM7094 in 1965; loss of accuracy quite possible in particular for $\alpha = 1$).

The appearance of negative coefficients, apart from the first term, is important for our present note. For example, for $n = 6$ there is a negative at the sixth term (coefficient of $\alpha^{11}$), and at the coefficient of $\alpha^9$ when $n = 5$. Large $n$ seems to avoid negative signs after the first.

## A2. A New Solution with Interesting Consequences

A solution to (A1.3), the fundamental recursive form, is

$$D_n(q) = A\theta_1^n + B\theta_2^n \tag{A1.5}$$

where

$$\theta_1 = (1 + \alpha^2 + 2q + \sqrt{\Delta})/2,$$
$$\theta_2 = (1 + \alpha^2 + 2q - \sqrt{\Delta})/2,$$
$$\Delta = (1 + \alpha^2 + 2q)^2 - 4\alpha^2. \qquad (\theta_1\theta_2 = \alpha^2)$$

Since $D_0 = D_1 = 1$, we have

$$D_n = A\theta_1^n + (1 - A)\alpha^{2n}/\theta_1^n$$

with $A\theta_1 + (1 - A)\alpha^2/\theta_1 = 1$. Thus

$$D_n = \left[\frac{1 - \alpha^2 x}{x^{n-1}} + (1 - x)x^n\alpha^{2n}\right] / (1 - \alpha^2 x^2), \qquad (n = 0, 1, 2, \cdots) \tag{A1.6}$$

where

$$x = 1/\theta_1; \quad \text{i.e.} \quad x = 2/\{1 + \alpha^2 + 2q + \sqrt{\Delta}\}. \tag{A1.7}$$

Moreover

$$\begin{cases} q = 0, & (x = 1) \\ q = \infty, & (x = 0) \\ \dfrac{dx}{dq} = -\dfrac{2x}{\sqrt{\Delta}} = -\dfrac{2x^2}{(1 - \alpha^2 x^2)}. \end{cases} \tag{A1.8}$$

Hence,

$$E(\hat{\alpha} - \alpha) = \frac{1}{2}\int_0^1 x^{\frac{n-5}{2}} \frac{d}{d\alpha}\left[\frac{(1 - \alpha^2 x^2)^{3/2}}{\{1 - \alpha^2 x + (1 - x)x^{2n-1}\alpha^{2n}\}^{1/2}}\right] dx.$$
$$(n = 2, 3, \cdots; \alpha^2 \leqslant 1) \tag{A1.9}$$

For quadrature purposes,

$$E(\hat{\alpha} - \alpha) = -\frac{3}{2}\alpha\int_0^1 x^{\frac{n-1}{2}}\left\{\frac{1 - \alpha^2 x^2}{1 - \alpha^2 x + (1 - x)x^{2n-1}\alpha^{2n}}\right\}^{1/2} dx$$
$$+ \frac{1}{2}\int_0^1 x^{\frac{n-5}{2}} \frac{(1 - \alpha^2 x^2)^{3/2}\{\alpha x - n(1 - x)(\alpha x)^{2n-1}\}}{\{1 - \alpha^2 x + (1 - x)x^{2n-1}\alpha^{2n}\}^{3/2}} dx. \tag{A1.10}$$

Forms of this type could be constructed, using for example, (A1.2b) for $E(\hat{\alpha}-\alpha)^2$, for higher moments. Note that the interchange of derivative and integral operation in (A1.10) is permissible for $n \geqslant 4$. From (A1.10) then, we have

$$E(\hat{\alpha}-\alpha) = \frac{d}{d\alpha}\{\Phi_0(n,\alpha) + \alpha^{2n}\Phi_1(n,\alpha) + \alpha^{4n}\Phi_2(n,\alpha) + \cdots\} \tag{A1.11}$$

where $\Phi_0$, $\Phi_1$, etc. are infinite power series in $\alpha^2$. In fact,

$$E(\hat{\alpha}-\alpha) = \frac{1}{2}\frac{d}{d\alpha}\sum_{s=0}^{\infty}\frac{(-1)^s(2s)!}{(2^s s!)^2} \times \int_0^1 x^{\frac{n-5}{2}}\frac{(1-\alpha^2x^2)^{3/2}}{(1-\alpha^2x)^{1/2}}\left\{\frac{(1-x)x^{2n-1}\alpha^{2n}}{(1-\alpha^2x)}\right\}^s dx\,. \tag{A1.12a}$$

In particular,

$$E(\hat{\alpha}-\alpha) = \Psi_0(n,\alpha) + \alpha^{2n-1}\Psi_1(n,\alpha) + \cdots$$

where

$$\Psi_0 = -\frac{2(n-2)\alpha}{(n+1)^{[2]}} + \sum_{s=2}^{\infty}\frac{6s(n+4s-6)^{[s-2]}\alpha^{2s-1}}{(n+4s-3)^{[s+1]}}\,, \tag{A1.12b}$$

$$(n^{[s]} \equiv n(n-2)\cdots(n-2s+2))$$

$$\begin{aligned}\alpha^{2n-1}\Psi_1 = {} & -\frac{2n\,\alpha^{2n-1}}{(5n-3)^{[2]}} - \frac{12(n+1)\alpha^{2n+1}}{(5n+1)^{[3]}} - \frac{6(n+2)(10n+13)\alpha^{2n+3}}{(5n+5)^{[4]}} \\ & - \frac{12(n+3)(5n+8)(5n+11)\alpha^{2n+5}}{(5n+9)^{[5]}} \\ & - \frac{15(n+2)(n+4)(10n+24)(10n+31)\alpha^{2n+7}}{(5n+13)^{[6]}} \\ & - \frac{3(n+5)}{32}\left\{\frac{231}{(5n+9)^{[3]}} - \frac{84}{(5n+11)^{[3]}} - \frac{14}{(5n+13)^{[3]}} - \frac{4}{(5n+15)^{[3]}} - \frac{1}{(5n+17)^{[3]}}\right\}\alpha^{2n+9} + \cdots\,;\end{aligned} \tag{A1.12c}$$

$$\alpha^{4n-1}\Psi_2 = \frac{12n\,\alpha^{4n-1}}{(9n-3)^{[3]}} + \cdots\,,$$

and the dominant term in general

$$\alpha^{2ns-1}\Psi_s \;=\; \frac{(-1)^s n\,(2s)!\alpha^{2ns-1}}{(s-1)!2^{s-1}(4ns+n-3)^{[s+1]}}\,. \tag{A1.13}$$

We have been unable to discover a simple structure similar to the series for $\Psi_0$ for $\Psi_1$, $\Psi_2$, $\cdots$. Note that in terms of gamma functions

$$\Psi_0(n,\alpha) \;=\; -\frac{2(n-2)\alpha}{(n+1)^{[2]}} + \frac{3}{4}\sum_{s=2}^{\infty}\frac{s\,\Gamma(n/2+2s-1)\Gamma(n/2+s-3/2)\alpha^{2s-1}}{\Gamma(n/2+2s-½)\Gamma(n/2+s+1)} \tag{A1.14}$$

and the series converges for $\alpha^2 \leqslant 1$ as fast (or slow) as the one-signed series

$$\sum \alpha^{2s-1}/s^2\,.$$

## A3. The Surges in the Moments

The basic series in $\alpha$ for $E(\hat{\alpha}-\alpha)$ given in (A1.4) and (A1.12b) consists of non-negative terms; the terms in the first surge will be negative and surges thereafter alternate in sign. This is an unusual phenomenon and contrasts with one-signed, or alternating signs in a series. The structure of the series in $\alpha$ for the moments is clearly quite complicated; notice that the evaluation of a few terms of $E(\hat{\alpha}-\alpha)$ might lead one to infer a one-signed series (ignoring the first term). Interpretating sign patterns as showing analytical properties of the function studied becomes difficult whenever the sign pattern periodicity exceeds two. We remind the reader that our interest lies in detecting properties from a series; the converse problem is not so complicated as in the present case, but for the most part an easily understood function definition is not available. It may be surprising to recall (Bowman and Shenton, 1988, Table 1.8) that the mean value of the standard deviation $(E\sqrt{m_2})$ in normal sampling as a series in $n^{-1}$ has sign periodicity of four.

As an illustration of surges for $E(\hat{\alpha}-\alpha)$ take $n=5$, $\alpha=1$. The basic coefficient of $\alpha^9$ is $30{\cdot}21^{[3]}/22^{[6]} = 0.009558$. The first surge contributes $-10/22^{[2]} = -0.022727$, the combined coefficient being -0.013169. Note from Table A1.1 that is this case the coefficient of $\alpha^{17}$ is negative; the contributing terms are

$$\frac{54{\cdot}37^{[7]}}{38^{[10]}} - \frac{15{\cdot}7{\cdot}9{\cdot}74{\cdot}81}{38^{[6]}} \;=\; -0.000451\,.$$

Similarly the coefficient of $\alpha^{19}$ has the three components

$$\frac{60{\cdot}41^{[8]}}{42^{[11]}} - 0.004259 + 0.000940 \;=\; 0.000132\,.$$

We think some readers may like to juggle with elementary mathematics. Take $n = 2$, $\alpha = 1$. Then

$$E(\hat{\alpha} - \alpha) = -\frac{3}{2}\int_0^1 \frac{\sqrt{x}}{\sqrt{(1 - x + x^2)}}dx + \frac{1}{2}\int_0^1 \frac{\{1 - 2(1 - x)x^2\}}{\sqrt{x}(1 - x + x^2)^{3/2}}dx$$

$$= -1 + \int_0^1 \frac{(1 - t^4)}{(1 - t^2 + t^4)^{3/2}}dt \; . \qquad \text{(A1.15)}$$

Thus since the integral is unity, (hint, use the transformation $x - (1 - t^2)/(1 + t^2)$), $E(\hat{\alpha}) = 1$. But $\hat{\alpha} = 1 + uv/u^2$, where $u, v$ are independent normals $\epsilon N(0, 1)$. Are the results reconcilable? More generally in this case

$$E(\hat{\alpha} - \alpha) = -\alpha + \alpha\int_0^1 \frac{(1 - \alpha^2 t^4)dt}{(1 - \alpha^2 t^2 + \alpha^2 t^4)^{3/2}}$$

so that

$$E(\hat{\alpha} - \alpha) = -\alpha + \alpha\int_0^1 (1 - \alpha^2 t^4)\sum_{s=0}^{\infty} \frac{(2s + 2)!\alpha^{2s} t^{2s}(1 - t^2)^s}{2^{2s+1}s!(s + 1)!}dt$$

$$= \alpha\int_0^1 \sum_{s=1}^{\infty} \alpha^{2s} k_s(t)dt \; ,$$

where $k_s(t) = \dfrac{(2s)!t^{2s}(1 - t^2)^{s-1}}{2^{2s}(s!)^2}\left\{2s + 1 - (4s + 1)t^2\right\}$.

Clearly $\displaystyle\int_0^1 k_s(t)dt = 0$.

## A4. Higher Moments

Consider $Var(\hat{\alpha})$ for which $E(\hat{\alpha} - \alpha)^2$ is required. From (A1.2b) the basic need concerns

$$\int_0^{\infty} D_{\alpha}^2 \frac{q}{\sqrt{D_n}}dq \; . \qquad \text{(A1.16)}$$

But $(\theta_1 + \theta_2 - 1 - \alpha^2)/2 = q$. Thus

$$q = ½(1/x + \alpha^2 x - 1 - \alpha^2)$$

in the notation of section A2, and (A1.16) becomes

$$-\frac{1}{4}\int_0^1 D_\alpha^2\left(\frac{1}{x} + \alpha^2 x - 1 - \alpha^2\right)x^{\frac{n-5}{2}} \frac{(1-\alpha^2 x^2)^{3/2}}{\{1-\alpha^2 x + (1-x)x^{2n-1}\alpha^2\}^{1/2}}dx \,. \tag{A1.17}$$

If we use series, then differentiation may be used after integration, and series similar to that for $E(\hat{\alpha} - \alpha)$ will arise, with rather more complicated surges. Exact quadrature will demand the second derivative in the integrand for given $n$ and $\alpha$.

Although there are some complications, there is no reason to doubt exact results for four or more moments can be set up in this way.

## A5. A COETS Approach

Computer oriented extended Taylor Series (COETS) (Bowman, 1986) may be used in the present case. We start from

$$\hat{\alpha} - \alpha = \frac{x_1\epsilon_2 + x_2\epsilon_3 + \cdots + x_{n-1}\epsilon_n}{x_1^2 + x_2^2 + \cdots + x_{n-1}^2}, \qquad (\epsilon_j \in N(0,1)) \tag{A1.18}$$

and express as sums of polynomials in products of the epsilons. For example

$$\begin{aligned} x_1 &= \alpha x_0 + \epsilon_1, \\ x_2 &= \alpha\epsilon_1 + \epsilon_2, \\ x_3 &= \alpha^2\epsilon_1 + \alpha\epsilon_2 + \epsilon_3, \\ &\cdot \\ &\cdot \\ &\cdot \\ x_{n-1} &= \alpha^{n-2}\epsilon_1 + \alpha^{n-3}\epsilon_2 + \cdots + \epsilon_{n-1}. \end{aligned}$$

The dominant term in the denominator is

$$\epsilon_1^2 + \epsilon_2^2 + \cdots + \epsilon_{n-1}^2,$$

a $\chi^2$-variate. An illustration will be sufficient to indicate the general case, directed towards computer implementation.

$$n = 4: \quad \hat{\alpha} - \alpha = \frac{A_0 + A_1\alpha + A_2\alpha^2}{B_0 + B_1\alpha + B_2\alpha^2 + B_3\alpha^3 + B_4\alpha^4}, \tag{A1.19a}$$

where

$$A_0 = \epsilon_1\epsilon_2 + \epsilon_2\epsilon_3 + \epsilon_3\epsilon_4, \quad A_1 = \epsilon_1\epsilon_3 + \epsilon_2\epsilon_4, \quad A_2 = \epsilon_1\epsilon_4;$$
$$B_0 = \epsilon_1^2 + \epsilon_2^2 + \epsilon_3^2, \quad B_1 = 2(\epsilon_1\epsilon_2 + \epsilon_2\epsilon_3),$$
$$B_2 = \epsilon_1^2 + \epsilon_2^2 + 2\epsilon_1\epsilon_3, \quad B_3 = 2\epsilon_1\epsilon_2, \quad B_4 = \epsilon_1^2.$$

Then

$$E(\hat{\alpha} - \alpha) = E\left\{\frac{A_0 + A_1\alpha + A_2\alpha^2}{B_0}\right\}\{-G_\alpha + G_\alpha^2 - G_\alpha^3 + \cdots\} \tag{A1.19b}$$

where

$$G_\alpha = (B_1\alpha + B_2\alpha^2 + B_3\alpha^3 + B_4\alpha^4)/B_0.$$

For the coefficients we have

$$\alpha:\quad E\left\{-\frac{A_0B_1}{B_0^2} + \frac{A_1}{B_0}\right\} = -\frac{4}{15},$$

$$\alpha^2:\quad E\left\{\frac{A_2}{B_0} - \frac{A_1B_1}{B_0^2} + \frac{A_0B_1^2}{B_0^3} - \frac{A_0B_2}{B_0^2}\right\} = 0,$$

$$\alpha^4:\quad E\left\{\frac{A_0}{B_0}\left(-\frac{B_1^3}{B_0^3} + \frac{2B_1B_2}{B_0^2} - \frac{B_3}{B_0}\right) + \frac{A_1}{B_0}\left(-\frac{B_2}{B_0} + \frac{B_1^2}{B_0^2}\right) - \frac{A_2B_1}{B_0B_1}\right\} = \frac{12}{5\cdot7\cdot9}.$$

It is easy seen that the weight of coefficients in every term is the same for numerator and denominator. Hence the basic problem is in general a term such as

$$\nu_{r_1r_2\cdots r_k} = E\left\{\frac{\epsilon_1^{r_1}\epsilon_2^{r_2}\cdots\epsilon_k^{r_k}}{(\epsilon_1^2 + \epsilon_2^2 + \cdots + \epsilon_{n-1}^2)^R}\right\}, \tag{A1.20}$$

$$(R = \tfrac{1}{2}(r_1 + r_2 + \cdots + r_k))$$

noting that $\sum r$ must be even. But the moment in (A1.20) is scale free. Hence

$$\nu_{r_1,r_2,\cdots r_k} = \{\prod_{s=1}^{r} E(\epsilon_s^{r_s})\}/\{E(\sum_{s=1}^{n-1}\epsilon_s^2)^R\} \tag{A1.21}$$

and from the $\chi^2$-distribution the denominator is $(n-1)(n+1)\cdots(n+2R-3)$. In the numerator we use $E(\epsilon^{2m}) = 1, 3, \cdots (2m-1)$, and $E(\epsilon^{2m+1}) = 0$.

For the expansion for the basic series in (A1.12b), use

$$\frac{(1-px^2)^{3/2}}{(1-px)^{1/2}} = A_0 + A_1 p + \cdots, \qquad (p = \alpha^2)$$

with

$$A_s = \frac{1}{2^s s!}\left\{\frac{(2s)!x^s}{2^s s!} - 3\binom{s}{1}\frac{(2s-2)!x^{s+1}}{2^{s-1}(s-1)!} + 3\binom{s}{2}\frac{(2s-4)!x^{s+2}}{2^{s-2}(s-2)!} - \cdots\right\}. \qquad (s = 1, 2, \cdots; A_0 = 1)$$

Identify the integrated form with the partial fraction developments of terms in (A1.12b).

## APPENDIX II

## THE DERIVATIVE OF A CONTINUED FRACTION WHITTAKER'S APPROACH

The question posed is given a c.f.

$$y = \frac{b_0}{x+a_1-}\;\frac{b_1}{x+a_2-}\;\cdots$$

how do we set up $dy/dx$. One approach would be to develop it as a power series, differentiate, and return to a c.f. But another approach is to use 2nd (or higher) order c.f.s which in fact are pivoted on the reciprocal matrix process. Perhaps we should make it clear here that Shenton's papers (*Proc. Edinb. Math. Soc.*) in the early 1950's do not mention Whittaker's approach to c.f.s - none of the referees either as far as their comments go, mentioned Whittaker, and one or two of them were Edinburgh based. Our first contact with the Whittaker paper was purely by chance, and a short note was written about it (Shenton and Bowman, 1978). Recall also that theories making use of *mere* determinants have not been fashionable in recent years, and most (if not all) modern treatments on c.f.s take the route of sequences of transformations. Lastly, note that Aitken's letter to Shenton was dated 1959, and it took some time to coordinate the ideas.

In the Whittaker Memorial Number (*Proc. Edinb. Math. Soc.*, 11, 1958), here are some relevant comments from Aitken's contribution.

> The introductory example of the paper of 1916 is a good one, namely the c.f.
>
> $$\frac{1}{2x-}\;\frac{1}{2x-}\;\frac{1}{2x-}\;\cdots$$
>
> as compared with the power series
>
> $$\frac{1}{2x}+\frac{1}{2!\,2^2x^3}+\frac{1\cdot 3^*}{3!\,2^3x^5}+\frac{1\cdot 3\cdot 5^*}{4!\,2^4x^7}+\cdots,$$ (* omitted in the article)

both being representations of the function $y = x - (x^2 - 1)^{1/2}$ . Whereas the series converges only outside the unit circle in the complex plane, the c.f. converges everywhere except along the segment of the real axis given by $-1 < x < 1$.

··· The manipulative disadvantage of continued fractions (this is the chief deterrent to their more general use) is that they do not lend themselves to addition, multiplication or differentiation - fundamental processes. Whittaker's aim was in part to remedy this, in respect of differentiation at least. His first result is that the c.f.

$$\frac{1}{b_0+x-}\;\frac{a_1}{b_1+x-}\;\frac{a_2}{b_2+x-}\;\cdots\;\frac{a_n}{b_n+x}$$

admits an expansion in powers of $x^{-1}$ generalizing the corresponding expansion of $1/(b+x)$; it is in fact

$$\frac{1}{x} - \frac{b^{(1)}}{x^2} + \frac{b^{(2)}}{x^3} - \frac{b^{(3)}}{x^4} + \cdots ,$$

where $b^{(p)}$ is the leading element in the $p$ th power of the continuant matrix

$$B = \begin{vmatrix} b_0 & c_1 & & & & \\ d_1 & b_1 & c_2 & & & \\ & d_2 & b_2 & \cdot & & \\ & & \cdot & \cdot & \cdot & \\ & & & \cdot & c_{n-1} & \\ & & & \cdot & b_{n-1} & c_n \\ & & & & d_n & b_n \end{vmatrix},$$

$c_i$ and $d_i$ being any numbers such that $c_i d_i = a_i$. (As is pointed out, a natural choice would be to take all $c_i = 1$, or to ensure symmetry of $B$ by making $c_i = d_i = a_i^{1/2}$ ). The theorem becomes more interesting when expressed (as at p.246) in matrix notation. It is then that the c.f. in question is the leading element in the reciprocal matrix $(B + xI)^{-1}$, connection being thus established with the theorem that a continued fraction is the quotient of two determinants, the numerator being a leading diagonal cofactor in the denominator. The rest of the paper is concerned with the relation of the result to Stieltjes' expansion of a continued fraction in a power series.

The differentiation of the c.f., let us say S, follows from the above result; for $-dS/dx$ must be the leading element in $(B + xI)^{-2}$. The rest of the paper is devoted to obtaining in explicit form $-dS/dx$ as the quotient of two determinants, the numerator of order $2n$, the denominator of order $2n+2$. In the notation of partitioned matrices we may write this quotient as

$$\begin{vmatrix} 0 & B_0 + xI \\ B_0 + xI & -I \end{vmatrix} \div \begin{vmatrix} 0 & B + xI \\ B + xI & -I \end{vmatrix}$$

where $B_0$ is $B$ with the first row and column removed, in fact the submatrix complementary to the leading element $b_0$ of $B$ and 0 denotes a block of zero elements.

For the 2nd order c.f., and Whittaker's example

$$y(x) = x - \sqrt{(x^2-1)} = \frac{1}{2x -}\ \frac{1}{2x -}\ \frac{1}{2x -}\ \cdots,$$

(3.39) gives the sequences $\chi_s$ for $dy/dx = 1 - x/\{\sqrt{(x^2-1)}\}$. For example, in terms of elements of $(B + xI)^{-2}$

$$\chi_1 = \frac{1}{2}\begin{vmatrix} 0 & 1 \\ 1 & x^2 + 1/4 \end{vmatrix} \div (x^2 + 1/4),$$

$$\chi_2 = \frac{1}{2}\begin{vmatrix} 0 & 1 & 0 \\ 1 & x^2 + 1/4 & x \\ 0 & x & x^2 + 1/2 \end{vmatrix} \div \begin{vmatrix} x^2 + 1/4 & x \\ x & x^2 + 1/2 \end{vmatrix},$$

$$\chi_3 = \frac{1}{2}\begin{vmatrix} 0 & 1 & 0 & 0 \\ 1 & x^2 + 1/4 & x & 1/4 \\ 0 & x & x^2 + ½ & x \\ 0 & 1/4 & x & x^2 + ½ \end{vmatrix}$$

$$\div \begin{vmatrix} x^2 + 1/4 & x & 1/4 \\ x & x^2 + ½ & x \\ 1/4 & x & x^2 + ½ \end{vmatrix},$$

and in general

$$\chi_s = -\frac{1}{2}\frac{K_{s-1}(\gamma_1, \beta_1, \alpha_1)}{K_s(\gamma_0, \beta_0, \alpha_0)}, \qquad (s = 1, 2, \cdots)$$

where

$$\gamma_0 = x^2 + 1/4, \qquad \gamma_s = x^2 + ½, \qquad s \geqslant 1$$
$$\beta_s = x, \qquad \alpha_s = 1/4.$$

For example,

$$\chi_1 = -(1/2)/(x^2 + 1/4),$$
$$\chi_2 = -(x^2/2 + 1/4)/(x^4 - x^2/4 + 1/8),$$
$$\chi_3 = -(x^4/2 + 1/8)/(x^6 - 3x^4/4 + 3x^2/16 + 1/32),$$

and so on. If

$$\chi_s = -T_3^*/T_s,$$

then

$$T_s^* = x^2 T_{s-1}^* - \frac{(4x^2-1)T_{s-2}^*}{8} + \frac{x^2 T_{s-3}^*}{16} - \frac{T_{s-4}^*}{256} + \frac{1}{4^{s-1}},$$

$$T_s = x^2 T_{s-1} - \frac{(4x^2-1)T_{s-2}}{8} + \frac{x^2 T_{s-3}}{16} - \frac{T_{s-4}}{256}, \qquad (s \geqslant 4)$$

with

$$T_0^* = 0, \; T_1^* = \frac{1}{2}, \qquad T_2^* = \frac{(2x^2+1)}{4}, \qquad T_3^* = \frac{x^4}{2} + \frac{1}{8};$$

$$T_0 = 1, \; T_1 = x^2 + \frac{1}{4}, \; T_2 = x^4 - \frac{x^2}{4} + \frac{1}{8}, \; T_3 = x^6 - \frac{3x^4}{4} + \frac{3x^2}{16} + \frac{1}{32}.$$

It is readily shown that

$$y(x) = \frac{1}{\pi}\int_{-1}^{1} \frac{\sqrt{(1-\omega^2)}d\omega}{x+\omega}$$

so that $(-\chi_s)$ provides an increasing convergent sequence to $dy/dx$ for real $x$ not in $-1 < x < 1$.

Numerical: $x = 2$, $\quad dy/dx = -(2-\sqrt{3})/\sqrt{3}$.

| $s$ | $T_s^*$ | $T_s$ | $T_s^*/T_s$ |
|---|---|---|---|
| 0 | 0 | 1 | 0 |
| 1 | 1/2 | 17/4 | 2/17 |
| 2 | 9/4 | 121/8 | 0.148760 |
| 3 | 130/16 | 1689/32 | 0.153937 |
| 4 | 1819/64 | 47059/256 | 0.154614 |
| 5 | 50697/512 | 655459/1024 | 0.154692 |
| 6 | 353077·5/1024 | 2282345/1024 | 0.1546994 |

For higher derivatives such as $(d^2y/dx^2)/2$, use $C_3(t) = (t+x)^3$ in a 3rd order c.f. with

$$\frac{1}{2}\frac{d^2y}{dx^2} = \frac{1}{\pi}\int_{-1}^{1} \frac{\sqrt{(1-\omega^2)}d\omega}{(x+\omega)^3}.$$

An increasing sequence is now

$$|V_r^{(0)}, U_{r+1}^{(2)}, U_{r+2}^{(1)}| \div |U_r^{(2)}, U_{r+1}^{(1)}, U_{r+2}^{(0)}|$$

with the recurrences in (4.19) and

$$p_1 = 3x\,, \qquad p_2 = 3x^2, \qquad p_3 = x^3.$$

Further derivatives merely increase the number of components.

In general the first derivative of a Stieltjes c.f. can be expressed as a 2nd order c.f. (similarly for higher derivatives and higher orders). The polygamma functions provide an interesting set. For example for $J(x)$ (p.234), the series part of $\ln\Gamma(x)$, we can use

$$J(\sqrt{x})/\sqrt{x} \;=\; \frac{\frac{1}{12}}{x+}\;\frac{\frac{1}{30}}{1+}\;\frac{\frac{53}{210}}{x+}\;\frac{\frac{195}{371}}{1+}\;\cdots \qquad (\sqrt{x} > 0)$$

and set up a 2nd order c.f. for its derivative with respect to $x$. However note for the integral form of the derivative, that in some cases integration by parts may provide a lower order c.f. (see Example 3.10). The integral forms for the polygamma functions (Bowman and Shenton, 1988, p.30) are examples; any one of this class, using its c.f. component, could be used to set up a 2nd order c.f. for its derivative with respect to $x^2$.

## APPENDIX III

## HENRICI BOUNDS

For the c.f.

$$\frac{\alpha_1}{z+}\ \frac{\alpha_2}{1+}\ \frac{\alpha_3}{z+}\ \cdots \qquad (\alpha_1, \alpha_2, \alpha_3, \cdots > 0)$$

with $z > 0$, Henrici (1977, p602) gives three assessments for the difference of successive convergents. We quote two of them:

$$|C_s - C_{s-1}| \leqslant K \prod_{k=2}^{s} (1 + 2\sigma\alpha_k^{-1/2})^{-1/2}, \tag{A3.1}$$

$$|C_{s+1} - C_s| \leqslant K \left\{ 1 + 2\sigma \left( \frac{\mu_0'}{\mu_s'} \right)^{1/2s} \right\}^{-s/2}, \tag{A3.2}$$

where $\frac{\alpha_1}{z+}\ \frac{\alpha_2}{1+}\ \cdots \sim \frac{\mu_0'}{z} - \frac{\mu_1'}{z^2} + \cdots$, $K = \frac{\alpha_1}{z}\sqrt{\left(\frac{1+\sqrt{5}}{2}\right)}$, and $\sigma = \sqrt{z} > 0$. An application is given on p160; see also assessments of the partial numerators in the applications on p.97 and p.101. Since in the present context the components in the products in (A3) are positive, bounds for the c.f. will tend to close when the series $\sum 1/\sqrt{\alpha_k}$ indicates divergence, and fail to close when the series converges. Compare this aspect with Carleman's criteria for the double-factorial series for which $1/(\mu_s')^{1/2s} \sim s^{-1}$ (the convergence critical barrier). Note also for the form

$$\frac{1}{\alpha_1^* z +}\ \frac{1}{\alpha_2^* +}\ \frac{1}{\alpha_3^* z +}\ \cdots$$

that the corresponding moment problem is determined if the parameters $\alpha_1^*, \alpha_2^*, \cdots$, are positive and $\sum \alpha_k^*$ diverges; thus broadly speaking the equivalence transformation maps the divergence of $\sum \alpha_k^*$ to the divergence of $\sum 1/\alpha_k^{1/2}$. A few example of interest are given in Bowman and Shenton (1988, Appendix A, VI).

## ANNOTATED REFERENCES

Aitken, A. C. (1946), *Determinants and Matrices*, Oliver and Boyd.

A fascinating account of the main results described by using simple algebraic or numerical illustrations. It is consistent with a remark he made in one of his lectures that for the most part the basic theorems in the subject could be written on one side of a postcard.

For our present subject we state his account of Schweinsian expansions for determinant quotients.

$$\text{(i)} \quad \frac{|h_1 b_2 c_3 d_4|}{|a_1 b_2 c_3 d_4|} = \frac{h_1}{a_1} + \frac{|h_1 a_2| b_1}{a_1 |a_1 b_2|} + \frac{|h_1 a_2 b_3| \, |b_1 c_2|}{|a_1 b_2| \, |a_1 b_2 c_3|} + \frac{|h_1 a_2 b_3 c_4| \, |b_1 c_2 d_3|}{|a_1 b_2 c_3| \, |a_1 b_2 c_3 d_4|};$$

$$\text{(ii)} \quad \frac{|a_1 b_2 c_3 d_4|}{|b_2 c_3 d_4|} = a_1 - \frac{a_2 b_1}{b_2} - \frac{|a_2 b_3| \, |b_1 c_2|}{b_2 |b_2 c_3|} - \frac{|a_2 b_3 c_4| \, |b_1 c_2 d_3|}{|b_2 c_3| \, |b_2 c_3 d_4|}.$$

In our application to c.f.s we have formally

$$\int_a^b \frac{\{A(x)\}^2 W(x) dx}{C(x)} = \sum_{s=0}^{\infty} \frac{|\alpha_0, \gamma_{01}, \gamma_{12}, \cdots, \gamma_{s-1,s}|^2}{\Delta_{s-1}\Delta_s}$$

$$= -\lim_{s \to \infty} \begin{bmatrix} 0 & \alpha_0 & \alpha_1 & \cdots & \alpha_s \\ \alpha_0 & \gamma_{00} & \gamma_{01} & \cdots & \gamma_{0s} \\ \alpha_1 & \gamma_{10} & \gamma_{11} & \cdots & \gamma_{1s} \\ \cdot & \cdot & \cdot & \cdot & \cdot \\ \cdot & \cdot & \cdot & \cdot & \cdot \\ \cdot & \cdot & \cdot & \cdot & \cdot \\ \alpha_s & \gamma_{s0} & \gamma_{s1} & \cdots & \gamma_{ss} \end{bmatrix} \div \Delta_s$$

where $\Delta_r = |\gamma_{00}, \gamma_{11}, \cdots, \gamma_{rr}|$, a case of (ii).

A further generalization is that the $n$ th order determinant

$$\left| \int_a^b \frac{A_j(x)A_k(x)W(x)dx}{C(x)} \right|_n \qquad (j,k = 1,2,\cdots,n)$$

$$= (-)^n \lim_{s \to \infty} \begin{vmatrix} 0 & 0 & \cdots & 0 & {}_1\alpha_0 & {}_1\alpha_1 & \cdots & {}_1\alpha_s \\ 0 & 0 & \cdots & 0 & {}_2\alpha_0 & {}_2\alpha_1 & \cdots & {}_2\alpha_s \\ \cdot & \cdot & \cdot & \cdot & \cdot & \cdot & \cdot & \cdot \\ \cdot & \cdot & \cdot & \cdot & \cdot & \cdot & \cdot & \cdot \\ \cdot & \cdot & \cdot & \cdot & \cdot & \cdot & \cdot & \cdot \\ 0 & 0 & \cdots & 0 & {}_n\alpha_0 & {}_n\alpha_1 & \cdots & {}_n\alpha_s \\ {}_1\alpha_0 & {}_2\alpha_0 & \cdots & {}_n\alpha_0 & \gamma_{00} & \gamma_{01} & \cdots & \gamma_{0s} \\ {}_1\alpha_1 & {}_2\alpha_1 & \cdots & {}_n\alpha_1 & \gamma_{10} & \gamma_{11} & \cdots & \gamma_{1s} \\ \cdot & \cdot & \cdot & \cdot & \cdot & \cdot & \cdot & \cdot \\ \cdot & \cdot & \cdot & \cdot & \cdot & \cdot & \cdot & \cdot \\ \cdot & \cdot & \cdot & \cdot & \cdot & \cdot & \cdot & \cdot \\ {}_1\alpha_s & {}_2\alpha_s & \cdots & {}_n\alpha_s & \gamma_{s0} & \gamma_{s1} & \cdots & \gamma_{ss} \end{vmatrix} \div \Delta_s$$

Details are given in Shenton (1953).

Bowman, K. O. and Shenton, L. R. (1975), *Tables of Moments of the Skewness and Kurtosis Statistics in Non-Normal Sampling*, UCCND-CSD-8, Oak Ridge National Laboratory.

Contains series in descending powers of $n$, the sample size, for the sample skewness $\sqrt{b_1} = m_3/m_2^{3/2}$, kurtosis $b_2 = m_4/m_2^2$, and $2b_2 - 3b_1 - 6$ (the Type III boundary when zero). The first 8 moments of $\sqrt{b_1}$ to order $n^{-8}$, and the first 6 moments of the other two statistics to order $n^{-6}$ are given. Populations sampled are from the Pearson system, Gram-Charlier Type A, and Normal Mixtures. Only a few cases for the Pearson system have $2\beta_2 - 3\beta_1 - 6 > 0$, for when computing was done in these cases overflow and other machine problems arose. More than 150 populations involving more than 4000 series are listed. The tabulations could be used to test out summation algorithms, alternative assessments being readily available in many cases using simulation. The report also contain a brief description of Bowman's "Computer Oriented Extended Taylor Series" (COETS) algorithm, a cornerstone of much of the later development of the subject.

Bowman, K. O. and Shenton, L. R. (1981), *Estimation Problems Associated with the Weibull Distribution*, ORNL/CSD-79, Oak Ridge National Laboratory. _____(1983), Moments Series for Moment Estimators of the Parameters of a Weibull Density, *Proc. of Computer Science and Statistics: 14th Symposium on the Interface*, Springer-Verlag, New York, 174-186.

Series in descending powers of the sample size are developed for the moments of the coefficient of variation $v^*$ for the Weibull Distribution $F(t) = 1 - \exp\{-(t/b)^c\}$. A similar series for the moments of the estimator $c^*$ of the shape parameter $c$ is derived from these. Comparisons are made with basic asymptotic assessments for the means and variances.

From the first four moments, approximations are given to the distribution of $v^*$ and $c^*$. In addition we give an almost unbiased estimator $\overline{c}$ of $c$ when a sample is provided with the value of $v^*$. Comments are given on the validity of the asymptotically normal assessments of the distributions.

This study involves series based on an implicit defining equation. Thus for the d.f. $F(t) = \exp\{-(t/b)^c\}$, the fundamental equations are

$$\sqrt{\{\Gamma(1+2/c^*)/\Gamma^2(1+1/c^*)\}} = \sqrt{m_2}/m_1',$$
$$b^*\Gamma(1+1/c^*) = m_1'$$

for the moment estimator $c^*$ of $c$, and $b^*$ of $b$. Terms to $n^{-12}$ in the sample size are given for the first four moments of $c^*$. Terms to $n^{-24}$ are also given for the first four moments of the coefficient of variation $v^* = \sqrt{m_2}/m_1'$. Since sampling studies of the Weibull population are straightforward, summation algorithm can be tested out (especially for $v^*$) against simulation studies.

Christoffel, E. B. (1981), *The Influence of his Work on Mathematics and the Physical Sciences*, Birkhäuser Verlag, Basel, Boston, Stuttgart.

Chapter II has an article by Walter Gautschi and concerns the Gauss-Christoffel quadrature formulae. The subject is intimately connected with continued fractions; in particular see pp.76-78 on the Gauss c.f.

Chapter III is devoted to orthogonal polynomials, continued fractions and Padé approximation. The introductory comments in the article by Peter Wynn (The work of E. B. Christoffel on Continued Fractions pp. 190-202) on the place of c.f.s in the general framework of mathematics are of special interest.

de Bruijn, N. G. (1961), *Asymptotic Methods in Analysis*, North Holland Publishing Co., Amsterdam.

Mainly theoretical development of the methods of the subject illustrated by many simple (and not so simple) examples. The author sheds light on why the subject is so difficult to reduce to a few simple theorems, and stresses techniques. His comments on the order of magnitude symbols $o$ and $O$ and their relevance to numerical work should be noted (see pp. 18-19)).

Henrici, Peter (1977), *Applied and Computational Complex Analysis*, Vol 2, John Wiley.

Chapter 12 gives an original and stimulating account of continued fractions and a complex variable. His remainder theorems and comments on rapidity of convergence (pp. 602-603, 624-626, 629) should be noted. A general account of acceleration processes is given in Brezinski (1985).

Jones, W. B. and Thron, W. J. (1980), *Continued Fractions, Analytic Theory and Application*,

It is described in the forward as the first systematic treatment of the theory of continued fractions in book form for over two decades, updating and supplementing to a large degree the Perron classic.

Khovanskii, A. N. (1963), *The Application of Continued Fractions and Their Generalizations to Problems in Approximation Theory*, (translated by Peter Wynn), P. Noordkoff, Groningen.

Gives some new results from an elementary point of view with many illustrations.

Lam, Hing-Kam (1978), *The Distribution of the Standard Deviation and Student's t from Non-normal Universes*, Thesis for Doctor of Philosophy, University of Georgia.

In sampling from the negative exponential ($f(x) = \exp(-x)$, $x > 0$) Lam shows that:

(a) $$E(\sqrt{m_2}) = 1/2, \quad n = 2;$$
$$= \sqrt{2}/3 + \{\ln(2+\sqrt{3})\}/(3\sqrt{6}), \quad n = 3;$$
$$= \sqrt{3}/4 - \pi/72 + 5\{\ln(\sqrt{2}+\sqrt{3})\}/(12\sqrt{2}), \quad n = 4.$$

Also

(b) $$E(m_2^{3/2}) = 3/4, \quad n = 2;$$
$$= \{11 + \sqrt{3}\ln(2+\sqrt{3})/2\}/(9\sqrt{2}), \quad n = 3;$$
$$= 31\sqrt{3}/64 - \pi/288 + 47\ln(\sqrt{2}+\sqrt{3})/(192\sqrt{2}),$$
$$n = 4.$$

For

(c) $t^* = (m_1' - \mu_1')/\sqrt{m_2}$,

$$E(t^*) = -\sqrt{2}\ln(2+\sqrt{3})/\sqrt{3}, \quad n = 3;$$

$$= \pi/12 - \ln(\sqrt{2}+\sqrt{3})/\sqrt{2}, \quad n = 4;$$

$$E(t^{*2}) = \pi\sqrt{3}\ln(3)/2 - 9\sqrt{2}\int_{\sin^{-1}(1/\sqrt{3})}^{\pi/2} \frac{\ln(\sin\theta)d\theta}{2+\cos^2\theta}, \quad n = 4.$$

Muir, T. (1930), *Contributions to the History of Determinants, 1900-1920*, Blackie and Son, Ltd., London and Glasgow.

There are four previous volumes. Interest lies in his chapters on continuant determinants. Papers by Muir on c.f.s are listed in Brezinski, *A Bibliography on Padé Approximation and Related Subject*, (1977).

Perron, O. (1957), *Die Lehre von den Kettenbrüchen*, Teubner, Stuttgart.

A classical mathematical treatise containing many examples from Gauss, Stieltjes, Laguerre, Padé and others. It apparently has not been translated from the German.

Shenton, L. R. (1953), A Determinantal Expansion for a Class of Definite Integral, Part 1, *Proc. Edinb. Math. Soc.*, 9(2), 44-52.

Gives a determinantal quotient for generalized c.f.s based on a modified Stieltjes transform. Traditional c.f.s turn out to be a special case. The formula (21) has the factor $(-1)^s$ omitted.

____________ (1954a), Inequalities for the Normal Integral Including a New Continued Fraction, *Biometrika*, 41(1&2), 177-189.

Considers asymptotic properties of the Mills' ratio with assessments of the rate of convergence of the associated continued fractions. The new continued fraction was new to the author but a rediscovery.

Birnbaum's inequality for $R(t)$ is generalized; but the generalizations have been overlooked in the literature.

____________ (1954b), A Determinantal Expansion for a Class of Definite Integral, Part 2, *Proc. Edinb. Math. Soc.*, 10(2), 78-91.

Exploits Parseval's theorem to settle some convergence problems.

____________(1956), A Determinantal Expansion for a Class of Definite Integral, Part 3, *Proc. Edinb. Math. Soc.*, 10(2), 134-140.

The case when the denominator of the Stieltjes transform is a polynomial of degree $n$ is studied.

____________(1957), A Determinantal Expansion for a Class of Definite Integral, Part 4, *Proc. Edinb. Math. Soc.*, 10(2), 152-160.

A fundamental formula (28) is given as a determinantal ratio for a generalized convergent. This formula is used in another form in a later paper.

____________(1957), A Determinantal Expansion for a Class of Definite Integral, Part 5, *Proc. Edinb. Math. Soc.*, 10(2), 166-188.

Recurrence relations are developed for 2nd order c.f.s. The author failed, although encouraged by a referee, to derive the recurrence for 3rd and higher orders - the third order case involves continuant-type determinants with elements in seven diagonals, zeros elsewhere.

____________ (1958), Generalized Algebraic Continued Fractions Related to Definite Integrals, *Proc. Edinb. Math. Soc.*, 9, 170-182.

Introduces new component functions based on the equation (28) of Part 4. Theoretically, approximating sequences can now be set up for c.f.s of any finite order.

Ursell, H. D. (1958), Simultaneous Linear Recurrence Relations with Variable Coefficients, *Proc. Edinb. Math. Soc.*, 9(2), 183-206.

Ursell was the referee for Part 5 of Shenton's study "A Determinantal Expansion for a Class of Definite Integral". He was able to determine the order of recurrences for generalized continuants, but not the actual form.

# BIBLIOGRAPHY


Abramowitz, M. and Stegun, I. A. (1964), *Handbook of Mathematical Functions*, National Bureau of Standard, Applied Math. Series 55, U.S. Department of Commerce.
[21,238,244,257]

Aitken, A. C. (1932), On the Graduation of Data by the Orthogonal Polynomials of Least Squares, *Proc. Roy. Soc. Edinb.*, 53, 54-78.
[28]

Aitken, A. C. (1946), *Determinants and Matrices*, Oliver and Boyd.
[218,304]

Aitken, A. C. (1958), The Contributions of E. T. Whittaker to Algebra and Numerical Analysis, *Proc. Edinb. Math. Soc.*, 11(1), 31-38.
[298]

Aitken, A. C. and Gonin, (1934), On Fourfold Sampling with or without Replacement, *Proc. Roy. Soc. Edinb.*, 55, 114-125.
[30,31]

Allen, G. D., Chui, C. K., Madych, W. R., Marcowich, F. J. and Smith, P. W. (1975), Padé Approximation of Stieltjes Series, *J. Approx. Theory*, 14, 302-316.
[33]

Baker, George A. Jr. (1975), *Essentials of Padé Approximants*, Academic Press, New York.
[13,42,53,111]

Baker, George A. Jr. and Gammel, J.L. ed (1970), *The Padé Approximant in Theoretical Physics*, Academic Press.
[13]

Baker, G. A. Jr., Gammel, J. L. and Wills, J. G. (1961), An Investigation of the Applicability of the Padé Approximant Method, *J. Math. Anal. Appl.*, 2, 405-418.
[53]

Baker, George A. Jr. and Graves-Morris, P. R. (1981), *Padé Approximants: Part i, Basic Theory, Part ii: Extensions and Applications.* Addison-Wesley.
[13]

Baker, George A. Jr. and Hunter, D. L. (1973), Methods of Series Analysis II. Generalized and Extended Methods with Application to the Ising Model, *Physical Review B*, 7(7), 3377-3392.
[54]

Barbeau, E. J. (1979), Euler Subdues a Very Obstreperous Series, *American Math. Month.*, 86, 356-372.
[5]

Bender, C. M. and Wu, T. T. (1969), Anharmonic Oscillator, *Physical Review*, 184(5), 1231-1260.
[42,94,258]

Borel, Émile. (1928), *Lecons sur les Sé*ries Divergentes. Gauthier-Villars, Paris. (Translated by Charles L. Critchfield and Anna Vakar (1975), Los Alamos Scientific Laboratory).
[12]

Bowman, K. O. (1986), *Small Sample Properties of Statistics - Computer Algorithmic Approach*, Dissertation requirement of the Doctor of Engineering Degree, University of Tokyo.
[95,295]

Bowman, K. O. and Shenton, L. R. (1975), *Tables of Moments of the Skewness and Kurtosis Statistics in Non-Normal Sampling*, UCCND-CSD-8, Oak Ridge National Laboratory.
[95,316]

Bowman, K. O. and Shenton, L. R. (1976), A New Algorithm for Summing Divergent Series Part 2: A Two-Component Borel Summability Model, *Journal of Computational and Applied Mathematics*, 2 (4), 259-266.
[256]

Bowman, K. O. and Shenton, L. R. (1978), Asymptotic Series and Stieltjes Continued Fractions for a Gamma Function Ratio, *Journal of Computation and Applied Mathematics*, 4, 105-111.
[106]

Bowman, K. O. and Shenton, L. R. (1981), *Estimation Problems Associated with the Weibull Distribution*, ORNL/CSD-79, Oak Ridge National Laboratory.
[305]

Bowman, K. O. and Shenton, L. R. (1983a), Moments Series for Moment Estimators of the Parameters of a Weibull Density, *Proc. of Computer Science and Statistics: 14th Symposium on the Interface*, Springer Verlag, New York, 174-186.
[305]

Bowman, K. O. and Shenton, L. R. (1983b), Continued Fractions and the Polygamma Functions, *Journal of Computational and Applied Mathematics*, 9(1), 29-39.
[229]

Bowman, K. O. and Shenton, L. R. (1985), The Distribution of a Moment Estimator for a Parameter of the Generalized Poisson Distribution, *Commun. Statist. Computa. Simula.*, 14(4), 867-893.
[141]

Bowman. K. O. and Shenton, L. R. (1988), *Properties of Estimators for the Gamma Distribution*, Marcel Dekker, Inc.
[58,95,98,105,121,140,160,228,245,287,293,302,303]

Brezinski, C. (1977), *A Bibliography on Padé Approximation and Related Subjects*. du Laboratoire de Calcul de l'Université des Sciences et Techniques de Lille.
[13,54]

Brezinski, C. (1978), A Bibliography on Padé Approximation and Related Subjects, Addendum 1.
[13,54]

Brezinski, C. (1979), *Programmes FORTRAN pour Transformations de Suites et de séries*, Publication ANO-3, du Laboratoire de Calcul de l'Univ. des Sciences et Techniques de Lille.
[58]

Brezinski, C. (1980a), A Bibliography on Padé Approximation and Related Subjects, Addendum 2.
[13,54]

Brezinski, C. (1980b), *Padé Type approximation and General Orthogonal Polynomials*, Birkhäuser Verlag.
[58]

Brezinski, C. (1981), A Bibliography on Padé Approximation and Related Subjects, Addendum 3.
[13,54]

Brezinski, Claude (1985), Convergence Acceleration Methods: The Past Decade. *Journal of Computational and Applied Mathematics*, 12 & 13, 19-36.
[13,307]

Cabannes, H. ed. (1976), *Padé* Approximants Method and its Application to Mechanics, edited, Lecture Note in Physics, 47, Springer Verlag.
[54]

Char, B. W. (1980), On Stieltjes' Continued Fraction for the Gamma Function, *Math. Comp.*, 34, 547-552.
[140]

Christoffel, E. B. (1981), *The Influence of his Work on Mathematics and the Physical Sciences*, Birkhäuser Verlaag, Basel, Borton, Stuttgart. [306]

Craig, C. C. (1929), Sampling when the Parent Population is a Pearson Type III, *Biometrika*, 21, 287-293. [9]

de Bruijn, N. G. (1961), *Asymptotic Methods in Analysis*, North Holland Publishing Co., Amsterdam. [94,306]

Dusenberry, W. E. and Bowman, K. O. (1977), The Moment Estimator for the Shape Parameter of the Gamma Distribution, *Comm. Stat. Part B., Simulation and Computation*, 6, 1-19. [95]

Erdelyi, A. (1956), *Asymptotic Expansions*, Dover Publication. [94]

Gaunt, D. S. and Guttman, A. J. (1973), Series Expansions: Analysis of Coefficients, Chapter in Phase Transitions and Critical Phenomena, 3, C. Domb and M. S. Green, Eds., Academic Pross, New York and London, 181-243. [42,54,111]

Gautschi, W. (1964), Error Function and Fresnel Integral, Chapter 7, NBS [106,208,211,213]

Geary, R. C. (1936), Moments of the Ratio of the Mean Deviation to the Standard Deviation for Normal Samples, *Biometrika*, 28, 295-307. [149]

Gokhale, D. V. (1962), On an Inequality for Gamma Function, *Skandinavisk Aktuarietidskrift*, 45, 213-215. [106]

Graffi, S., Grecchi, V. and Turchetti, G. (1971), Summation Methods for the Perturbation Series of the Generalized Anharmonic Oscillator, , *Il Nuovo Cimento*, 4B, No. 2, 313-340. [266,269]

Graves -Morris, P. R. ed (1973), *Padé Approximants and Their Applications*, Academic Press, New York. [13,54]

Gurland, J. (1956), An Inequality Satisfied by the Gamma Function, *Skandinavisk Aktuarietidskrift*, 39, 171-172. [106]

Hardy, G. H. (1949), *Divergent Series*, Oxford, at Clarendon. [94]

Henrici, Peter (1977), *Applied and Computational Complex analysis*, Vol 2, John Wiley. [54,160,303,307]

Hunter, D. L. and Baker, George A. Jr. (1973), Methods of Series Analysis, Part 1 Comparison of Current Methods used in the Series of Critical Phenomena, *Physical Review*, B7, 3346.
[54]

Jones, W. B. and Thron, W. J. (1980), *Continued Fractions, Analytic Theory and Application*, Addison-Wesley Publishing Co.
[54,307]

Karamata, J. (1960), Sur Quelques Problems Poses par Ramanujan, *J. Indian Math. Soc.(NS)*, 24,343-365.
[148]

Kausler, C. F. (1803), *Die Lehre von den Continuirlichen Brüchen*, Stuttgart.
[58]

Khovanskii, A. N. (1963), *The Application of Continued Fractions and Their Generalizations to Problems in Approximation Theory*, Noordhoff, Groniengen.
[33,54,55,307]

Krawtchouk, M. (1929), Sur Une Gènèralisation des Polynomes de Hermite, *Comptes Rendus de l'Acadèmie des Sciences*, Paris 189, 620-622.
[30]

Knopp, K. (1928), *The Theory and Application od Infinite Series*, Blackie and Son, Ltd., London and Glasgow.
[44]

Lam, H. K. (1978), *The Distribution of the Standard Deviation and Student's t from Non-Normal Universes*, Thesis in fulfillment of Doctor of Philosophy degree, University of Georgia.
[95,307]

Lam, H. K. (1980), Remarks on the Distribution of the Sample Variance in Exponential Sampling, *Commun. Statist. Simula. Computa.*, 9(16), 639-647.
[95,96]

Levin, D. (1973), Development of Non-Linear Transformations for Improving Convergence of Sequences. *Internat. J. Computer Math.*, B3, 371-388.
[10,245]

Luke, Y. L. (1975), *Mathematical Function and Their Applications*, Academic Press, New York.
[148]

McCabe, J. H. (1983), On an Asymptotic Series and Corresponding Continued Fraction for a Gamma Function Ratio, *Journal of Computational and Applied Mathematics*, 9, 125-130.
[104,106 ]

Mitrinović, D. S. (1970), *Analytic Inequalities*, Springer Verlag.
[106,148]

Muir, T. (1884), Note on the Condensation of a Special Continuant, *Proc. Edinb. Math. Soc.*, 11, 16-18.
[177]

Muir, T. (1920), *The Theory of Determinant in the Historical Order of Development*, V3, Macmillan Co., Ltd., London.
[176]

Muir, T. (1930), *Contributions to the History of Determinants, 1900-1920*, Blackie and Son, Ltd., London and Glasgow.
[6,308]

Padé, H. (1892), Sur la Representation Approchèe d'une Function des Fractions Rationnelles, *Ann. Sci. Scol. Norm. Sup. Suppl.*, (3), 9, 1-93.
[50]

Pearson, E. S. (1929), Note on Dr Craig's Paper. *Biometrika*, 21, 294-302.
[10]

Pearson, E. S. (1930), A Further Development of Tests for Normality, *Biometrika*, 22, 232-249.
[51]

Perron, O. (1910), Über eine Spezielle Klasse von Kettenbrüchen, *Rend. Pal.*, 29.
[23,160]

Perron, O. (1957), *Die Lehre von den Kettenbrüchen*, Teubner, Stuttgart.
[13,32,54,58,61,78,106,170,243,308]

Reddall, W. F. (1972), *The Asymptotic Trajectory of a Strong Planer Shock Wave arising from a Non-Self-Similar Piston Motion*, Ph D. Dissertation, Stanford University, California.
[94]

Rietz, H. L. (1931), Note on the Distribution of the Standard Deviation of Sets of Three Variables drawn at Random from a Rectangular Distribution, *Biometrika*, 23, 424-426.
[110]

Rogers, L. J. (1907a), On the Representation of Certain Asymptotic Series as Continued Fractions, *Proc. Lnd. Math. Soc.*, 4(2), 393-395.
[158]

Rogers, L. J. (1907b), Supplementary Note on the Representation of Certain Asymptotic Series as Continued Fractions, *Proc. Lond. Math. Soc.*, 4(2), 393-395.
[158]

Saff, E. B. and Varga, R. S. ed (1977), *Padé* and Related Approximation Theory and Application, Academic Press, New York.
[13]

Shanks, D. (1955), Non-Linear Transformations of Divergent and Slowly Convergent Sequences, *J. Math. Phys.*, 34, 1-42.
[10,150,245]

Shenton, L. R. (1949), On the Efficiency of the Method of Moments and Neyman's Type A Distribution, *Biometrika*, 36, 450-454.
[63]

Shenton, L. R. (1953), A Determinantal Expansion for a Class of Definite Integral, Part 1, *Proc. Edinb. Math. Soc.*, 9(2), 44-52.
[298,308]

Shenton, L. R. (1954a), Inequalities for the Normal Integral Including a New Continued Fraction, *Biometrika*, 41(1&2), 177-189.
[36,54]

Shenton, L. R. (1954b), Efficiency of the Method of Moments and the Gram-Charlier Type A Distribution, *Biometrika*, 38, 58-73.
[1,36]

Shenton, L. R. (1954c), A Determinant Expansion for a Class of Definite Integral, Part 2, *Proc. Edinb. Math. Soc.*, 10(2), 78-91.
[298,308]

Shenton, L. R. (1956), A Determinantal Expansion for a Class of Definite Integral, Part 3, *Proc. Edinb. Math. Soc.*, 10(2), 134-140.
[298,308]

Shenton, L. R. (1957a), A Determinantal Expansion for a Class of Definite Integral, Part 4, *Proc. Edinb. Math. Soc.*, 10(2), 153-166.
[175,298,308]

Shenton, L. R. (1957b), A Determinantal Expansion for a Class of Definite Integral, Part 5, *Proc. Edinb. Math. Soc.*, 10(2), 167-188.
[298,309]

Shenton, L. R. (1957c), Generalized Continued Fractions; Identities and Recurrence Relations, unpublished, 1-28.
[217]

Shenton, L. R. (1958), Generalized Algebraic Continued Fractions Related to Definite Integrals, *Proc. Edinb. Math. Soc.*, 9, 170-182.
[309]

Shenton, L. R. (1963), A Note on Bounds for the Asymptotic Sampling Variance of the Maximum Likelihood Estimator of a Parameter in the Negative Binomial Distribution, *Annals of the Institute of Statistical Mathematics*, 15, 145-151.
[54,298]

Shenton, L. R. and Bowman, K. O. (1971), Continued Fractions for the Psi Function and Its Derivatives, *SIAM Jour. Appl. Math.*, 20, 547-554.
[54,139,229,232]

Shenton, L. R. and Bowman, K. O. (1975), The Development of Techniques for the Evaluation of Sampling Moments, *Internat. Statist. Rev.*, 43(3), 317-334.
[95]

Shenton, L. R. and Bowman, K. O. (1976), A New Algorithm for Summing Divergent Series Part 1. Basic Theory and Illustrations, *Journal of Computational and Applied Mathematics*, 2(3), 151-167.
[256]

Shenton, L. R. and Bowman, K. O. (1977a), *Maximum Likelihood Estimation in Small Samples*. Charles Griffin & Co., Ltd., England. (Macmillan Publishing Co., New York)
[42]

Shenton, L. R. and Bowman, K. O. (1977b), A New Algorithm for Summing Divergent Series Part III: Application, *Journal of Computational and Applied Mathematics*, 3, 35-51.
[253]

Shenton, L. R. and Bowman, K. O. (1978), Generalized Continued Fractions and Whittaker's Approach, *Journal of computational and Applied Mathematics*, 4(1), 3-5..br [298]

Shenton, L. R., Bowman, K. O. and Lam, H. K. (1979), Comments on a Paper by Geary, *Biometrika*, 66, 400-401.
[150]

Shenton, L. R., Bowman, K. O. and Sheehan, D. (1971), Sampling Moments of Moments Associated with Univariate Distribution, *J.R.S.S.(B)*, 33, 444-457.
[9]

Shenton, L. R. and Carpenter, J. A. (1964), The Mills' Ratio and the Probability Integral for a Pearson Type IV Distribution, *Biometrika*, 52(1&2), 119-126.
[50, 54]

Shenton, L. R. and Johnson, W. (1965), Moments of a Serial Correlation Coefficient, *J.R.S.S.(B)*. 27(2), 308-320.
[289]

Shenton, L. R. and Kemp, A. W.(1989), An S-Fraction and $\ln^2(1+x)$, *Journal of Computational and Applied Mathematics*.
[146]

Shohat, J. A. and Tamarkin, J. D. (1943), *The Problem of Moments*. American Mathematical Society, New York.
[40,161,172,181,193]

Sidi, A. (1979), Convergence Properties of Some Sequence Transformations, *Mathematics of Computation*, 33, 30-47.
[246]

Stieltjes, T. J. (1905), *Correspondance D'Hermite et de Stieltjes*, Gauthier-Villars, Paris.
[11,84,102,183]

Stieltjes, T. J. (1918), *Oeuvres Complètes*. Tome 2, P. Noordhoff, Groningen.
[3,11,28,43,59,75,92,232]

Szegö, G. (1939), *Orthogonal Polynomials*. American Math. Soc. Colloq. Publ., XXIII, Amer. Math. Soc., New York.
[19,23,25,26,31,33,38,79,172,184]

Turnbull, H. W. (1932), Matrices and Continued Fractions, *Proc. Roy. Soc. Edinb.*, 53, 151-163.
[164]

Uppuluri, V. R. R. (1966), On a Stronger Version of Wallis' Formula, *Pacific Journal of Mathematics*, 19(1), 183-187.
[106]

Ursell, H. D. (1958), Simultaneous Linear Recurrence Relations with Variable Coefficients, *Proc. Edinb. Math. Soc.*, 9(4), 183-206.
[8,192,309]

Van Dyke, M. (1974), Analysis and Improvement of Perturbation Series, *Q. Jl. Mech. Appl. Math.* XXVII, Pt. 4, 423-450.
[42,94,153]

Van Dyke, M. (1975), Computer Extension of Perturbation Series in Fluid Mechanics, *SIAM J. Appl. Math.*, 28(3), 720-734.
[42,94]

Wall, H. S. (1929), On the Padé Approximants Associated with the Continued Fraction and Series of Stieltjes, *Thesis, Trans. Amer. Math. Soc.*, 33, 511-532.
[217]

Wall, H. S. (1948), *Analytic Theory of Continued Fractions*, D. Van Nostrand, New York.
[13,21,22,28,35,50,54,58,183,186,193,242]

Watson, G. N. (1962), *A Treaties on the Theory of Bessel Function*, Cambridge University Press.
[48]

Whittaker, E. T. (1916), On the Theory of Continued Fractions, *Proc. Roy. Soc. Edinb.*, 36(4), 243-255.
[163]

Whittaker, E. T. and Robinson, G. (1924), *Calculus of Observations*, Blackie and Son, Ltd., Glasgow.
[1]

## NOMENCLATURE

| | |
|---|---|
| Continued Fractions | c.f.s |
| Densities: | |
| Gamma | $e^{-x}x^{a-1}/\Gamma(a) \qquad (a>0)$ |
| Uniform ($U(0,1)$) | $\begin{cases} 1, & 0<x<1 \\ 0, & \text{elsewhere.} \end{cases} \quad (\mu_1' = 1/2 \ , \ \sigma^2 = 1/12)$ |
| Normal | $e^{-x^2/2}/(2\pi), \qquad (-\infty<x<\infty)$ |
| Half Normal | $\surd(2/\pi)e^{-x^2/2}, \qquad (x>0)$ |
| Logistic | $e^{-x}/(1+e^{-x})^2. \quad (-\infty<x<\infty)$ |
| Determinant Notation | $\begin{bmatrix} a & b \\ c & d \end{bmatrix}$ is used (same as the notation used for matrix, but from the context the meaning should be clear.) |
| Diagonal Notation for Determinant | $\lvert a_1 b_2 c_3 \rvert = \begin{bmatrix} a_1 & a_2 & a_3 \\ b_1 & b_2 & b_3 \\ c_1 & c_2 & c_3 \end{bmatrix}$ |
| Euler's Constant | $\int_0^\infty \left( \frac{1}{1+x} - e^{-x} \right) \frac{dx}{x} = 0.5772 \cdots$ |
| Exponential Integral | $E_1(x) = \int_x^\infty \frac{e^{-t}}{t} dt = e^{-x} \int_0^\infty \frac{e^{-t}\,dt}{x+t}$ |
| | $E_i(x) = \int_{-\infty}^x \frac{e^{-t}}{t} dt, \qquad (x>0)$ |
| Factorial Series | $\sum_{s=0}^\infty (-1)^s \frac{(ms)!}{n^s} \qquad (m=1,2,\cdots)$ |

Gamma Function $\Gamma(x)\Gamma(1-x)=\dfrac{\pi}{\sin \pi x}, \qquad (0<\mathrm{Re}(x)<1)$

Geary's

Absolute Deviation $\omega_n'=(n+1)^{-1/2}\dfrac{\sum |x_j-\bar{x}|}{\sqrt{\sum(x_j-\bar{x})^2}}$

$d_n$ $d_n=\sum_{i=1}^{n+1}|z_i|/(n+1)=d\ .$

$(z_i=x_i-\bar{x}\ ,\ x_i \in N(0,1))$

Hamburger moment problem Hmp

Hypergeometric Function $F(a,b;c;z)$

$$=\frac{\Gamma(c)}{\Gamma(a)\Gamma(b)}\sum_{\lambda=0}^{\infty}\frac{\Gamma(a+\lambda)\Gamma(b+\lambda)}{\Gamma(c+\lambda)\lambda!}z^{\lambda}$$

$$=\frac{\Gamma(c)}{\Gamma(b)\Gamma(c-b)}\int_0^1\frac{x^{b-1}(1-x)^{c-b-1}dx}{(1-xz)^a}$$

$(\mathrm{Re}(c)>\mathrm{Re}(b)>0)$

Integer part of $x$ $[x]$ Algebraically greatest integer not exceeding $x$.

J Fraction J.f. $F(x)=\dfrac{b_0}{x+a_1-}\ \dfrac{b_1}{x+a_2-}\ \cdots$

Matrix Notation $\begin{bmatrix} a & b \\ c & d \end{bmatrix}$

Maximum Likelihood m.l.

One-Component Borel 1cB

Padé Approximation $P(n\,|\,m)=\pi_n(x)/\pi_m(x)$
($\pi_s$ is a polynomial of degree $s$)

Population Moments $\mu_s'=E(X^s)$

$\mu_s=E(X-E(X))^s$

$\mu_2=\mu_2'-\mu_1'^2$

$\mu_3=\mu_3'-3\mu_2'\mu_1'+2\mu_1'^3$

$\mu_4=\mu_4'-4\mu_1'\mu_3'+6\mu_2'\mu_1'^2-3\mu_1'^4$

Population Skewness $\alpha_3 = \sqrt{\beta_1} = \mu_3 / \mu_2^{3/2}$

Kurtosis $\alpha_4 = \beta_2 = \mu_4 / \mu_2^2$

Psi Function $\Psi(x+1) = \Psi(x) + 1/x$

Random Sample $\underset{\sim}{x}_n \equiv (x_1, x_2, \cdots, x_n)$

Recurring Decimals $0.\dot{x}\dot{y}$

Sample Moments $m_s' = \sum_1^n x_j^s / n \qquad (m_1' = \text{mean} = \bar{x})$

$$m_s = \sum_1^n (x_j - m_1')^s / n$$

$$s_x^2 = \sum_1^n (x_j - \bar{x})^2 / (n-1)$$

$$\text{Variance} = m_2' - m_1'^2$$

$$E(m_1') = \mu_1'$$

$$E(m_2) = (1 - 1/n)\mu_2 \qquad (\mu_2 = \sigma^2)$$

Sample Skewness $\sqrt{b_1} = m_3 / m_2^{3/2}$

Kurtosis $b_2 = m_4 / m_2^2 \qquad (b_2 - b_1 \geqslant 1)$

S Fraction S.f. $F(x) = \frac{b_0}{x-}\ \frac{b_1}{x-}\ \cdots$

Standard Deviation s.d.

Stieltjes Fraction $F(x) = \frac{q_0}{x+}\ \frac{p_1}{1+}\ \frac{q_1}{x+}\ \frac{p_2}{1+}\ \cdots$

or

$$= \frac{1}{\alpha_1 x +}\ \frac{1}{\alpha_2 +}\ \frac{1}{\alpha_3 x +}\ \cdots$$

Even part $\frac{q_0}{x+p_1-}\ \frac{p_1 q_1}{x+p_2+q_1-}\ \frac{p_2 q_2}{x+p_3+q_2-}\ \cdots$

Odd part $\frac{q_0}{x} - \frac{q_0}{x}\left\{\frac{p_1}{x+p_1+q_1-}\ \frac{q_1 p_2}{x+p_2+q_2-}\ \frac{q_2 p_3}{x+p_3+q_3-}\ \cdots\right\}$

Stieltjes moment problem Smp

Stieltjes Transform $F(z) = \int_0^\infty \frac{d\,\sigma(t)}{z+t}$

| | |
|---|---|
| | ($d\,\sigma(t)$ corresponds to Gamma, Half-normal, $U(0, 1)$; or stepfunction such as Binomial) |
| Student's Modified Statistic | $t = (m_1' - \mu_1')/ s_x$ <br> $t^* = (m_1' - \mu_1')/ \sqrt{m_2}$ |
| Three Component Borel | 3cB |
| Two Component Borel | 2cB |

# GUIDE TO CONTINUED FRACTIONS AND SERIES

## CONTINUED FRACTIONS

SERIES

# INDEX